Image Operators

Image Operators

Image Processing in Python

Jason M. Kinser

CRC Press
Taylor & Francis Group
Boca Raton London New York

CRC Press is an imprint of the
Taylor & Francis Group, an **informa** business

CRC Press
Taylor & Francis Group
6000 Broken Sound Parkway NW, Suite 300
Boca Raton, FL 33487-2742

Printed on acid-free paper

International Standard Book Number-13: 978-1-4987-9618-7 (Hardback)

Library of Congress Cataloging-in-Publication Data

Names: Kinser, Jason M., 1962- author.
Title: Image operators: image processing in Python/Jason M. Kinser.
Description: First edition. | Boca Raton, FL: CRC Press/Taylor & Francis Group, [2019] |
"A CRC title, part of the Taylor & Francis imprint, a member of the Taylor & Francis Group, the academic division of T&F Informa plc." | Includes bibliographical references and index.
Identifiers: LCCN 2018017140 (print) | LCCN 2018017764 (ebook) | ISBN 9780429451188 (eBook) |
ISBN 9780429835940 (Adobe PDF) | ISBN 9780429835933 (ePUB) |
ISBN 9780429835926 (Mobipocket) | ISBN 9781498796187 (hardback: acid-free paper)
Subjects: LCSH: Image processing—Digital techniques. | Python (Computer program language)
Classification: LCC TA1637 (ebook) | LCC TA1637 .K48 2019 (print) | DDC 006.6—dc23
LC record available at https://lccn.loc.gov/2018017140

Visit the Taylor & Francis Web site at
http://www.taylorandfrancis.com

and the CRC Press Web site at
http://www.crcpress.com

This book is dedicated to my beautiful wife – Sue Ellen.

Contents

PART IV Texture and Shape

Python Codes

Preface

Image processing and analysis is a burgeoning field that is gaining renewed interest in recent years. The need for image analysis tools is ever increasing. Along with this is also the need to be able to efficiently and explicitly describe processes used in analyzing images. Unfortunately, the current state of publications is that each author has their own way of describing processes. Two different authors describing the same process will often provide vastly different ways of communicating their proposed process.

The recent development of high-powered scripting languages such as Python compounds the issue. Publications can consume more real estate in explaining the process than it takes to write the Python script to execute the process. Furthermore, the descriptions can be imprecise, because some authors prefer to describe their processes through textual descriptions. Readers attempting to replicate their results may find it a difficult process as not all of the steps are clearly explained.

The purpose of this text is to provide a unified mathematical language that coincides with Python scripting. Image operators represent processes in a image analysis sequence, and these are associated with Python scripts. Thus, a concise mathematical description of a process is easily translated into Python scripts through this correlation. The conversion of Python scripts to image operators is nearly as easy. Thus, this text introduces the initial set of image operators, complete with associated Python scripts and examples.

Jason Kinser, D.Sc.
George Mason University
Fairfax, VA, USA
jkinser@gmu.edu

Software and Data

Software and data used in this text are available at:
`https://jmkinser49.wixsite.com/imageoperators`

Author

Jason M Kinser, DSc, has been an associate professor at George Mason University for more than 18 years teaching courses in physics, computational science, bioinformatics and forensic science. Recently, he converted the traditional university physics course into an active learning technology environment at GMU. His research interests include modern teaching techniques, more effective methods in text-based education, image operators and analysis, pulse image processing and multi-domain data analysis. This book was born from a desire to engage students in physics education and to find ways of reducing the external costs that both students and institutions incur within the traditional education framework.

Part I

Image Operators

1 Introduction

Analysis of images has been a growing field of science and application for several decades. As computers become more powerful, the ability to process and analyze images increases. Furthermore, the process of collecting digital data has become trivial as the proliferation of smart phones to almost every society provides individuals with the means to quickly gather gigabytes of digital image information.

While the ability to analyze images has been increasing, the ability to communicate these processes has been stagnate. Traditional sciences, such as physics and chemistry, have a standardized language. Physicists from one part of the world can quickly scan equations written by physicists in another part of the world and understand the message, often without requiring definitions. Independent of their institute of education, physicists understand that

$$\Delta t' = \Delta t \gamma \tag{1.1}$$

represents time dilation in a relativistic system. Furthermore, the language is so universally standardized that they also understand that

$$\gamma = \frac{1}{\sqrt{1 - v^2/c^2}}. \tag{1.2}$$

The language of the field is known worldwide, and hence communication of the concept does not require the recipient to dig into a text to understand what each variable represents. Because they are knowledgeable of the field, the communication is compact and precise.

The field of image analysis does not enjoy this efficiency. Papers in the same journal will display different mathematics to describe similar phenomena. The reader is thus required to reset their understanding of the mathematics to read about similar concepts. Often, authors resort to textual descriptions of their protocols instead of equations, leading to imprecise descriptions of their methods. Replication of their published process may require some guesswork since the descriptions lack complete clarity.

Compounding this issue is the growing collection of scripting languages and tools. Several decades ago, researchers would have to write several lines of code to perform a certain process because libraries were scant. Today, freely available languages come with powerful toolkits. Programmers now need only a few lines to perform those same image analysis tasks. The issue that arises is that quite often, the amount of real estate required to write the program is much less than the amount of real estate used to describe the same process. In the 1990s, Wilson and Ritter developed a mathematical language for image processing that was coincident with a C++ set of tools. This culminated in an image algebra [34]. New languages and toolkits, though, have surfaced since then, and the need for a common language is even greater than it was before.

The main goal of this book is to introduce a mathematical language for image analysis processes that is precise, compact, and matches well with modern scripting languages. Of course, several protocols and examples will be used to describe the language. There are also two appendices that contain organized descriptions of the operators to use as references.

1.1 SCRIPTING IN PYTHON

Multiple scripting and programming languages exist that are suitable for image processing. One of the most popular languages is Python, which is also freely available. So, it is the language used in this text. The rest of this chapter reviews elements of Python that are useful for the subsequent chapters; however, by no means is this chapter intended on being a comprehensive description of Python.

1.2 INSTALLATION

Python has evolved as several contributors provide useful libraries. Now distributions exist that contain suites of these useful packages and powerful editors. One such distribution is Anaconda [13], which contains multiple editors (Spyder, IDLE, and Jupyter). Anaconda provides all the tools used in this text except that are written by the author. Readers will not be required to install third-party software.

1.2.1 EXAMPLE CODES

Examples will be provided for many topics throughout the chapters such as seen in Code 1.1. Line 1 begins with the comment character (#) and the name of the file that contains the following function. This is used for software provided by the authors. Functions that are provided with the Python installation do not display the name of the file (or module) from which the functions are maintained. Lines 2 through 5 contain a function that is stored in the file listed in line 1. Line 6 starts the Python prompt >>>. These lines are to be typed directly in the Python console. Users of IPython (the default interface in Spyder) will have a different prompt such as `In [1]:`. However, the commands that follow >>> in the listed Codes will work in the same manner if they are typed into the IPython interface.

The text also includes the following conventions to assist in presenting the materials:

- Computer directories and subdirectories are presented in italic font.
- Python module names are presented in italic font.
- Python function names are presented in bold font.
- Python script and variables are presented in an evenly spaced font.
- Math variables are presented in italic font.
- Matrices and tensors are presented in bold font with uppercase letters.
- Image operators are represented by uppercase letters in italic font.
- The vector $\vec{w}$ is reserved for the frame size (number of pixels) of the image.

1.2.2 ESTABLISHING A WORK SPACE

Once Python is installed, the user should establish working directories in order to maintain organization. For example, a directory named *imageops* could be created that will house all of the work performed in the following chapters. The next step would be to create several subdirectories inside *imageops* that will contain specific types of files. For example, one subdirectory should contain only

Code 1.1 Positioning Python to the user's directory

```
1 # myfile.py
2 def NewFunction( a ):
3     b = a + 5
4     return b
5 
6 >>> NewFunction( 6 )
7 11
```

Python code, another should contain only data, another should contain only reports that are written, and so on. This text will assume that the users have a working directory and that there are the following subdirectories:

- pysrc (to contain Python scripts)
- data (to contain generic data)
- results (to receive results that are computed)

When the IDLE environment is started, the Python editor is positioned within a certain directory that is not usually a good place to put user files. For example, the Anaconda distribution on Windows will start Python in the *C:/ProgramData/Anaconda*. The Spyder interface will start Python in the user's main directory. In either case, these are not directories where the user shall place a lot of files.

So, the first item of business is to move the Python to the proper working directory that the user has created. Code 1.2 shows an example of this is accomplished. Line 1 imports standard modules that come with Python. Section 3.6 reviews the concept of modules. Line 2 changes the directory for the Python interpreter. The string */Users/myhome/imageops* is merely an example, and the user should replace this string with their own directory structure. Line 3 adds the user's *pysrc* directory to those directories that are searched when Python loads a module.

In line 1, the command is to import two different modules which are basically Python files that came with the installation. They reside in a directory, and when the **import** command is used, Python searches a list of directories for the specified module. Line 3 adds the user's directory to that list so that they can import modules that they have written.

1.2.3 THE SPYDER INTERFACE

Anaconda is the recommended Python installation. This comes with the Spyder interface which is shown in Figure 1.1. The lower right panel is the console where the user will type into commands. Responses to those commands will appear in this panel. At the bottom of the panel, there are tabs of which the "IPython" tab is active. The "Python console" tab offers an IDLE-like interface for the user. Spyder will allow multiple sessions of Python to be available to the user.

The left panel is the editor. Here, the user can enter scripts and create modules. Code written in a window in this panel can be saved and entered into the console by clicking on the green arrow just below the "Run" pulldown menu. The upper right panel will contain information about variables and files. Users can select the presentation through tabs below this panel.

1.2.4 INTENT OF THE TEXT

The purpose of this textbook is several fold. First, it presents a unified mathematical description for use in the field of image processing and image analysis. Second, it connects this language to the Python scripting language, so that the conversions between description and execution are straightforward and effective. Third, this text reviews several methods of image analysis as examples. The provided Python scripts are intended to be educational, and thus, some of the codes may not be the most efficient implementation. Highly efficient codes can sometimes be more difficult to read which is in opposition to the intent of this text. Problems are provided at the end of each chapter (except this one), so this book can also be used in a classroom setting.

Code 1.2 Positioning Python to the user's directory

```
1 >>> import os, sys
2 >>> os.chdir( '/Users/myhome/imageops')
3 >>> sys.path.append( 'pysrc' )
```

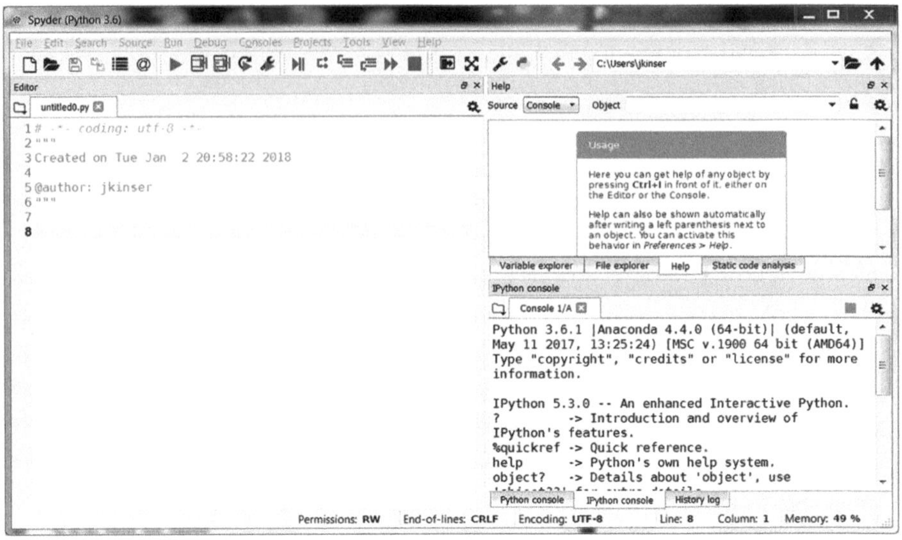

Figure 1.1 The Spyder interface.

This book is divided into major categories:

- Presentations of image operators,
- Image space manipulations,
- Frequency space manipulations,
- Texture and shape,
- Image basis, and
- Appendices contain tables and descriptions of the image operators.

There are occasions where two (or more topics) are interrelated. For example, topic A relies on topic B, but topic B relies on topic A. One must be presented before the other, and thus forward referencing is required. In this situation, components of topic B will be used in the presentation of topic A, but a thorough understanding of topic B is not required. Basically, the Python of topic B is used with a description of what the code is doing. Thus, in studying the concepts of topic A, the script of topic B can be used, without alteration, to complete the example. There will be a forward reference to topic B, but knowledge of that section of the text will not be required in order to understand the concepts in topic A.

This text is designed to be an educational tool for those entering into the field of image processing, by providing a mathematical framework to explicitly and efficiently describe their protocols.

2 Operator Nomenclature

Image analysis is a very lively field of research, but it lacks a unified language. Thus, technical publications describe their protocols in vastly different manners and even rely on inefficient and inexact textual descriptions. The development of image analysis libraries has expanded the ability to construct short scripts that perform complicated tasks. Thus, the amount of real estate in publications required to precisely describe the algorithm is often much larger than the space required to write the computer script.

The intent of this text is to present a unified operator notation that offers highly efficient and accurate operator notation that can easily be translated into modern Python scripts. Furthermore, the operators are presented with extensive examples to facilitate the correlation between operator notation and protocol realization. The operator notation is precise so that the programmer will not have to guess at the parameters required to replicate results. The operators are efficient in that a single operator is directly related to a single Python function.

The library of operators will take the form of

$$\mathbf{b}[\vec{x}] = A_m B_n C_p \mathbf{a}[\vec{x}], \tag{2.1}$$

where $\mathbf{a}[\vec{x}]$ is the input image and $\mathbf{b}[\vec{x}]$ is the output image. Each operator, i.e., A_m, B_n, and C_p, defined a specific function with the subscripts defining parameters of that function. For example, $R_{\vec{v},\theta}$ is the Rotation operator with $\vec{v}$ being the center of rotation and θ being the angle of rotation. Many functions have optional arguments. If the rotation is about the center of the frame then $\vec{v}$ is not required in the Rotation operator. The goal is that each operator also maps to a Python function from either the standard packages or from modules provided in this text. The generic operators in Equation (2.1) are translated it Python script of the form shown in Code 2.1 where each function in the script corresponds directly with an operator in the mathematical description.

In this example, $\mathbf{a}[\vec{x}]$ is the original image, which is represented by `adata` in the code. The first operator applied to the image is C_p, since it is nearest to the image. Operators B_n and A_m are applied to previous results in that order. The output is $\mathbf{b}[\vec{x}]$, which is represented by `bdata` in the code.

2.1 IMAGE NOTATION

An image is usually conceived as a two-dimensional array of pixel values. However, the number of dimensions increases with the inclusion of color, motion, or more spatial dimensions. Thus, the definition of an image begins with the declaration that the space in which the image exists is $\mathbf{X}$ and a vector $\vec{x}$ spans this space. The pixels (or voxels) locations are defined by this scanning vector $\vec{x} \in \mathbf{X}$. An image is therefore represented in bold as, $\mathbf{a}[\vec{x}]$. This represents any type of image from a grayscale image to a hyperspectral image.

Several operations will transform the image from one type of space into another. For example, a Fourier transform will create a new image in frequency space instead of the original image space. Thus, the output space is defined with a different notation, such as $\mathbf{Y}$, which is spanned by $\vec{y}$. Images $\mathbf{a}[\vec{x}]$ and $\mathbf{b}[\vec{y}]$ are in different types of spaces.

Code 2.1 Corresponding Python outline

```
>>> t1 = C( p, adata )
>>> t2 = B( n, t1 )
>>> bdata = A( m, t2 )
```

A set of images is enclosed by braces and is represented by $\{\mathbf{a}[\vec{x}]\}$ or by a subscript $\mathbf{a}_i[\vec{x}]$ where $i = 1, ..., N$ and N is the number of images in the set.

2.2 OPERATORS

Images are manipulated through the application of operators which are organized into several categories. This section will present just a few of these operators to demonstrate how the notation functions. The entire list of operators are shown in Appendices A and B. Operators will be detailed in the chapters in which they are first.

The operator categories are as follows:

- Creation operators. These create an image or sets of images.
- Channel operators. These extract information from specific channels or combine multiple channels into one.
- Information operators. These extract information from images but do not alter the images.
- Intensity operators. These modify the intensity of the pixels in an image without changing the shapes (or content) within the image.
- Geometric operators. These move content within the image but do not change the intensity of the content.
- Transformation operators move the information into a completely different coordinate system or representation.
- Expansion operators. These convert the image information into an expanded space.

A few Python examples are shown so readers can replicate the results. However, these appear before the Python chapter. Readers can thus replicate the results without being require to understand every command at the onset.

2.2.1 CREATION OPERATORS

The Creation operators are used to create images or sets of images. These are the only operators that do not act on existing images. For example, the Rectangle operator, **r**, creates an image that has binary valued pixels within a rectangular region. In a two-dimensional image, there are six parameters that are required for this operator with two defining the size of the image, two defining the location of one corner of the rectangle, and two more defining the size of the rectangle. An image with more dimensions requires more parameters. In order to be consistent for any number of dimensions, the parameters for this operator are described by three vectors: $\vec{w}$, $\vec{v}_1$, and $\vec{v}_2$. The vector $\vec{w}$ defines the size of the image, and the other two vectors define opposing corners of the rectangle, thus defining its location and size.

The creation of a rectangle for any number of dimensions is then represented as

$$\mathbf{a}[\vec{x}] = \mathbf{r}_{\vec{w};\vec{v}_1,\vec{v}_2}[\vec{x}]. \tag{2.2}$$

The vector $\vec{w}$ may be omitted if the frame size of the image is implied through the application in which it is being used. The result is that $\mathbf{a}[\vec{x}]$ is an image with a solid rectangle filled with pixels that have an intensity of 1, on a black background that have pixels with intensity of 0.

The Circle operator, **c**, creates an image with a binary circle or sphere. Other operators create a Kaiser window, a set of Gabor filters, or a set of Zernike filters. These will be explained in more detail in the chapters where they are first employed.

Images are loaded from a file using the File operator, Y, that receives a file name.

$$\mathbf{a}[\vec{x}] = Y(\texttt{fileName}). \tag{2.3}$$

Table 2.1
The Creation Operators

Symbol	Name	Description
$\mathbf{z}_{\vec{w}}$	Zeros	Returns an image of size $\vec{w}$ with all pixels set to 0.
$\mathbf{r}_{\vec{w};\vec{v}_1,\vec{v}_2}[\vec{x}]$	Rectangle	Returns an image with a solid rectangle located at $\vec{v}_1$ and size $\vec{v}_2$.
$\mathbf{o}_{\vec{w};\vec{v},r}[\vec{x}]$	Circle	Returns an image with a solid circle of radius r located at position $\vec{v}$.
$\mathbf{q}_{\vec{w}}[\vec{x}]$	Random	Returns an image with random numbers between 0 and 1.
$\mathfrak{G}_{\vec{w},\vec{f},\vec{t},sw}[\vec{x}]$	Gabor	Returns a set of images that are Gabor filters.
$\mathcal{Z}_{\vec{w};r,m,n}[\vec{x}]$	Zernike	Returns a set of images that are Zernike functions.
$k_{\vec{w};r_1,r_2}[\vec{x}]$	Kaiser	Returns an image which contains a Kaiser mask with an inner radius of r_1 and an outer radius of r_2.
$Y(\texttt{fileName})$	Load	Loads from a file.

Table 2.1 shows the Creation operators. The Zero operator, **z**, creates an array of a specified size with all pixel values set to 0. The operators **r**, **o**, and **q** create an image with a solid rectangle, a solid circle, or random values, respectively. The rest of the operators create an image or image set according to the specified algorithm. Each function is listed in the appendices with a brief description.

2.2.2 CHANNEL OPERATORS

Channel operators are presented in Chapter 5 and perform functions on individual channels. Most commonly, these are applied to images that have color channels but can be used for other types of applications as needed. For example, a medical scan will produce several 2D images, and thus, each slice is considered as a channel.

Consider an image $\mathbf{a}[\vec{x}]$ which contains three color channels with the common color format of red, green, and blue channels. The channels are denoted as separate entries inside a stack encased by curly braces. In this example, the color channels can be separated by

$$\begin{Bmatrix} \mathbf{r}[\vec{x}] \\ \mathbf{g}[\vec{x}] \\ \mathbf{b}[\vec{x}] \end{Bmatrix} = \mathcal{L}_{\mathrm{RGB}}\mathbf{a}[\vec{x}], \tag{2.4}$$

where $\mathcal{L}$, as described in Section 5.1, is the Color Conversion operator that translates the image $\mathbf{a}[\vec{x}]$ into the RGB format. The result of this operation is the creation of three grayscale images, $\mathbf{r}[\vec{x}]$, $\mathbf{g}[\vec{x}]$, and $\mathbf{b}[\vec{x}]$, which depict the amount of red, green, and blue present in the original image.

In the case of a medical scan, there are too many channels to place in a vertical array. Brackets are used to indicate which channels are being used. So,

$$\mathbf{b}[\vec{x}] = \begin{Bmatrix} \varnothing \\ < 12 > 1 \\ \varnothing \end{Bmatrix} \mathbf{a}[\vec{x}] \tag{2.5}$$

will extract just channel 12 and multiply it by 1. The output $\mathbf{b}[\vec{x}]$ is a grayscale image of just that one channel. The index inside - the brackets does not have to be numeric. So,

$$\mathbf{b}[\vec{x}] = \begin{Bmatrix} \varnothing \\ < \texttt{blue} > 1 \end{Bmatrix} \mathbf{a}[\vec{x}] \tag{2.6}$$

extracts just the blue channel. The $\varnothing$ symbols are used to indicate that this is a Channel operator, as the curly braces with a single entity may be confused with an image set. The $\varnothing$ symbol is used to block a channel from passing through an operation. Thus, the isolation of the red information from a color image is described as

$$\mathbf{r}[\vec{x}] = \begin{Bmatrix} 1 \\ \varnothing \\ \varnothing \end{Bmatrix} \mathbf{a}[\vec{x}], \tag{2.7}$$

where $\mathbf{r}[\vec{x}]$ is the grayscale image depicting the red information contained in $\mathbf{a}[\vec{x}]$. The $\bowtie$ symbol allows a channel to pass through the operation without being altered. So, to divide the red channel by 2 and leave the other channels unchanged, the notation is

$$\mathbf{b}[\vec{x}] = \begin{Bmatrix} 0.5 \\ \bowtie \\ \bowtie \end{Bmatrix} \mathbf{a}[\vec{x}]. \tag{2.8}$$

Finally, a scalar value indicates that all pixels in that channel should have the scalar value. The operation creates an image $\mathbf{b}[\vec{x}]$ from

$$\mathbf{b}[\vec{x}] = \begin{Bmatrix} 0.5\mathbf{f}[\vec{x}] \\ 0 \\ \mathbf{g}[\vec{x}] \end{Bmatrix}. \tag{2.9}$$

In this operation, all of the values in the red channel are half of the values from a grayscale image $\mathbf{f}[\vec{x}]$, all of the values in the green channel are set to 0, and all of the values in the blue channel are taken from some other grayscale image denoted as $\mathbf{g}[\vec{x}]$.

Consider the example is shown in Figure 2.1 which shows the original RGB image and a second image in which the information in the color channels have been swapped. In this case, the information from the red channel is place in the blue channel, the information in the green channel is placed in the red channel, and the information from the blue channel is placed in the green channel. The notation for splitting the information is Equation (2.4), and the notation for putting these channels in a different order to create a new image is

$$\mathbf{c}[\vec{x}] = \begin{Bmatrix} \mathbf{g}[\vec{x}] \\ \mathbf{b}[\vec{x}] \\ \mathbf{r}[\vec{x}] \end{Bmatrix}. \tag{2.10}$$

Code 2.2 shows two methods for accomplishing this color shift. Lines 1 through 4 use PIL (Python Image Library) to load an image and split it into three grayscale images that represent the intensities

(a) (b)

Figure 2.1 (a) An original image and (b) the same image with the color channels in a different order.

Code 2.2 Swapping the color channels

```
1 >>> from PIL import Image
2 >>> mg = Image.open( 'data/bird.jpg')
3 >>> r,g,b = mg.split()
4 >>> mg2 = Image.merge('RGB', (g,b,r) )
5 #
6 >>> import imageio
7 >>> import scipy.ndimage as nd
8 >>> adata = imageio.imread( 'data/bird.jpg')
9 >>> bdata = nd.shift( adata, (0,0,1), mode='wrap' )
```

of each color band. In line 4, the data is merged back into an RGB image but the color information is arranged differently as prescribed. The green information is placed in the red channel and so on.

The second method uses the **ndimage.shift** function. If this function was applied to the first axis of the data, then it would shift it vertically. The Wrap option places the rows of data that are shifted off of the frame to the other end of the image. If the image is being shifted upwards, then the rows that shift outside of the frame are placed at the bottom of the image. A shift along the second axis would move the image horizontally. Line 8 loads the image into a tensor of which the first two axes are the vertical and horizontal dimensions. The third axis is the color channels, and these are eligible to be shifted as well. Line 9 uses the argument `(0,0,1)` to shift only the third channel and turns on the Wrap option. Thus, the red information is shifted into the blue channel, the blue information is shifted into the green channel, and the green information is shifted into the red channel. The result is the same as in Figure 2.1b.

Table 2.2 shows the channel-related operators. Access to individual channels uses the curly braces, color model conversion uses $\mathcal{L}$, the $\varnothing$ blocks a channel while $\bowtie$ passes a channel through to the next operation. Merging of the channels can be accomplished either by summation or multiplication using the $\sum$ or $\prod$ operators, respectively.

A simple example of the summation function is to convert a color image to grayscale by

$$\mathbf{b}[\vec{x}] = \sum_{\mathcal{L}} \left\{ \begin{matrix} 0.5 \\ 0.75 \\ 0.25 \end{matrix} \right\} \mathbf{a}[\vec{x}]. \tag{2.11}$$

Code 2.3 shows the step in line 2. This conversion favors the green information at the expense of the blue information, which is common for grayscale conversions.

Table 2.2
The Channel Operators

Symbol	Name	Description
$\left\{ \begin{matrix} \cdot \\ \cdot \\ \cdot \end{matrix} \right\}$	Channel isolation	Accesses individual channels in an image.
$\mathcal{L}_m$	Color model conversion	Converts an image into the specified color model m.
$\varnothing$	Block	Prevents information from one channel from participating in the computation.
$\bowtie$	Pass	Allows a channel to pass through without alteration.
$\sum_{\mathcal{L}}$	Summation	Creates a new image from the addition of the different color channels.
$\prod_{\mathcal{L}}$	Product	Creates a new image from the multiplication of the different color channels.

Code 2.3 Converting an RGB image to a grayscale image

```
>>> adata = imageio.imread( 'data/bird.jpg')
>>> bdata = 0.5 *adata[:,:,0] + 0.75*adata[:,:,1] + 0.25*adata[:,:,2]
```

2.2.3 INFORMATIONAL OPERATORS

An Informational operator returns information about an image without altering the image, and these are shown in Table 2.3. The first example is the Size operator, Z, which returns the size of the image as a vector. The bird image shown in Figure 2.1(a) has 519 pixels in the vertical dimension and 653 pixels in the horizontal dimension; thus, the size vector is $\vec{v} = (519, 653)$. In the case of the color bird image, $\vec{v} = (519, 653, 3)$. The Python script is shown in lines 1 through 4 in Code 2.4, and the operator notation is

$$\vec{v} = Z\mathbf{a}[\vec{x}]. \tag{2.12}$$

The second example, is the operator, ⊠, that returns the location of the center of mass. This is applied to a binary valued image with a single contiguous region of ON pixels. It also returns a vector and is

$$\vec{v} = \boxtimes\mathbf{a}[\vec{x}]. \tag{2.13}$$

This is shown in lines 5 through 7 of Code 2.4. Line 6 creates a binary-valued image with a filled circle centered at (200,300). The **center_of_mass** function is called in line 7.

The third example is to determine the average pixel value for those pixels within a defined region that are greater than a threshold value γ. This process is defined as

$$t = \frac{\sum_{\vec{x}} \Gamma_{p,>\gamma} \square_A \mathbf{a}[\vec{x}]}{\sum_{\vec{x}} \Gamma_{>\gamma} \square_A \mathbf{a}[\vec{x}]}. \tag{2.14}$$

Table 2.3
The Informational Operators

Symbol	Name	Description
$\mathbf{a}^{\dagger}[\vec{x}]$	Complex conjugate	Returns the complex conjugate of the image.
$\Re\mathbf{a}[\vec{x}]$	Real component	Returns the real component of the image.
$\Im\mathbf{a}[\vec{x}]$	Complex component	Returns the imaginary component of the image.
$\mathbf{a}[\vec{x}] \stackrel{?}{=} \mathbf{b}[\vec{x}]$	Is equal	Returns a binary valued image with pixels set to 1 if the same pixels have the same value in the input images.
$>, <, \geq, \leq$	Comparisons	Used just as the previous operator to determine the relationships between pixel values.
$\mathbf{a}[\vec{x}] \cdot \mathbf{b}[\vec{x}]$	Inner product	Returns a scalar value for the inner product.
$\boxtimes\mathbf{a}[\vec{x}]$	Center of mass	Returns a vector that is the center of mass of the image.
$\bigvee, \bigwedge$	Max and min	Returns the maximum or minimum value in an image.
$A_{\vee}, A_{\wedge}$	Max and min locations	Returns the locations of the max or min.
$\mathfrak{D}(a,b)$	Distance	Returns the Euclidean distance between entities a and b.
$\mathcal{E}\mathbf{a}[\vec{x}]$	Energy	Returns the energy of an image.
$\mathcal{M}\mathbf{a}[\vec{x}]$	Average	Returns the average pixel value.
$N\mathbf{a}[\vec{x}]$	Count	Returns the number of pixels.
$\mathcal{N}\mathbf{a}[\vec{x}]$	Nonzero	Returns the locations of the nonzero elements.
$\mathcal{O}_{\downarrow,C}$	Sort	Returns a sort order of the data. The direction of the arrow indicates if the return is incremental or decremental. Condition C may be applied if necessary.
$\mathcal{T}_A\mathbf{a}[\vec{x}]$	Regional standard deviation	Returns an image in which are the local standard deviation values over a region of size A for every pixel in the image.
$V\mathbf{a}[\vec{x}]$	Covariance	Returns the covariance matrix.
$Z\mathbf{a}[\vec{x}]$	Dimension	Returns the dimensions of the image.

Code 2.4 A few informational operations

```
>>> bdata.shape
(519, 653)
>>> data.shape
(519, 653, 3)
>>> import mgcreate as mgc
>>> cdata = mgc.Circle( (512,512), (200,300), 25 )
>>> nd.center_of_mass( cdata )
(200.0, 300.0)
```

The $\square_A$ operator will extract part of the image as defined by A. Usually, this is in the forms of two vectors which define the upper-left and lower-right corners of a rectangular region, and the result is a smaller image which is the information within this rectangle. The $\Gamma_{p,>\gamma}$ is a passive threshold, and so the pixels that are less than γ are set to 0, while the other pixels maintain their current value. One common use of this operator is to convert nearly dark backgrounds to completely dark backgrounds using a low value for γ. The denominator uses $\Gamma_{>\gamma}$ which converts those pixels that pass the threshold to a value of 1. Thus, the sum of these is the number of pixels that pass the threshold. The ratio, then, is the average value of the pixels within a defined region that pass a threshold. The Python code that matches the operator notation is shown in Code 2.5, but it is not very efficient since it performs the same comparison twice. An operator notation that produces more efficient Python code is shown in three steps:

$$\mathbf{b}[\vec{x}] = \square_A \mathbf{a}[\vec{x}], \tag{2.15}$$

and

$$t = \frac{\sum_{\vec{x}} \left((\Gamma_{p,>\gamma} \mathbf{b}[\vec{x}]) \times \mathbf{b}[\vec{x}] \right)}{\sum_{\vec{x}} \Gamma_{>\gamma} \mathbf{b}[\vec{x}]}. \tag{2.16}$$

Here, $\mathbf{c}[\vec{x}]$ is a mask of all of the pixels above a threshold. In this notation, computations are not repeated. There is an Average operator, $\mathcal{M}$, but that cannot be used here as it would include the pixels that did not pass threshold in the computation of the average.

The final example uses the Energy operator, $\mathcal{E}$, which returns the energy contained within an image. This is accomplished by summing the squares of the pixel values and dividing by the number of pixels. The energy of a red channel is

$$f_r = \frac{\sum_{\vec{x}} \left(\begin{Bmatrix} 1 \\ \varnothing \\ \varnothing \end{Bmatrix} \mathbf{a}[\vec{x}] \right)^2}{N \mathbf{a}[\vec{x}]}, \tag{2.17}$$

where N is the Count operator that returns the number of pixels. The value f_r is the energy of the red channel of image $\mathbf{a}[\vec{x}]$. The total energy is the sum of the energies of the three channels and is defined as

Code 2.5 Computing the average of selected pixels

```
>>> gamma = 200
>>> num = ((adata[100:200, 100:300] > gamma)*adata[100:200, 100:300]).sum()
>>> den = (adata[100:200, 100:300] > gamma).sum()
>>> num/den
229.95188875669245
```

$$f = \mathcal{E}\mathbf{a}[\vec{x}] \equiv \frac{\sum_{\text{RGB}} \sum_{\vec{x}} (\mathbf{a}[\vec{x}])^2}{N\mathbf{a}[\vec{x}]}. \tag{2.18}$$

The task in this final example is to determine which of the color channels has the most energy. The notation is

$$t = A_\vee \left\{ \mathcal{E} \left\{ \begin{matrix} 1 \\ \varnothing \\ \varnothing \end{matrix} \right\} \mathbf{a}[\vec{x}], \mathcal{E} \left\{ \begin{matrix} \varnothing \\ 1 \\ \varnothing \end{matrix} \right\} \mathbf{a}[\vec{x}], \mathcal{E} \left\{ \begin{matrix} \varnothing \\ \varnothing \\ 1 \end{matrix} \right\} \mathbf{a}[\vec{x}] \right\}. \tag{2.19}$$

Here, the curly braces have two purposes with the first being the separation of the color channels. The second is the outer layer of curly braces which collects the three energy values into a set. The operator $A_\vee$ then returns the location of the largest value, which in this case is the identifier of which image in the set had the largest value.

The Python script is shown in Code 2.6. Line 3 performs several steps. The **lambda** function is used to compute the energy for a single channel, and the **map** function is used to repeat this computation for all three channels. The output is `nrg` which is a list that contains the energy of the three channels. Line 4 converts this into a vector, and the **argmax** function determines which channel had the highest energy. The three color channels are represented numerically by 0, 1, and 2, and so the answer in line 6 indicates that the blue channel has the most energy.

2.2.4 INTENSITY OPERATORS

Table 2.4 shows the Intensity operators. These create a new image by modifying the pixel intensities of an original image. Such an operator can brighten an image but does not move the content such as a shift or rotation.

One example is a Threshold operator, Γ, which compares all values in an image to a threshold. This operator has several variants. The first is to simply determine which pixels are greater than (or less than) some scalar value γ as in

$$\mathbf{b}[\vec{x}] = \Gamma_{>\gamma}\mathbf{a}[\vec{x}]. \tag{2.20}$$

The output $\mathbf{b}[\vec{x}]$ is a binary valued image in which a pixel therein is set to 1 if the same pixel in $\mathbf{a}[\vec{x}]$ is greater than γ. Otherwise, the pixel is set to 0. The operator $\Gamma_{a<b}$ would set pixels to 1 if they are greater than a and less than b. The subscript p is used to describe a passive threshold. Thus, $\Gamma_{p>4}$ would set all pixels less than 4 to a value of 0 but allow the other pixels to maintain their current value.

As an example, consider the task of isolating the cane pixels in the image in Figure 2.2(a). The cane for the most part has a uniform color and so the isolation process relies on color thresholds. This process will convert the RGB to the YIQ color model and then pixels within a specified range of values will be set to 1. The YIQ color model separates pixel intensity from hue, and it will be discussed in Chapter 5. Once again, the main goal here is merely to introduce the philosophy inherent in the operator notation instead of thoroughly understanding all of the details.

The hue information is maintained in the I and Q channels, and these are shown in Figure 2.2(b) and (c). As seen, the cane has a fairly uniform intensity in each channel. Furthermore, this uniform

Code 2.6 Determining which channel has the most energy

```
1 >>> adata = imageio.imread( 'data/bird.jpg')
2 >>> V,H,N = adata.shape
3 >>> nrgs = map( lambda x: (adata[:,:,x]**2).sum()/(V*H), [0,1,2] )
4 >>> t = np.array( list(nrgs) ).argmax()
5 >>> t
6 2
```

Table 2.4
The Intensity Operators

Symbol	Name	Description
$f\mathbf{a}[\vec{x}]$	Scaling	Multiplies value f to all the pixels.
$\mathbf{a}[\vec{x}]\&\mathbf{b}[\vec{x}]$	Binary AND	Returns a binary-valued image that it is the binary AND of the two input images.
$\mathbf{a}[\vec{x}]\Vert\mathbf{b}[\vec{x}]$	Binary OR	Returns a binary-valued image that it is the binary OR of the two input images.
$\mathbf{a}[\vec{x}]+\mathbf{b}[\vec{x}]$	Addition	Returns the addition of two images.
$\mathbf{a}[\vec{x}]-\mathbf{b}[\vec{x}]$	Subtraction	Returns the subtraction of two images.
$\mathbf{a}[\vec{x}]\times\mathbf{b}[\vec{x}]$	Multiplication	Returns the elemental multiplication of two images.
$\bigtriangledown_n, \triangle_n$	Lo, Hi bits	Returns an image that passes the n lo or hi bits.
$\lhd_n, \rhd_n$	Erosion and Dilation	Returns an image after applying the Erosion or Dilation operators for n iterations.
$A_f\mathbf{a}[\vec{x}]$	Value location	Returns the location of the pixels that have a value of f.
$\mathcal{B}\mathbf{a}[\vec{x}]$	Binary fill holes	Returns an image with the holes filled in.
$E_m\mathbf{a}[\vec{x}]$	Edge enhancement	Applies an edge enhancement to an image. The type of edge enhancement is defined by m.
$\Gamma_n\mathbf{a}[\vec{x}]$	Threshold	Returns an image after applying a threshold defined by n.
$\mathcal{H}\mathbf{a}[\vec{x}]$	Harris	Applies the Harris corner detection algorithm.
$\mathcal{S}\mathbf{a}[\vec{x}]$	Smooth	Applies a smoothing algorithm.

(a) (b) (c) (d)

Figure 2.2 (a) An original image, (b) the I channel from the YIQ color model, (c) the Q channel, and (d) pixels that passed both thresholds.

intensity is different than background surrounding the cane, and so isolation of the cane by color is not too difficult. The process applies thresholds to both of these channels and any pixel that survives all of the thresholds is set to 1.

The protocol for this process is:

1. Convert the image to the YIQ color model,
2. Apply a threshold to the I channel that sets to 1 only those pixels which have a value between 18 and 23,
3. Apply a threshold to the Q channel which sets to 1 only those pixels which have a value between 2 and 8, and
4. Multiply the resultant channels to form the final image.

The operator notation is

$$\mathbf{b}[\vec{x}] = \prod_{\mathcal{L}} \begin{Bmatrix} \varnothing \\ \Gamma_{18<23} \\ \Gamma_{2<8} \end{Bmatrix} \mathcal{L}_{\text{YIQ}}\mathbf{a}[\vec{x}]. \tag{2.21}$$

The Python script is shown in Code 2.7, where the image $\mathbf{a}[\vec{x}]$ is loaded in line 1. It is converted to the YIQ color model in line 3 represented by the operator $\mathcal{L}_{\text{YIQ}}$. The Channel operator (curly braces) separates the Y, I, and Q channels. The Y channel contains the intensity information which

Code 2.7 Isolating the man's cane

```
1 >>> a = imageio.imread('data/man.png')
2 >>> rgb = a.transpose( 2,0,1 )
3 >>> yiq = colorsys.rgb_to_yiq( rgb[0], rgb[1], rgb[2] )
4 >>> c1 = (yiq[1]>8) * (yiq[1]<23)
5 >>> c2 = (yiq[2]>2) * (yiq[2]<8)
6 >>> c = c1 * c2
```

is not used in this procedure, and therefore, it is blocked by $\varnothing$. A threshold, $\Gamma_{18<23}$, is applied to the I channel in line 4. Another threshold, $\Gamma_{2<8}$, is applied to the Q channel in line 5. The results from these two channels are multiplied by the $\prod_{\mathcal{L}}$ operator in line 6. The $\varnothing$ blocks the channel, but it does not set the values to zero. Otherwise, the $\prod_{\mathcal{L}}$ would result in an image with 0 at every pixel. The $\varnothing$ simply blocks the channel from participating in the calculation. The final result is shown in Figure 2.2(d). While pixels other than the cane survive this process, the cane is clearly isolated from its immediate background that accomplishes the goal.

Table 2.4 displays the Intensity operators which includes standard arithmetic and binary operators. There are also operators to manipulate the individual bits of an integer pixel. More advanced operators include the erosion, dilation, filling holes, edge enhancement, smoothing, and even the Harris operator which is a corner enhancement algorithm. Again, these will be explained in detail in their first uses in the following chapters.

2.2.5 GEOMETRIC OPERATORS

The Geometric operators are those that move pixels rather than changing the intensity. One simple example is the Rotation operator, *R*, which rotates the image an angle θ about a defined point $\vec{v}$:

$$\mathbf{b}[\vec{x}] = R_{\theta,\vec{v}}\mathbf{a}[\vec{x}]. \tag{2.22}$$

The Scaling operator changes the size of the image and is presented slightly different than other operators. The factor α governs the scaling with $\alpha > 1$, increasing the size of the image. The notation is has two forms,

$$\mathbf{b}[\vec{x}] = \mathbf{a}[\alpha\vec{x}] = S_\alpha\mathbf{a}[\vec{x}].$$

Table 2.5 displays the Geometric operators. Some operators are used to extract subportions of the image such as windowing or down sampling. Some operators change the shape of the content through scaling, affine transformations, bending, shifting, rotating, or swapping. The Concatenation operator creates a new larger image by abutting smaller images. The Warp and Morph operators are used to change shape into a specified grid. Actually, the Morph operator breaks the rule of altering the pixel values of the content but this is through a merger of the set of input images rather than through a user-defined operation. The Plop operator places a smaller image in a bigger frame, and the Reshape operator converts an image into a very long vector or any other geometry desired by the user.

2.2.6 TRANSFORMATION OPERATORS

The Transformation operators create a new image or matrix representing the data in a very different coordinate system. The most popular of these is a Fourier transform which converts data from an image space to a frequency space. This transformation is written as

$$\mathbf{b}[\vec{y}] = \mathfrak{F}\mathbf{a}[\vec{x}]. \tag{2.23}$$

Table 2.5
The Geometric Operators

Symbol	Name	Description
$\mathbf{a}[\alpha\vec{x}]$	Scaling	Returns an image that is a different size.
$\mathbf{a}[\mathbf{M}\vec{x}]$	Affine	Multiplies the matrix **M** by each coordinate vector.
$\square_{\vec{v}_1,\vec{v}_2}$	Window	Extracts a subimage.
$\Downarrow_n$	Downsample	Extracts pixels according to prescription n.
B_β	Bending	Applies a barrel or pincushion transformation.
C	Coordinate map	Moves the pixel according to the user-defined map.
$\mathcal{C}_a$	Concatenation	Creates a larger image from the concatenation of a images from an image set.
$D_{\vec{v}}$	Shift	Shifts the image.
M_α	Morph	Creates a set of new images from the morph of at least two input images.
$R_{\alpha,\vec{v}}$	Radial coordinate	Radial coordinate transformation.
$\mathcal{R}_{\theta,\vec{v}}$	Rotation	Rotates an image through an angle θ centered at $\vec{v}$.
$U_{\vec{v}}$	Plop	Places a smaller image in the center of a larger frame.
$\mathcal{V}_{\vec{w}}$	Reshape	Reshapes the array.
W_G	Warp	Warps an image to grid G.
X	Quadrant swap	Exchanges quadrants I and IV. Exchanges quadrants II and III.
$\mathfrak{L}_m$	Flip	Flips an image about a specified axis.

Table 2.6
The Transformation Operators

Symbol	Name	Description
$\mathbf{a}[\vec{x}] \otimes \mathbf{b}[\vec{x}]$	Correlation	Returns the correlation of two images.
$\mathfrak{F}$	Fourier	Returns the Fourier transform.
H	Hough	Returns the Hough transform.
P	Polar	Returns the polar coordinates.
P^{-1}	Inverse polar	Converts polar coordinates to rectilinear.
$\mathcal{P}_{\vec{v}}$	Radial polar	Returns the image as a radial-polar transformation.
$\mathcal{P}_{\vec{v}}^{-1}$	Inverse radial polar	Returns the image as the inverse radial-polar transformation.
$\mathfrak{P}_n$	PCA	Returns the data in a new space defined by the principal components.
$\mathcal{W}$	Wavelet decomposition	Returns the wavelet decomposition image.

The output space is defined by $y \in \mathbf{Y}$ as it is a different space than $\mathbf{X}$. In this case, $\mathbf{X}$ represents spatial coordinates and $\mathbf{Y}$ represents frequency coordinates.

Table 2.6 shows the Transformation operators. All of these will be explained in their first uses in other chapters.

2.2.7 EXPANSION OPERATORS

The Expansion operators expand the dimensionality of an image usually in an attempt to decrease the complexity of the information contained therein. These operators can increase the number of dimensions or more commonly create a set of images from a single image. One example is empirical mode decomposition (EMD) which creates several images isolating information in limited frequency bands. An image set is enclosed by curly braces with an index parameter. The EMD operator is $\mathfrak{E}$, and thus, the operator notation is

$$\{\mathbf{b}[\vec{x}]\} = \mathfrak{E}\mathbf{a}[\vec{x}].$$

In this example, the image set is $\{\mathbf{b}[\vec{x}]\} = \mathbf{b}_1[\vec{x}], \mathbf{b}_2[\vec{x}], \ldots, \mathbf{b}_N[\vec{x}]$ where N is the number of images generated by the operator.

Table 2.7 shows the Expansion operators.

Table 2.7
The Expansion Operators

Symbol	Name	Description
$\mathfrak{E}$	EMD	Returns a set of images from EMD decomposition.
I_n	ICM	Returns a set of pulse images from the ICM process.
$\mathcal{I}$	Isolation	Returns a set of images that isolate the contiguous regions in the input image.
J_n	PCNN	Returns a set of pulse images from the PCNN process.
T_n	Eigenimages	Returns a set of eigenimages.

2.3 COMBINATIONS AND REDUCED NOTATION

Most image processing tasks require several steps. Consider again the task of identifying the presence of the cane in the image in Figure 2.2(a). In the previous example, the cane was isolated but there was no automated detection. Here, the process is completed by detecting the presence of a long line using a Hough transformation. The protocol is

1. Convert the image to YIQ space,
2. Apply thresholds on the I and Q channels,
3. Combine the channels,
4. Apply a Hough transformation, and
5. Detect the presence of a peak.

The output of the Hough transform is $\mathbf{b}[\vec{y}]$ and the presence of a large peak indicates the presence of a strong line in the original image. Furthermore, the location of the peak indicates the location and orientation of the strong line in the input. Thus, there are three operations that are used. The first creates the image $\mathbf{b}[\vec{y}]$ while the other two extract the intensity of the peak and the location of the peak. The image $\mathbf{b}[\vec{x}]$ is generated by applying the thresholds to the I and Q channels as before and then applying the Hough transform:

$$\mathbf{b}[\vec{y}] = H \left\{ \begin{matrix} \varnothing \\ \Gamma_{18<23} \\ \Gamma_{2<8} \end{matrix} \right\} \mathcal{L}_{\mathrm{YIQ}} \mathbf{a}[\vec{x}]. \tag{2.24}$$

The detection of a peak is accomplished by finding the maximum value of $\mathbf{b}[\vec{x}]$ with the $\bigvee$ operator and comparing this to the threshold value of γ:

$$t = \left(\bigvee \mathbf{b}[\vec{y}] \right) > \gamma. \tag{2.25}$$

If the value of t is 1, then the peak had a sufficient height and then it becomes prudent to find the location of the peak as in

$$\vec{v} = A_{\vee} \mathbf{b}[\vec{y}], \tag{2.26}$$

where $A_{\vee}$ returns the location of the maximum value.

Curly braces on the outside of an operator indicates that the operator produced a set of images as in

$$\{F \mathbf{a}[\vec{x}]\}, \tag{2.27}$$

where F is the generic operator. The notation $F\{\mathbf{a}[\vec{x}]\}$ would imply that the operator was applied to the entire set rather than individual images.

The operator subscripts can get tedious and so a shorthand notation can be defined for each application. Consider a case in which the original image needs to be separated into several subimages. The Window operator is used to create these subimages, which requires at least two vectors to define the opposing corners of the bounding box and so the operator appears as $\square_{\vec{v}_1,\vec{v}_2}$, and a second bounding box would appear as $\square_{\vec{w}_1,\vec{w}_2}$. The user can shorten the notation by defining

$$\square_1 \equiv \square_{\vec{v}_1,\vec{v}_2}$$

and

$$\square_2 \equiv \square_{\vec{w}_1,\vec{w}_2}.$$

Thus, repeated uses of these operators will appear cleaner. An example case is to create a single image which is the concatenation of the same region extracted from three original images. The operator notation for this protocol is

$$\mathbf{b}[\vec{x}] = \mathcal{C}_H \left\{ \square_1 \mathbf{a}_1[\vec{x}], \square_1 \mathbf{a}_2[\vec{x}], \square_1 \mathbf{a}_3[\vec{x}] \right\}. \tag{2.28}$$

Using the notation that applies the same operator to all images in a set, this notation can be shortened to

$$\mathbf{b}[\vec{x}] = \mathcal{C}_H \left\{ \square_1 \mathbf{a}[\vec{x}] \right\}. \tag{2.29}$$

An operator that is applied multiple times is denoted by a superscript. Thus,

$$\mathbf{b}[\vec{x}] = R_\theta^n \mathbf{a}[\vec{x}] \tag{2.30}$$

indicates that the Rotation operator of an angle θ is applied n times. If the subscript were a vector (e.g., $\vec{\theta}$), then successive uses of the operator would sequentially use the elements in the vector. For example, if $\vec{\theta} = (20, 10, 15)$, then $R_{\vec{\theta}}^3$ would apply rotations of 20, 10, and 15 degrees in succession to the image. Repetition of a sequence of operators requires the use of parenthesis to group the iteration, as in

$$\mathbf{b}[\vec{x}] = (AB)^n \mathbf{a}[\vec{x}], \tag{2.31}$$

where A and B are generic operators.

2.4 SUMMARY

Operators are an efficient method of accurately depicting common image processing protocols. These operators are organized into several operational categories. Creation operators are used to generate simple image frames. The Channel operators isolate color channels, and the Informational operators report information about an image without alteration. There are four classes that create new images from operations performed on input images or sets of images. The Intensity operators are designed to alter the intensity but not the shape of the content. The Geometric operators are just the opposite and are designed to move pixels to new locations but not to alter the intensities of the content. The Transformation operators rearrange the information into a very different coordinate system. Finally, the Expansion operators create a set of images from a single image in an attempt to simplify the information within an image.

The purpose of the operator notation is to mimic the scripting implementation of the protocol. In an ideal situation, each operator should warrant only a single line of Python script. As this is accomplished, the conversion between theory and implementation becomes more amenable.

PROBLEMS

1. Given $\vec{w} = (200, 200)$, $\vec{v}_1 = (10, 100)$ and $\vec{v}_2 = (50, 150)$, how many pixels are set to 1 in

$$\mathbf{a}[\vec{x}] = \mathbf{r}_{\vec{w},\vec{v}_1,\vec{v}_2}$$

2. What is the value of t in,
$$t = \sum_{\vec{x}} \mathbf{o}_{\vec{w},\vec{v},r}$$
where $\vec{w} = (200,200)$, $\vec{v} = (100,100)$ and $r = 50$?
3. What is the value of t in,
$$t = \sum_{\vec{x}} \mathbf{o}_{\vec{w},\vec{v},r}$$
where $\vec{w} = (200,200)$, $\vec{v} = (100,100)$ and $r = 150$?
4. Write the notation for making a solid annular ring with an inner radius of 60 and an outer radius of 75. This ring should be centered in a frame of size 200×200.
5. Write the operator notation for the absolute value of the difference between the values in the red and green channels of an RGB image $\mathbf{a}[\vec{x}]$.
6. Given an image as shown. The background pixels have a value of 0 and the target pixels (inside the small boxes) the pixels have a value of 1. What does the following notation return?
$$\{\vec{v}_i\} = \{\boxtimes I \mathbf{a}[\vec{x}]\}$$

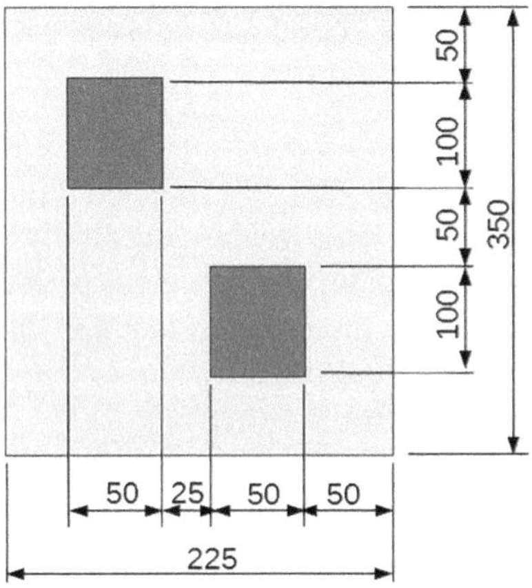

7. Is the following statement true (since there is no vector specified for the rotation the center is assumed to be the center of the frame)?
$$\mathbf{a}[\vec{x}] = \mathcal{S}_{-\vec{v}} \mathcal{R}_{-\theta} \mathcal{S}_{\vec{v}} \mathcal{R}_{\theta} \mathbf{a}[\vec{x}]$$

3 Scripting in Python

Multiple scripting and programming languages exist that are suitable for image processing. Python is the language selected for this text because it has powerful, efficient tools, it is high demand by employers, and it is free. The rest of this chapter reviews elements of Python that are useful for the subsequent chapters; however, by no means is this chapter intended on being a comprehensive description of Python. Readers are highly encouraged to collect and frequently consult Python references.

3.1 BASIC PYTHON SKILLS

This section will review Python skills and in some cases present the corresponding image operator notation. Readers who are familiar with Python may wish to scan sections to see the relationships between Python and the image operator notation.

3.1.1 VARIABLES

Python provides standard variables such as integers and floats, but it also offers a complex variable. Code 3.1 shows an example of each. Line 1 creates an integer, line 4 creates a float, and line 5 creates a complex number. Lines 6 and 8 extract the real and imaginary parts of the variable. Line 2 uses the **type** function to return the data type of variable a. As seen in line 3, the variable is an integer.

Python obeys the algebraic hierarchy. The order of operations is as follows:

1. ( ) : Parenthesis
2. ** : Power
3. * / % : Multiplication, division, modulus
4. + − : Addition and subtraction

Math functions use standard notation, similar to other languages. Code 3.2 shows commands to perform some of these. Line 1 is addition, line 3 is subtraction, and line 5 is multiplication. Line 7 shows the command for division, and even though both the dividend and the divisor are integers, this function returns a float answer. This is different from the behavior of Python 2.7 and earlier which returned an integer. In Python 3, the integer division is performed by two slashes as seen in line 7. The remainder is returned by line 9. Line 13 uses the **divmod** function performs both of the operations in lines 7 and 9.

Code 3.1 Creating an integer

```
1 >>> a = 63426
2 >>> type(a)
3 <class 'int'>
4 >>> c = 63426.
5 >>> b = 23.3 + 9.342j
6 >>> b.imag
7 9.342
8 >>> b.real
9 23.3
```

Code 3.2 Simple math functions

```
>>> 5+6
11
>>> 6-5
1
>>> 4*5
20
>>> 9/4
2.25
>>> 9 // 4
2
>>> 9%4
1
>>> divmod( 9, 4 )
(2, 1)
```

Code 3.3 Type casting

```
1 >>> xyz = 6.5
2 >>> abc = int(xyz)
3 >>> abc
4 6
5 >>> abc + 0.
6 6.0
```

Type casting converts a variable from one data type to another. Casting is performed through functions such as **int**, **float**, and **complex**. Line 1 in Code 3.3 creates a float and line 2 converts that to an integer. In doing so, the decimal portion of the value is removed as shown in line 4. Line 5 may look a bit odd but basically it is adding an integer to a float and thus the answer is a float without changing the numerical value.

3.1.2 STRINGS

Two strings are created as shown in either of the first two lines in Code 3.4. There is no distinction between the use of single or double quotes with the single caveat that they cannot be mixed. Thus, lines 1 and 2 perform the same assignment. Two strings can be easily combined to create a larger string using the plus sign as shown in line 3.

Code 3.4 Creating strings

```
1 >>> astr = 'hello'
2 >>> bstr = "howdy"
3 >>> cstr = astr + bstr
4 >>> cstr
5 hellohowdy
```

3.1.3 TYPE CONVERSIONS WITH STRINGS

Strings consisting of only numerical characters can be converted to numerical data types and numerical data can be converted to a string. Several examples are shown in Code 3.5. Line 1 creates a string which has the characters nine, point, and five. This is not a number but a text that looks like a number. So, it is not possible to apply math operations on `astr`. Line 2 converts this string to a floating point numeral which can be used in mathematical operations. Line 3 uses the **str** function to convert a numeral into a string. Line 2 uses the **float** function but there are equivalents for all data types. As seen in line 4, a string is converted to an integer with the **int** function. However, the string represents an integer and not a float. The conversion of a string that represents a decimal value must first be converted to a float and then to an integer. The **int** function will remove the decimal value.

Consider a case in which it is necessary to store the data into many different files with incrementing file names such as *file1.txt*, *file2.txt*, and *file3.txt*. Before the file can be stored, the filename must be created which is a concatenation of a string, an integer, and a string. An example is shown in Code 3.6, where the integer is defined in line 1, and line 2 builds the filename. Code 3.25, in a later section, will demonstrate the construction of sequential filenames using the **str** function.

3.2 TUPLES, LIST, DICTIONARIES, AND SETS

One of the most convenient aspects of Python is its ability to collect data of varying types. There are four main methods in which data is coalesced into a collection with later versions of Python offering even more convenient management functions. For this text, only the four main collections are reviewed.

3.2.1 TUPLE

A *tuple* is a collection of data of varying types which is not to be altered once established. The data within the tuple can be of any type such as integer, float, and strings. In fact, a tuple can contain other tuples, lists, dictionaries, sets, and any user-defined data type. A tuple is surrounded by parenthesis as shown in Code 3.7, which defines the variable `atup` as a tuple with an integer, a float, and a string.

3.2.2 SLICING

Slicing is the ability to extract part of the entity, or when allowed, to change to part of the entity. For example, a task may need to use only the first item in a tuple, and thus access to that single item is

Code 3.5 Type conversions

```
1  >>> astr = "9.5"
2  >>> f = float( astr )
3  >>> bstr = str( f )
4  >>> d = int( "9" )
5  >>> d
6  9
7  >>> g = int(float(astr))
```

Code 3.6 Building a name

```
1  >>> n = 1
2  >>> fname = 'file' + str(n) + '.txt'
```

Code 3.7 Tuple

```
>>> atup = ( 5, 5.6, 'a string' )
```

needed. Python uses square brackets to access an item in a tuple, list, or other collection. Like C and Java, Python starts indexes at 0 instead of 1. Code 3.8 shows the retrieval of individual components of the `atup` tuple. Lines 1, 3, and 5 retrieve the first three items in the tuple. Line 7 uses a -1 as the index which is the method of extracting the last item in the tuple. Using a -2 would retrieve the next to the last item.

It is possible to retrieve multiple elements with slicing elements. Consider the examples in Code 3.9, which starts with the creation of a tuple in line 1. In line 2, two arguments are given to indicate which elements are to be obtained. This particular notation includes the first index but excludes the second index. Thus, the return is the items at locations 3, 4, 5, and 6, but not 7. There are some logical reasons for this. First, the difference between the values of the given indexes is $7-3=4$, and four elements are returned.

The second logical reason is shown in lines 4 and 6 which extract the first part of the tuple and the last part of the tuple. There are two items of note in these lines. To extract the first seven elements, the notation could be `btup[0:7]`, but the first index is optional. Line 4 extracts the elements from the beginning (or the 0 index) up to index 7. Line 6 extracts the elements from index 7 to the end. Note that both lines 4 and 6 use 7 as the index, but only line 7 contains `btup[7]`.

Line 8 shows a method of reversing the data. This starts at element 12 and goes down to, but not including, element 0. The third argument is the step size, and since it is -1, this extraction is reversed. Line 10 shows the extraction from end, to beginning, in reverse order.

Code 3.8 Extraction

```
1 >>> atup[0]
2 5
3 >>> atup[1]
4 5.6
5 >>> atup[2]
6 'a string'
7 >>> atup[-1]
8 'a string'
```

Code 3.9 Slicing selected elements

```
1  >>> btup = (1,2,3,4,3,21,2,3,4,3,2,1)
2  >>> btup[3:7]
3  (4, 3, 21, 2)
4  >>> btup[:7]
5  (1, 2, 3, 4, 3, 21, 2)
6  >>> btup[7:]
7  (3, 4, 3, 2, 1)
8  >>> btup[12:0:-1]
9  (1, 2, 3, 4, 3, 2, 21, 3, 4, 3, 2)
10 >>> btup[::-1]
11 (1, 2, 3, 4, 3, 2, 21, 3, 4, 3, 2, 1)
```

Slicing methods are applied to tuples, lists, strings, and even user-defined collections. The rules are the same for each application.

3.2.3 LISTS

A *list* is similar to a tuple in that it is a collection of any type of data. However, a list is changeable. Code 3.10 shows some simple list manipulations. Line 1 creates a list with three elements by enclosing them in square brackets. Line 2 appends an integer at the end of the list as shown in line 4. This is extremely useful if the number of elements that are to be collected is not known. The list simply collects the items, and the size of the list grows as necessary.

Line 5 inserts a string into position 1. Line 8 removes the item at position 1 and places that into variable `b`. Line 13 removes the first instance of the data that has a value of 78. This is not the index but the actual data in the argument. This is just a sampling of the commands that are available with a list.

3.2.4 DICTIONARIES

A *dictionary* is also a collection of data, storing data in a manner that facilitates efficient recall. There are two components to a dictionary. The first part is the *key* which replaces the idea of an index. The second part is the *value* which is the associated data to the key. A word dictionary is an excellent example to demonstrate these concepts, where the key is the word being looked up, and the value is the definitions that are associated with that word. Much like a word dictionary, a Python dictionary can only search on the key.

The advantage of a dictionary is that it provides a significantly faster search than do tuples or lists. So, if the data set is large, the use of dictionaries can provide a significant speed advantage.

Line 1 in Code 3.11 creates an empty dictionary using curly braces. Line 2 establishes the first entry which has an index of 0 and a value which is a string ‘`a string`’. The fact that the first key is 0 is purely coincidental as shown in line 3 where the second entry uses a key of 5. In fact, keys do not have to be integers but can be many different types including a tuple. Line 4 creates an entry in which the key is a string and the value is a tuple which contains two integers and a list.

Code 3.10 Using a list

```
1  >>> alist = [1, 5.4, 'hello']
2  >>> alist.append( 78)
3  >>> alist
4  [1, 5.4, 'hello', 78]
5  >>> alist.insert( 1, 'here' )
6  >>> alist
7  [1, 'here', 5.4, 'hello', 78]
8  >>> b = alist.pop( 1 )
9  >>> b
10 'here'
11 >>> alist
12 [1, 5.4, 'hello', 78]
13 >>> b = alist.remove( 78 )
14 >>> alist
15 [1, 5.4, 'hello']
```

Code 3.11 Using a dictionary

```
1 >>> dct = { }
2 >>> dct[0] = 'a string'
3 >>> dct[5] = 'more data'
4 >>> dct['a'] = (5, 7, [8.9, 3.4] )
5 >>> dct
6 {0: 'a string', 'a': (5, 7, [8.9, 3.4]), 5: 'more data'}
7 >>> dct[0]
8 'a string'
9 >>> dct[5]
10 'more data'
```

To optimize searching, Python will order a dictionary as it sees fit. As shown in line 6, the three items in the dictionary are not in the same order as entered. A dictionary is only searched via its keys. Lines 7 and 9 extract two items in the dictionary by their keys, not by their position.

There are many functions that accompany a dictionary and three of the most common are shown in Code 3.12. Line 1 uses the **keys** function, which returns all of the keys in the dictionary. Likewise, line 3 returns all of the values. Both of these functions returned objects which can be converted to lists by applying the **list** function. Line 5 determines if a key is inside a dictionary. In this case, the query is to determine if the integer 11 is a key, and the result indicates that it is not.

3.2.5 SETS

A *set* is a collection of data that also has the standard functions that accompany a set such as union or intersection. Code 3.13 displays a few set functions with line 1 creating a set from a tuple. Line 4 attempts to create a set with duplicate data. One very useful property of a set is that it does not allow duplicates. Removal of duplicates from a list is performed by converting the list to a set and then back to a list. The conversion to a set removes the duplicates and the conversion back to a list returns the data to the original collection type. Three other functions are shown in lines 10, 12, and 14.

3.3 FLOW CONTROL

This section displays a few of the methods in which the flow of the program can be controlled. These commands are standard in procedural languages.

Code 3.12 Some dictionary functions

```
1 >>> dct.keys()
2 dict_keys([0, 5, 'a'])
3 >>> dct.values()
4 dict_values(['a string', 'more data', (5, 7, [8.9, 3.4])])
5 >>> 11 in dct
6 False
```

Code 3.13 Some set functions

```
>>> aset = set( (0,3,4,5))
>>> aset
{0, 3, 4, 5}
>>> aset = set( (0,3,4,5,5))
>>> aset
{0, 3, 4, 5}
>>> bset = set( (4,5,0,1,2))
>>> aset.union(bset)
{0, 1, 2, 3, 4, 5}
>>> aset.intersection( bset )
{0, 4, 5}
>>> aset.difference( bset )
{3}
>>> bset.difference( aset )
{1, 2}
```

3.3.1 THE IF COMMAND

The `if` command is used to make decisions and follows similar logical expressions as other languages but with its own unique syntax. The first example is shown in Code 3.14 where the value of `x` is set to 5. In line 2, the value of `x` is compared to a value of 3, and if the statement is true, then the next line is executed which does occur as seen in line 5.

Unlike other languages, Python does not use curly braces or BEGIN and END statements to group the commands that are executed. Instead, Python uses a colon and indentations. Line 2 ends with a colon, and line 3 is indented, and therefore belongs to the `if` statement. Another example is shown in Code 3.15, which shows two lines indented after the `if` statement. Both of these are executed because the statement is true.

A problem does occur if the user mixes different types of indentations. In this case, both indentations are achieved with space characters and not a TAB character. Some editors will use a TAB

Code 3.14 A simple `if` statement

```
1 >>> x = 5
2 >>> if x > 3:
3         print( 'Hi' )
4 
5 Hi
```

Code 3.15 A multiple line `if` statement

```
>>> if x > 3:
        print( 'hi' )
        print( x + 4 )

hi
9
```

instead of spaces. As long as all of the lines within the `if` statement are indented in the same manner, Python will function. It is not possible, however, to mix the indentations. So, it is not possible to use white spaces to indent line 2 and a TAB to indent line 3 in this case. This problem usually occurs when the code is edited with different editors.

Like other languages Python offers and `if-else` combination as shown in Code 3.16. The `else` statement is not indented but it does end with a colon. Line 2 is executed if line 1 is true, and line 4 is executed if line 1 is false.

Python also offers an else-if statement with the keyword `elif` as shown in Code 3.17. If line 1 is false, then line 3 will be considered. If line 3 is also false, then line 6 will be executed.

Compound statements use the keywords `and`, `or`, and `not`. The example shown in Code 3.18 shows the case where two conditions must be true before line 2 is executed. Line 6 shows a shorter manner in which the same logic is achieved.

3.3.2 THE WHILE COMMAND

The `while` loop is the first of two control loops that Python uses. Line 2 shows the beginning of the statement with the condition $x < 5$. As long as a condition is true, the commands within the `while` loop are executed. Code 3.19 shows a simple loop which executes lines 3 and 4, while the condition

Code 3.16 The `if-else` statement

```
1 >>> if x < 3:
2         print( 'hi' )
3 else:
4         print( 'lo' )
5
6 lo
```

Code 3.17 The `elif` statement

```
1 >>> if x < 3:
2         print( 'hi' )
3 elif x == 5:
4         print( 'lo' )
5 else:
6         print( 'nada' )
7
8 lo
```

Code 3.18 A compound `if` statement

```
1 >>> if x > 4 and x < 6:
2         print( 'yes' )
3
4 yes
5
6 >>> if 4 < x < 6:
7         print( 'cool' )
8
9 cool
```

Code 3.19 A `while` statement

```
1 >>> x = 0
2 >>> while x < 5:
3         print(x,end='')
4         x += 1
5
6 01234
```

in line 2 is true. Line 3 prints out the values of `x`, and the `end=''` indicates that the print statements should not include a newline.

3.3.3 BREAK AND CONTINUE

Two commands that can alter the flow of the program are `break` and `continue`. The `break` command can be used to exit from a loop prematurely as shown in Code 3.20. In this script, the loop should print line 3 six times. However, line 4 catches the loop when `x = 4` and prints line 5. Line 6 contains the `break` statement and this terminates the `while` loop immediately. As seen in the print, the value of `x` never becomes 5.

The `continue` statement is similar except that instead of terminating the rest of the loop, it only terminates the current iteration. In Code 3.21, the loop is again scheduled to run for six iterations, and line 5 again catches the progress when `x = 4`. The `continue` statement terminates the rest of the iteration, and as seen, line 8 is not executed for this iteration. However, the looping process does continue and considers the case of `x = 5`.

3.3.4 THE FOR LOOP

The `for` loop is similar to the `while` loop except that the iterator takes on values from a collection. Consider the example in Code 3.22 which shows the creation of a tuple in line 1. Line 2 establishes the `for` loop, and the iterator (the variable `i`) will progress through the values in the tuple as the iterations progress.

Code 3.20 Usgin `break` command

```
1  >>> x=0
2  >>> while x < 6:
3           print( 'a', x )
4           if x == 4:
5                   print( 'b' )
6                   break
7           x += 1
8
9  a 0
10 a 1
11 a 2
12 a 3
13 a 4
14 b
```

Code 3.21 Using the `continue` command

```
>>> x = 0
>>> while x < 6:
        print( 'a', x, end='')
        x += 1
        if x == 4:
                print( 'b' )
                continue
        print( 'c' )

a 0c
a 1c
a 2c
a 3b
a 4c
a 5c
```

Code 3.22 The `for` loop

```
>>> atup = (3,5,7)
>>> for i in atup:
        print( i,'', end='' )

3 5 7
```

In other languages, the iterator is often an integer, and the `for` loop instruction provides a starting value, an ending value, and a step size to the increment. Python uses the **range** to accomplish the same effect. Three examples of the range function are shown in Code 3.23. Line 1 creates a range of values from 0 up to 5 with a single argument in the **range** function. The result is converted to a list so that the values can be examined. Similar to the slicing rules, the **range** function considers value that includes the first number but excludes the second. Line 3 accepts two arguments with the first being the included start value and the second being the excluded stop value. Line 5 accepts three arguments with the first two being the start and stop values, and the third being the step size.

Thus, the **range** function can create the iteration object as the `for` loop is created as shown in Code 3.24.

Code 3.23 Using the `range` command

```
1 >>> list(range(5))
2 [0, 1, 2, 3, 4]
3 >>> list(range(3,5))
4 [3, 4]
5 >>> list(range(1,10,2))
6 [1, 3, 5, 7, 9]
```

Code 3.24 Using the `range` command in a `for` loop

```
>>> for i in range( 4 ):
        print( i,end=' ')

0 1 2 3
```

3.3.5 THE MAP AND LAMBDA FUNCTIONS

The combination of the **map** and **lambda** functions can replace a simple `for` loop containing a single line. The **lambda** command creates a simple function, and the **map** command provides the iterations. An example shown in Code 3.25 starts with the creation of a small list named `nlist`. Line 2 uses the **map lambda** combination to perform a string concatenation using each item in the list. The result is converted to a list and shown in line 4.

3.3.6 IMAGE OPERATORS AND CONTROL

Mathematically, an `if` statement is described by a case statement. For example, Code 3.26 presents two options which are described as

$$\mathbf{b}[\vec{x}] = \begin{cases} F_x \mathbf{a}[\vec{x}] & \texttt{if } x < 3 \\ F_{-x} \mathbf{a}[\vec{x}] & \texttt{otherwise} \end{cases}, \tag{3.1}$$

where $F_m \mathbf{a}[\vec{x}]$ refers to a generic Python function `Function(a,m)`.

Looping that involves the same operation applied to multiple images is described in operator notation by

$$\{\mathbf{b}[\vec{x}]\} = \{F\mathbf{a}[\vec{x}]\}. \tag{3.2}$$

The braces represent a set of images. The function F is inside the braces, and therefore, it is applied to each image. The output is also a set of images. The Python code is shown in Code 3.27 where both `amgs` and `bmgs` represent a list of images.

An operation that is applied to the whole set of images will have the operator on the outside of the braces, as in

$$\mathbf{b}[\vec{x}] = F\{\mathbf{a}[\vec{x}]\}. \tag{3.3}$$

In this case, the function is called only once, and the input is a set of images.

Code 3.25 Using the **map** and **lambda** functions

```
1 >>> nlist = [4,3,5]
2 >>> blist = list(map(lambda x: 'file'+str(x)+'.png', nlist ))
3 >>> blist
4 ['file4.png', 'file3.png', 'file5.png']
5 >>>
```

Code 3.26 Choosing parameters in a function

```
>>> if x < 3:
        bmg = Function( amg, x )
else:
        bmg = Function( amg, -x)
```

Code 3.27 Choosing parameters in a function

```
>>> bmgs = []
>>> for i in range( N ):
        bmgs.append( Function( amgs[i] )
```

3.4 INPUT AND OUTPUT

Python can easily read and write files to the hard drive. Files can be either text format or binary. The former saves data as text characters which editors and word processors can readily read. The binary format tends to be more compact, but also sensitive to the computer's operating system.

3.4.1 READING AND WRITING TEXT FILES

Code 3.28 shows a simple method to writing text data to a file. Line 1 creates the data that is to be written, and line 2 opens the file and assigns the file to the file pointer `fp`. The second argument 'w' indicates that this file is opened for writing. Line 3 writes the information and line 4 closes the file. This creates a simple text file that can be read by any text editor. If a file named *myfile.txt* already exists then line 2 will destroy it. There is no recovery and no forgiveness. Line 2 will destroy the existing file of the same name without remorse.

Reading the information from a text file is just as easy as shown in Code 3.29. Line 1 opens the file, but the second argument is missing. The default mode for the **open** function is to only read data. The absence of a second argument reverts the function to this mode. Line 2 reads the data and puts into a string named `bstring`, and line 3 closes the file.

The codes as shown read and write only text data. This method does not write other types of data such as integers and floats. Instead these need to be converted to a string using the **str** command before writing. When the data is read it is in a string and needs to be converted to the data type using commands such as **int** or **float**.

The process of reading and writing binary files is slightly different. The argument 'wb' is used to create a file for writing binary data, and the argument 'rb' is used to read from a binary file. In this text, the only use of reading and writing binary files is in the pickling process which is in the next section.

3.4.2 PICKLING FILES

The `pickle` module provides a very convenient method of quickly storing multiple types of data. The steps are shown in Code 3.30 where line 2 opens a standard file for writing as a binary file.

Code 3.28 Writing to a file

```
1 >>> astring = 'Write me to the file.'
2 >>> fp = open( 'myfile.txt', 'w' )
3 >>> fp.write( astring )
4 >>> fp.close()
```

Code 3.29 Reading to a file

```
1 >>> fp = open( 'myfile.txt' )
2 >>> bstring = fp.read()
3 >>> fp.close()
```

Code 3.30 Pickling a file

```
1 >>> import pickle
2 >>> fp = file( 'myfile.pickle', 'wb' )
3 >>> pickle.dump( data, fp )
4 >>> fp.close()
```

Code 3.31 Reading a pickle file

```
1 >>> fp = file( 'myfile.pickle', 'rb' )
2 >>> data = pickle.load( fp )
3 >>> fp.close()
```

Code 3.32 Reading a pickle file

```
>>> fp = file( 'myfile.txt' )
>>> data = pickle.load( fp, protocol=2)
>>> fp.close()
```

Line 3 uses the `pickle.dump()` function to store data into this file. The variable data can be almost anything including any concoction of lists, sets, tuples, variables, and dictionaries. Retrieving data from a pickled file is just as easy as shown in Code 3.31. The file is opened for reading in line 1 and the data is read in line 2.

There are significant changes in the Pickle module from Python 2.7 to Python 3. The default mode in Python 2.7 stored data as text, but the default mode for Python 3 is to store items as binary data. The protocols are assigned numbers with the first two numbers being associated with much older versions of Python. Code 3.32 shows the method by which Python 3 can read Python 2 pickle files by setting the protocol to 2.

The operator notation for reading an image file is

$$\mathbf{a}[\vec{x}] = Y(\texttt{Filename}). \tag{3.4}$$

There are several methods to create image data; thus, this operator exists to indicate that an image is retrieved from the disk. Once an image is created, saving the information to disk can be assumed, and thus, no operator exists to depict the process of saving an image.

3.5 DEFINING FUNCTIONS

A function is a collection of Python statements that can be invoked by calling the assigned function name. They are very useful in cases where the same sequence of commands are applied to multiple images or data.

3.5.1 FUNCTION COMPONENTS

Code 3.33 displays a simple function which starts with the keyword `def`. This is followed by the name of the function and optional input arguments. The parenthesis are required even if there are no input arguments. The definition is followed by a colon and then indentations for commands that are inside the function.

In this case, lines 2 through 4 are the commands inside the function. The definition of the function ends with line 4. The function is called in line 6 with the result returned in lines 7 and 8.

Code 3.33 Defining a function

```
>>> def MyFunction( indata ):
        print( indata )
        x = indata + 5
        print (x)

>>> MyFunction( 8  )
8
13
```

3.5.2 RETURNS

A function returns a value (or values) through the use of the `return` statement as shown in line 3 of Code 3.34. The function is called in line 5 and the variable oup becomes the value that was returned in line 3.

Languages like C and Java can only return a single variable. Technically, the same is true for Python, but that variable can be a tuple which itself contains multiple items. The function defined in Code 3.35 returns a single item which is a tuple that contains an integer, a float, and a string. Line 4 calls this function and as seen in lines 5 and 6, the return type is a tuple. Functions in Python can also return other types of collections like a list or dictionary.

Another call to the function is shown in line 7, where both sides of the equal signs are tuples. This call to the function works only if both sides have the same number of items. The items on the left side are assigned the values from the right side. Lines 8 and 9 show that the variable b becomes the float value. In this manner, Python behaves as though it is a function returning multiple variables which is a feature of the language that is unique and very useful.

Code 3.34 Return a value from a function

```
1 >>> def BFunction( inp ):
2         a = inp - 9
3         return a
4
5 >>> oup = BFunction( 17 )
6 >>> oup
7 8
```

Code 3.35 Return a tuple from a function

```
1 >>> def CFunction( ):
2         return 4, 5.6, 'string'
3
4 >>> answ = CFunction()
5 >>> type( answ )
6 <type 'tuple'>
7 >>> a,b,c = CFunction()
8 >>> b
9 5.6
```

3.5.3 DEFAULT ARGUMENTS

A function can receive multiple arguments which are separated by commas. Python also has the ability to receive default arguments as shown in Code 3.36. The function is defined in line 1 and receives four arguments. The last two have default definitions which are used unless the user overrides them. Line 4 calls the function with two arguments which are required. Line 2 will print the values for all four variables. This shows that `x` and `y` were defined by the input and `z` and `w` used the default values. Line 6 calls the function using a third argument which is changes the value of `z`. Line 8 calls the function with the user defining all four variables. The user can also specify that only a few default arguments are altered as shown in lines 10 and 12.

Default arguments have already shown in previous examples. Consider again Code 3.23 which displays the **range** function with differing number of arguments. These examples show that the **range** function must receive one input and can receive up to three inputs. Thus, there are two default inputs.

3.5.4 FUNCTION HELP

Adding lines of helpful comments to functions is achieved by surrounding the help text with triple-double quotes as shown in Code 3.37. When the left parenthesis is typed, as seen in Figure 3.1, then a balloon pops up with the help strings. In this case, it shows just the first two lines because line 3 is empty. All else after the empty line is not shown. However, the `help(DFunction)` command will print all of the strings to the console as shown in Code 3.38.

Code 3.36 Default arguments

```
1  >>> def FFunction( x, y, z = 5, w = 9 ):
2          print( x,y,z,w )
3
4  >>> FFunction( 1, 2 )
5  1 2 5 9
6  >>> FFunction( 1, 2, 3 )
7  1 2 3 9
8  >>> FFunction( 1, 2, 3, 4 )
9  1 2 3 4
10 >>> FFunction( 1, 2, z = 0 )
11 1 2 0 9
12 >>> FFunction( 1, 2, w = 0 )
13 1 2 5 0
```

Code 3.37 Function help

```
1  >>> def DFunction( inp ):
2          """inp = input
3          DFunction does very little
4          return: inp plus a little extra"""
5          return inp + 5
```

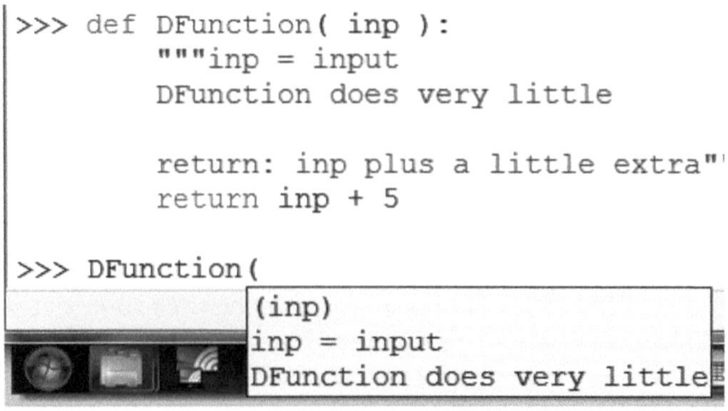

Figure 3.1 The help balloon.

Code 3.38 Showing help

```
>>> help(DFunction)
Help on function DFunction in module __main__:

DFunction(inp)
    inp = input
    DFunction does very little

    return: inp plus a little extra
```

3.6 MODULES

A module in Python is basically a file that contains definitions of variables, functions, and commands. The file can be stored in the directory of the user's choice. However, to be imported, Python must have this directory in its search path. Recall line 3 Code 1.2 which adds a directory to the system path which allows Python to search that directory for modules to import.

One simple way of to stay organized is to create a working directory for a project and inside that create a subdirectory to contain Python modules that are created by the user. When Python is started, the user then adds that subdirectory to the search path so that user written modules can be imported. An example module is shown in Figure 3.2 which is simply a file that contains comments, a command, and a function. In this example, the user moves Python to the *C:/science/courses/DBase* directory using the commands in lines 1 and 2 in Code 3.39. Line 3 then adds the subdirectory to the search path, and any Python module in that subdirectory can now be accessed.

Once a module is stored, it can be loaded into Python through several methods. The first is to use the `import` command which is shown in line 1 of Code 3.39. When this command is invoked, a *.pyc* file is created a subdirectory. This is pseudo-compiled Python code and is platform independent. If the user imports a module and then decides to change the module, it will have to be reloaded as shown in Code 3.40. The module is imported in line 1, and the variable `avar` is accessed in line 2.

Code 3.39 Initial commands

```
1 >>> import os, sys
2 >>> os.chdir( 'C:/science/courses/DBase')
3 >>> sys.path.append( 'pysrc' )
```

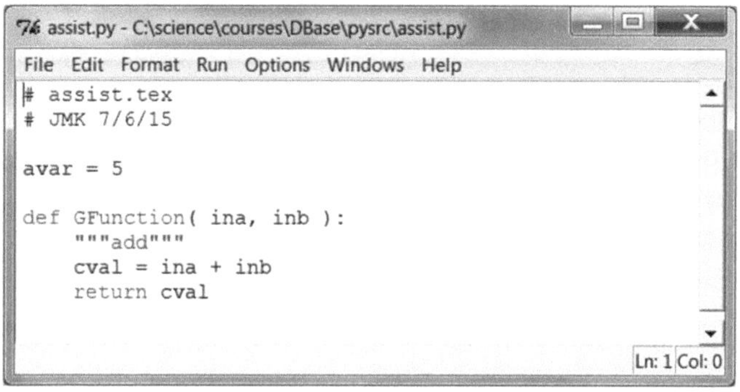

Figure 3.2 A module.

Code 3.40 Reading a module

```
1 >>> import assist
2 >>> assist.avar
3 5
4 >>> assist.Gfunction( 5,6 )
5 11
6 # user modifies assist.py
7 >>> import importlib
8 >>> importlib.reload( assist )
```

At line 6, the user modifies the contents of the module *assist.py*, and in order to use, these changes the module is reloaded in line 8.

Modules can adopt a new name as shown in line 1 of Code 3.41. This is convenience for the programmer. The name of `assist` is shortened to `at` and subsequent use is shown in line 2.

Using the import command requires that the user type in the module name each time a component is accessed. Another method of importing a module is to use the `from-import` protocol as shown in Code 3.42. Line 1 shows the command that imports everything from the module *assist.* Line 2 shows a method by which only specified components of the module are imported. When using one of these components, the module name is no longer required as shown in line 3. There are two caveats to usage. The first is that the module cannot be reloaded if it is modified. The second is that if there is a previous variable or function named `avar`, then this import would have overwritten this

Code 3.41 Shortcut name

```
1 >>> import assist as at
2 >>> at.avar
3 5
```

Code 3.42 From import

```
1 >>> from assist import *
2 >>> from assist import avar
3 >>> avar
4 5
```

Code 3.43 Executing commands in either version of Python

```
1 >>> execfile('pysrc/assist.py')              # Python 2.7
2 >>> exec(open('pysrc/assist.py').read()) # Python 3.x
```

previous definition. Perhaps, the user would be aware of another variable using the same name, and this simple awareness could prevent this problem from occurring. Line 1, however, loads everything from the module, and in this action, the programmer does not readily see the new definitions that are being loaded. Now, it is easy to destroy a previously defined variable or function. So, this method should be used with caution.

The third method is to execute the commands in a file. Line 1 in Code 3.43 shows a Python 2.7 command that basically runs all of the instructions in the file through the useful command `execfile`. Such convenience did not survive the evolution of Python 3, and so the set of commands in line 2 are used to achieve the same effect. If the user modifies the file, then this line can be called again to use the modified scripts. It is important to note that this command does not use `sys.path`, and therefore, the path structure must be included in the filename.

3.7 ERRORS

The construction of programs is rarely achieved without encountering code with errors. Common errors include syntax errors and execution errors (such as divide by 0), which are shown in Code 3.44. The Traceback shows the command and a statement as to which type of error that it is.

The Traceback provides the steps from the last command to the error as shown in Code 3.45. This uses the **assist.Gfunction()** defined in Figure 3.2, which works if the input arguments are two numerals or two strings. It does not work if the inputs are one number and one string as attempted in line 1. Line 4 shows that the error trace starts with the input from the command line, and line 6 indicates that the problem occurred in line 8 of the file *assist.py*. Line 7 shows the command, and line 8 indicates the nature of the error. The IDLE editor will allow users to right-click on the error and it will open the file and place the cursor on that line.

Code 3.44 Divide by 0 error

```
>>> 4/0
Traceback (most recent call last):
  File "<pyshell#0>", line 1, in <module>
    4/0
ZeroDivisionError: division by zero
```

Code 3.45 Traceback through a module

```
1 >>> assist.GFunction( 5, 'string' )
2
3 Traceback (most recent call last):
4   File "<pyshell#7>", line 1, in <module>
5     assist.GFunction( 5, 'string' )
6   File "pysrc\assist.py", line 8, in GFunction
7     cval = ina + inb
8 TypeError: unsupported operand type(s) for +: 'int' and 'str'
```

Code 3.46 Try-except

```
1 >>> try:
2         4/0
3 except ZeroDivisionError:
4         print( 'oops.  my bad')
5
6 oops.  my bad
```

An error is fatal and the Python program stops when an error is encountered. It is possible to catch an error and allow the Python script to continue using the `try-except` combination of commands. Code 3.46 shows a simple case in which the command on line 2 is attempted, and if a `ZeroDivisionError` occurs, then line 4 is executed. The most important aspect is that the Program is not terminated. Line 4 can be replaced with multiple command lines, and all types of errors can be used in line 3. Python even allows the user to define new errors, but that is far beyond the scope or necessity of this text.

3.8 NUMPY

Processing large collections of numerical values in Python can be very slow since it is an interpreted language. The *NumPy* package offers tools to create large arrays and to mathematically manipulate them with very good efficiency. In this text book, one dimensional arrays are called vectors, two dimensional arrays are called matrices, and higher dimensional arrays are called tensors. In Python nomenclature, all of these are called arrays.

3.8.1 CREATING ARRAYS

There are several different manners in which arrays can be constructed. For image applications, one method of creating an array is to load an image from a file. This process will be reviewed in Section 3.9.1. This section reviews a variety of other methods that can create an array.

3.8.1.1 Zeros and Ones

There are two functions that will create an array and initialize the elements to either 0 or 1. Both the **zeros** and the **ones** functions receive an argument defining the size of the array. An additional argument will also define the data type. Several examples are shown in Code 3.47. Line 2 creates an array of ten 0's, and by default these are all floats. Line 5 creates an array of 1's, and the second argument declares that these are to be integers. Line 8 creates an array with complex valued numbers.

Two dimensional arrays are constructed using the same functions, but the argument is a tuple instead of an integer. Code 3.48 shows the command for creating a two-dimensional array consisting of zeros. In this example, there are 128 elements in the vertical direction and 256 elements in the horizontal direction. Often these functions are the first step in a sequence of commands to construct an image containing definable geometric shapes. An example is shown in Code 3.52.

The operator notation for creating an array of 0's is

$$\mathbf{a}[\vec{x}] = z_{\vec{w}}[\vec{x}], \tag{3.5}$$

where $\vec{w}$ is the frame size for the image. A unique operator does not exist for the **ones** function since $z_{\vec{w}}[\vec{x}] + 1$ will suffice.

Code 3.47 Creation of vectors

```
>>> import numpy as np
>>> vec = np.zeros( 10 )
>>> vec
array([ 0.,  0.,  0.,  0.,  0.,  0.,  0.,  0.,  0.,  0.])
>>> vec = np.ones( 10, int )
>>> vec
array([1, 1, 1, 1, 1, 1, 1, 1, 1, 1])
>>> vec = np.zeros( 4, complex )
>>> vec
array([ 0.+0.j,  0.+0.j,  0.+0.j,  0.+0.j])
```

Code 3.48 Creating tensors

```
>>> tensor = np.zeros( (12,15,16), float )
```

Tensors are created in the same manner as matrices except that there are at least three elements in the tuple. However, in allocation of tuples, the user should be aware that memory can be consumed rather quickly. A 256×256 matrix with floating point elements consumes 0.5 megabytes. However, a tensor that is $256 \times 256 \times 256$ with floating point elements consumes 128 megabytes. Furthermore, if two tensors $c = a + b$ are added, then there needs to be space for four tensors (a, b, c, and $a + b$). Before tensors are allocated, the user needs to make sure that there is sufficient memory in the computer. Code 3.48 demonstrates the allocation of a much smaller tensor.

Accessing data in arrays is performed through slicing. However, for matrices and tensors, it is possible to access a single cell by passing a tuple to the slice. Code 3.49 demonstrates how to retrieve and change cells in a matrix. In line 1, the tuple (0,1) indicates that the cell is the first row and second column. The selection of rows, columns, subimages, and selected pixels is discussed in Section 3.8.2.8.

3.8.1.2 Random

The NumPy package contains a module to create several types of random arrays. In this text, the only function that is used creates random values, evenly distributed between the values of 0 and 1. There are other functions that create values with different types of distributions. Code 3.50 shows two methods of creating random arrays. Line 1 uses the **rand** function to create a vector with ten random numbers. Line 2 uses the **ranf** function to create a two dimensional array that has over a million elements. Since these random numbers are evenly distributed between 0 and 1, it is expected that the average of the values is close to 0.5. This is shown to be the case in lines 4 and 5.

The operator notation for creating an image or random values is

$$\mathbf{a}[\vec{x}] = \mathbf{q}_{\vec{w}}[\vec{x}], \tag{3.6}$$

Code 3.49 Accessing data in a matrix

```
1 >>> mat[0,1]
2 0.283461998819
3 >>> mat[1,1] = 4
```

Code 3.50 Creating random arrays

```
>>> vec = np.random.rand( 10 )
>>> mat = np.random.ranf( (1024,1024) )
>>> mat.mean()
0.49981380909534784
```

Code 3.51 Using a random seed

```
>>> np.random.seed( 5 )
>>> amat = np.random.ranf( (512,512))
```

where $\vec{w}$ defines the frame size. The creation of random values in a different range is represented by applying an shift, b, and bias, c, as in

$$\mathbf{a}[\vec{x}] = c\,\mathbf{q}_{\vec{w}}[\vec{x}] + b. \tag{3.7}$$

In some cases, it is desired that a random sequence can be repeatedly created. Some codes in this text use random numbers in an example. In order to replicate the example, it will be necessary to use the same random numbers. The **random.seed** function receives an integer argument and establishes the random sequence. Thus, Code 3.51 will always generate the same matrix `amat`.

3.8.1.3 Geometric Shapes

The rectangle and the circle are two geometric shapes that are commonly used in isolating regions in images. The Rectangle operator needs the frame size $\vec{w}$, the coordinates of the upper left corner $\vec{v}_1$, and the coordinates of the lower right corner $\vec{v}_2$. The operator is

$$\mathbf{a}[\vec{x}] = r_{\vec{w};\vec{v}_1,\vec{v}_2}. \tag{3.8}$$

The Python script to create a solid rectangle is straightforward as seen in Code 3.52. This creates a frame $\vec{w} = (256, 256)$. The upper left corner of the rectangle is $\vec{v}_1 = (100, 75)$, and the lower right corner is $\vec{v}_2 = (150, 125)$. All pixels in the interior have a value of 1, and all other pixels have a value of 0.

The operator for creating a solid circle is

$$\mathbf{a}[\vec{x}] = o_{\vec{w};\vec{v},r}, \tag{3.9}$$

where $\vec{v}$ is the location of the center of the circle and r is the radius. Creating this image in Python is not as straightforward, and so it is presented in Section 3.8.3 after another concept is reviewed.

3.8.1.4 Conversion of Numerical Data

Another method of creating arrays is to convert numerical data in another type of variable to an array. For example, an application extracts a single measurement from a set of N images, thus creating a list of N values. To facilitate further analysis, the user needs to convert these to a vector. This type of

Code 3.52 Creating a solid rectangle

```
>>> amg = np.zeros((256,256))
>>> amg[100:150,75:125] = 1
```

Code 3.53 Creating arrays from data

```
1 >>> vec = np.array( [1,2,3,4] )
2 >>> data = [ [1,2,3], [3,4,5] ]
3 >>> mat = np.array( data )
4 >>> mat
5 array([[1, 2, 3],
6        [3, 4, 5]])
```

conversion is performed by the **array** function as shown in Code 3.53 shows a couple of examples. Line 1 converts a list to an array. Lines 2 and 3 create a matrix using the same logic except that the input is a list of lists. Each of these internal lists must have the same number of elements. The output shown in lines 5 and 6 shows the arrangement of values in this newly formed matrix.

3.8.2 MANIPULATING ARRAYS

This section will review just a few of the methods by which numerical arrays are used to perform computations. The NumPy library has a large number of powerful tools, and thus, users are encouraged to peruse NumPy documentation to understand the breadth of operations that are available.

3.8.2.1 Display Option

Printing arrays to the console can consume a lot of real estate. One method of improving the presentation is to limit the number of digits in the decimal places that are printed. Code 3.54 shows the **set_printoptions** function with the `precision` set to a value of 3. This will then limit the number of printed decimals to 3. This command only alters the printed data. The values in the arrays are not changed and still have the original precision. Furthermore, this function applies only to arrays and does not apply to floats or values in a collection like a list.

3.8.2.2 Converting Arrays

Converting between a matrix and a vector is easily performed with the **resize** and **ravel** commands. Line 1 in Code 3.55 converts the 2×3 matrix into a 6 element vector. Line 4 converts `data` into a 3×2 matrix. In line 1, the answer is returned to a new variable, but in line 4, the original variable is changed. It is possible to convert a matrix of one size ($M \times N$) to a matrix of another size ($P \times Q$) using `resize` as long as $M \times N = P \times Q$. The `shape` command returns the dimensions of the array as seen in line 9.

3.8.2.3 Simple Math

The main reason that arrays are used is the speed of computation. Applying mathematical operations to an array is much easier to write and significantly faster to execute. Expressions for arrays are

Code 3.54 Setting the number of decimal places that are printed to the console

```
>>> np.set_printoptions( precision=3)
>>> vec1 = np.random.rand( 4 )
>>> vec1
array([ 0.938,  0.579,  0.867,  0.066])
```

Code 3.55 Converting between vectors and matrices

```
>>> data = mat.ravel()
>>> data
array([ 0.181,  0.283,  0.185,  0.978,  4.   ,  0.612])
>>> data.resize( (3,2) )
>>> data
array([[ 0.181,  0.283],
       [ 0.185,  0.978],
       [ 4.   ,  0.612]])
>>> data.shape
(3, 2)
```

Code 3.56 Math operations for vectors

```
1 >>> vec = np.ones( 4, int )
2 >>> vec + 2
3 array([3, 3, 3, 3])
4 >>> (vec + 12)/4
5 array([ 3.25,  3.25,  3.25,  3.25])
6 >>> vec1 = np.random.rand( 4 )
7 >>> vec2 = np.random.rand( 4 )
8 >>> vec1 + vec2
9 array([ 1.512,  0.906,  1.159,  0.716])
```

very similar to those used on scalar data. Code 3.56 shows a few examples. Line 2 adds a scalar to a vector, line 4 divides a vector by a scalar, and line 8 adds two vectors.

Operator notation for the addition of two arrays, such as seen in line 8, is simply

$$\mathbf{c}[\vec{x}] = \mathbf{a}[\vec{x}] + \mathbf{b}[\vec{x}]. \tag{3.10}$$

This is an elemental addition as in

$$c_i = a_i + b_i, \quad \forall i. \tag{3.11}$$

3.8.2.4 Multiplying Vectors

Addition of arrays is simple in concept. Multiplication of arrays, however, is not so straightforward. The operations for vectors and matrices are slightly different. There are, in fact, four different methods in which two vectors can be multiplied together. These are the elemental product, the inner product, the outer product, and the cross product.

The element-by-element product is similar in nature to the array addition as it is

$$c_i = a_i b_i, \quad \forall i. \tag{3.12}$$

The operator notation is

$$\vec{c} = \vec{a} \times \vec{b}. \tag{3.13}$$

Usually, the $\times$ symbol is used to represent the cross product, but since cross products are not used in this text, there should be no confusion.

The inner product (also known as the dot product) projects one array onto another and is described as

$$f = \sum_i a_i b_i. \tag{3.14}$$

The output is a scalar value and the operator notation is

$$f = \vec{a} \cdot \vec{b}. \tag{3.15}$$

The outer product creates a matrix, **M**, from two vectors, $\vec{x}$ and $\vec{y}$, as in

$$M_{i,j} = x_i y_j, \quad \forall i, j. \tag{3.16}$$

This operation is used in some neural network learning processes, but otherwise is not used in any significant manner in this text.

The final multiplication method is the cross product, in which two vectors are multiplied to create a third vector that is orthogonal to both of the original. This type of multiplication is not used here.

Code 3.57 shows the first two types of multiplications. Line 3 performs the elemental multiplication of two matrices, and line 4 performs the inner product of two matrices using the **dot** function.

3.8.2.5 Multiplying Matrices

There are three methods of multiplying matrices: elemental, inner product, and matrix-matrix. The first is

$$c_{i,j} = a_{i,j} b_{i,j}, \quad \forall i, j. \tag{3.17}$$

The second method is the inner product (dot product) and is similar to the vector version as in

$$f = \sum_i \sum_j a_{i,j} b_{i,j}. \tag{3.18}$$

The third method is matrix-matrix multiplication which is

$$c_{i,k} = \sum_j a_{i,j} b_{j,k}, \quad \forall i, k. \tag{3.19}$$

The **numpy.dot** function performs the inner product for two vectors, but it performs the matrix-matrix multiplication when applied to two matrices. While this type of multiplication is not used much in the analysis of images, it does appear in many other scientific applications. Thus, the inner product of two matrices is written as shown in Code 3.58.

3.8.2.6 Array Functions

There are many functions that can be applied to arrays, and a few of these are reviewed here in terms of image analysis. Consider a grayscale image, $\mathbf{a}[\vec{x}]$, in which the pixel values are stored in a NumPy

Code 3.57 Multiplication with vectors

```
1 >>> avec = np.random.rand( 5 )
2 >>> bvec = np.random.rand( 5 )
3 >>> cvec = avec * bvec
4 >>> f = avec.dot( bvec )
```

Code 3.58 The inner product of two matrices

```
>>> adata = np.random.ranf((2,3))
>>> bdata = np.random.ranf((2,3))
>>> f = (adata * bdata).sum()
```

array, thus allowing the use of NumPy functions on the image. Demonstration of the functions in Python requires that an image be loaded. The two steps for loading an image of a bird as gray scale is shown in the first two lines of Code 3.59, but the Python commands are discussed in Section 3.9.1. For now, it is sufficient to understand that the variable `amg` is a matrix that contains the grayscale values of the bird image.

The frame size is the number of pixels in the vertical and horizontal directions. For a color image, there is a third dimension with different color values. The frame size is represented by a vector and the operator notation is

$$\vec{w} = \mathbf{Za}[\vec{x}]. \tag{3.20}$$

In Python, this information is retrieved by the **shape** command as seen in lines 3 and 4.

The maximum value of the image is

$$f = \bigvee \mathbf{a}[\vec{x}]. \tag{3.21}$$

This is shown in lines 5 and 6. The maximum value is obtained by the **max** function. Likewise, the minimum value is obtained by the **min** function and is represented by the $\bigwedge$ operator.

The **max** function without arguments finds the maximum value of the entire array. Some applications require that the max of each column or each row be known. Each dimension in a Python array is called an axis, and Python starts counting at 0. Thus, the first axis (the vertical dimension) is the 0 axis. Likewise, the horizontal dimension is the 1 axis. Lines 7 and 8 use the axis argument in the **max** function to extract maximum values in the vertical dimension (line 7) or the horizontal dimension (line 8). Operator notation to extract maximum values according to a user described criteria, m, is

$$\vec{v} = \bigvee_m \mathbf{a}[\vec{x}]. \tag{3.22}$$

Several functions operate in the same manner. The sum over an entire array is

$$f = \sum_{\vec{x}} \mathbf{a}[\vec{x}], \tag{3.23}$$

and the sum over a user-specified criteria is

$$f = \sum_m \mathbf{a}[\vec{x}]. \tag{3.24}$$

Three operations are shown in Code 3.60 with the first being the sum over the entire image, and the others sum in the vertical and horizontal directions.

The average of the pixel values is computed by the **mean** function and is described by

$$\mu = \mathcal{M}_m \mathbf{a}[\vec{x}]. \tag{3.25}$$

Code 3.59 Maximum values in an image

```
1 >>> import imageio
2 >>> amg = imageio.imread( 'data/bird.jpg', as_gray=True)
3 >>> amg.shape
4 (519, 653)
5 >>> amg.max()
6 255.0
7 >>> vmax = amg.max(0)
8 >>> hmax = amg.max(1)
```

Code 3.60 Application of several functions

```
>>> sm = amg.sum()
>>> vsum = amg.sum(0)
>>> hsum = amg.sum(1)
```

The standard deviation is computed by the **std** function and is represented by

$$\sigma = \mathcal{T}_m \mathbf{a}[\vec{x}]. \tag{3.26}$$

Both Equations (3.25) and (3.26) create a scalar value. This is changed to a vector is the criteria m dictates that the output is a vector.

The **argmax** function is used to provide the location of the maximum. The notation is

$$\vec{v} = A_{\vee} \mathbf{a}[\vec{x}]. \tag{3.27}$$

The **argmax** function returns a single value which is element number of the location of the first occurrence of the maximum value. Line 2 in Code 3.61 shows the result. This location of the max is the 26,446 element in the array which has 653 elements in each row. Thus, the vertical-horizontal location of the pixel is obtained through the **divmod** function as shown in lines 3 and 4. This only is the first occurrence of the maximum value in the array. Lines 5 and 6 indicate that there are 28 such pixels as in

$$n = \sum_{\vec{x}} \Gamma_{=\vee \mathbf{a}[\vec{x}]} \mathbf{a}[\vec{x}]. \tag{3.28}$$

The **nonzero** function returns an array inside a tuple which at first may seem odd. Code 3.62 shows the application of this function. Line 1 creates a vector and line 3 compares this vector to a scalar threshold. The output is an array, `gvec`, that contains binary values as seen in line 5. The **nonzero** function is applied in line 6, and line 8 shows that there are two locations which were `True` in `gvec`. Line 8 shows the results as an array inside a tuple. This format is employed so that the output is congruent with the application of the **nonzero** function to other types of arrays. Consider the matrix, `mat`, which is shown in lines 12 through 15. Once again a comparison is performed (line 16), and the nonzero elements are extracted (line 17). Since there are two dimensions in `mat`, there are two arrays that are returned by the function. These are contained within the tuple `nz` generated in line 17. For a data array of N dimensions, the **nonzero** function will return N arrays inside a tuple. This is true for the case in lines 1 through 8 in which the data is only one dimension. The **nonzero** function still returns N arrays inside a tuple. The first item of this tuple is accessed by `nz[0]` as seen in lines 9 through 10.

The `nz` from line 17 is a tuple with two vectors. The first vector contains the vertical locations of the nonzeros, and the second vector contains the horizontal locations of the nonzeros. Thus, the four locations in which `gmat` is not zero are (0,1), (1,1), (2,0), and (2,1).

Code 3.61 Locating the maximum

```
1 >>> amg.argmax()
2 26446
3 >>> divmod(amg.argmax(),653 )
4 (40, 326)
5 >>> (amg==amg.max()).sum()
6 28
```

Code 3.62 Using the `nonzero` function

```
>>> vec
array([ 0.21 ,  0.268,  0.023,  0.998,  0.386])
>>> gvec = vec > 0.3
>>> gvec
array([False, False, False,  True,  True], dtype=bool)
>>> nz = gvec.nonzero()
>>> nz
(array([3, 4], dtype=int64),)
>>> nz[0]
array([3, 4], dtype=int64)

>>> mat
array([[ 0.17 ,  0.824,  0.414,  0.265],
       [ 0.018,  0.865,  0.112,  0.398],
       [ 0.721,  0.77 ,  0.083,  0.065]])
>>> gmat = mat > 0.5
>>> nz = gmat.nonzero()
>>> len( nz )
2
>>> nz[0]
array([0, 1, 2, 2])
>>> nz[1]
array([1, 1, 0, 1])
```

The operator for the `nonzero` function is

$$\mathbf{B} = \mathcal{N}_m \mathbf{a}, \tag{3.29}$$

where **B** is a matrix in which the rows are the locations of the nonzero values. The subscript m is optional and used to the describe restrictions on the operation. The output is a matrix in which each row is a location of a nonzero value.

3.8.2.7 Decisions

There are six comparison operators >, >=, <, <=, ==, and !=. Two different operator methods are used to describe a comparison. The first uses the comparison notation similar to Python as in

$$\mathbf{b}[\vec{x}] = \mathbf{a}[\vec{x}] > \gamma. \tag{3.30}$$

However, this notation is a bit awkward in cases where several operators are applied to an image in sequence. More amenable in these situations is the notation

$$\mathbf{b}[\vec{x}] = \Gamma_{>\gamma} \mathbf{a}[\vec{x}]. \tag{3.31}$$

Elemental comparisons of two images create a new image of binary values, as in

$$\mathbf{c}[\vec{x}] = \mathbf{a}[\vec{x}] > \mathbf{b}[\vec{x}] = \Gamma_{>\mathbf{b}[\vec{x}]}\, \mathbf{a}[\vec{x}]. \tag{3.32}$$

This is equivalent to

$$c_{i,j} = a_{i,j} > b_{i,j}, \quad \forall i, j. \tag{3.33}$$

To determine if all of the values in an array surpass a threshold, the product operator is also employed:

$$f = \prod_{\vec{x}} \Gamma_{>\gamma}\, \mathbf{a}[\vec{x}]. \tag{3.34}$$

The $\Gamma_{>\gamma}$ operator returns a binary-valued array and the $\prod_{\vec{x}}$ operator multiplies all of the values. If just one value is false (or 0), then f is zero. Only if all of the elements pass the threshold will f be equal to 1. The NumPy function for this is **alltrue**. Likewise, the **sometrue** function will return a 1 value if at least one of the elements pass the threshold. This is expressed as

$$f = \Gamma_{>0} \sum_{\vec{x}} \Gamma_{>\gamma}\, \mathbf{a}[\vec{x}]. \tag{3.35}$$

3.8.2.8 Advanced Slicing

Arrays can be sliced in manners similar to strings, tuples, etc. However, arrays offer advanced slicing techniques that are also quite useful in making efficient code. In Code 3.63, a vector of ten elements is created. In line 5, the first three elements are retrieved, and in line 7, every other element is retrieved. These methods behave the same as in the case of strings, tuples, lists, etc.

In line 9, a list of integers is created, and in line 10, this list is used as an index to the vector. The result is that the elements are extracted from the vector in the order prescribed by the list `n`. This allows the user to extract data in a specified order. Likewise, it is possible to set values in the array in a specified order as shown in line 12.

This same advanced technique applies to arrays with multiple dimensions. In Code 3.64, elements in a matrix are accessed using two index lists `v` and `h`. The first index represents element locations along the first axis (hence `v` represents the vertical dimension). The first element extracted is `M[1,1]`, the second element is `M[2,1]`, and the third element is `M[0,2]`.

Using this technique, the user can access elements in an array in any order without employing a Python `for` loop. This will dramatically increase execution time especially for large arrays.

3.8.2.9 Universal Functions

NumPy offers several mathematical functions that are applied to all elements of the array. Code 3.65 shows the square root function applied to all elements of an array. The popular functions are shown in Table 3.1.

Code 3.63 Advanced slicing for arrays

```
1  >>> V = np.random.rand( 10 )
2  >>> V
3  array([ 0.06 ,  0.796,  0.775,  0.592,  0.174,  0.764,  0.952,  0.871,
4          0.569,  0.545])
5  >>> V[:3]
6  array([ 0.06 ,  0.796,  0.775])
7  >>> V[::2]
8  array([ 0.06 ,  0.775,  0.174,  0.952,  0.569])
9  >>> n = [4,1,3,0]
10 >>> V[n]
11 array([ 0.174,  0.796,  0.592,  0.06 ])
12 >>> V[n] = -1, -2, -3, -4
13 >>> V
14 array([-4.   , -2.   ,  0.775, -3.   , -1.   ,  0.764,  0.952,  0.871,
15         0.569,  0.545])
```

Code 3.64 Advanced slicing for arrays with multiple dimensions

```
>>> M = np.random.ranf( (3,4) )
>>> M
array([[ 0.181,  0.663,  0.933,  0.791],
       [ 0.561,  0.145,  0.687,  0.877],
       [ 0.876,  0.881,  0.774,  0.347]])
>>> v = [1,2,0]
>>> h = [1,1,2]
>>> M[v,h]
array([ 0.145,  0.881,  0.933])
```

Table 3.1
Math Functions

sin	arcsin	sinh	arcsinh
cos	arccos	cosh	arccosh
tan	arctan	tanh	arctanh
exp	log	log2	log10
sqrt	conjugate	floor	ceil

Operator notation follows standard practices for these functions. The sine of all of the values is expressed as

$$\mathbf{b}[\vec{x}] = \sin(\mathbf{a}[\vec{x}]), \tag{3.36}$$

and the square root of all of the values is

$$\mathbf{b}[\vec{x}] = \sqrt{\mathbf{a}[\vec{x}]}. \tag{3.37}$$

3.8.2.10 Sorting

There are two manners in which information is sorted in this text. One is to rearrange the values in a vector to a particular criteria, and the other is to organize data entities such as images. Sorting values in a vector is expressed as

$$\vec{z} = \mathcal{O}_{\downarrow,C}\vec{v}, \tag{3.38}$$

Code 3.65 Mathematical functions for an array

```
>>> mat
array([[ 0.17 ,  0.824,  0.414,  0.265],
       [ 0.018,  0.865,  0.112,  0.398],
       [ 0.721,  0.77 ,  0.083,  0.065]])
>>> np.sqrt( mat )
array([[ 0.412,  0.908,  0.643,  0.514],
       [ 0.134,  0.93 ,  0.335,  0.631],
       [ 0.849,  0.878,  0.287,  0.256]])
```

Code 3.66 Sorting data in an array

```
>>> avec = np.random.rand(5 )
>>> avec
array([ 0.279,  0.562,  0.984,  0.906,  0.733])
>>> avec.sort()
>>> avec
array([ 0.279,  0.562,  0.733,  0.906,  0.984])
```

where the $\downarrow$ indicates that the result is in descending order, and the C is the optional user-defined criteria. The **sort** function from the NumPy module performs a similar operation. The Python function actually rearranges the data as shown in Code 3.66, whereas the operator notation creates a new vector.

A more useful function, in terms of image processing, is **argsort**. This function will return the indices of the data, but it does not rearrange the data. Consider an example in which there is a set of images, $\{\mathbf{a}[\vec{x}]\}$, and this data is to be sorted by a criteria C. Perhaps the criteria can be something as simple as counting the number of pixels that are greater than some value γ. The goal is to rearrange the images according to this metric. In this case, the sort order is determined by a measurement of each image, but it is the images that are rearranged. Thus, the sorting indices are extracted from one set of data (the measurements) and applied to another (the data). This example is expressed by

$$\{\mathbf{b}[\vec{x}]\} = \mathcal{O}_{\downarrow}\left\{\sum_{\vec{x}} \Gamma_{>\gamma}\mathbf{a}[\vec{x}]\right\}. \tag{3.39}$$

The expression $\{\sum_{\vec{x}} \Gamma_{>\gamma}\mathbf{a}[\vec{x}]\}$ applies a threshold to each image and the sum is the count of the number of pixels that pass threshold. The expression is completely inside the braces to indicate that the sum and threshold are applied to each image in the set. The $\mathcal{O}$ operator then sorts this data.

The Python script to accomplish this is shown in Code 3.67. The original data, `amgs`, is a three-dimensional array of the size $N \times V \times H$, where N is the number of images, and V and H are the dimensions of the images. In this example, all images have the same frame size. Line 1 applies the threshold to each image. The `x > 200` will create an array in which there are 1 values at the locations where the image passed the threshold, and the `sum` function will return the sum of this array which is also the number of pixels that pass the threshold. The output of the **map** function is converted to a list so that it can then be converted to a vector. Line 3 shows the values for this example. Line 4 uses **argsort** to determine the sort order. The variable `ag` is order of the indices for the sort. Thus, `amgs[6]` has been determined to be the image with the fewest pixels passing the threshold, and `amgs[4]` is the image with the most pixels passing the threshold. Line 7 creates a

Code 3.67 Sorting images according to a user-defined criteria

```
1 >>> v = np.array(list(map(lambda x: (x > 200).sum(), amgs )))
2 >>> v
3 array([3579, 3538, 3554, 3639, 3644, 3616, 3438, 3587, 3584, 3570])
4 >>> ag = v.argsort()
5 >>> ag
6 array([6, 1, 2, 9, 0, 8, 7, 5, 3, 4], dtype = int64)
7 >>> bmgs = amgs[ag]
8 >>> (bmgs[0] > 200).sum()
9 3438
```

new array of images in the sorted order. Lines 8 and 9 confirm that the first image in **bmgs** has the fewest pixels above threshold.

3.8.3 INDICES

The **indices** function returns an array in which the elements numerically increment along individual axes. The argument to the **indices** function in line 1 of Code 3.68 is a tuple (4,6). This will create two matrices. The first is of size 4×6 and the values increment downwards as show in lines 2 through 6. The second is a matrix of the same size that increments along the columns as shown in lines 7 through 11.

Consider a case in which a circular region of an image is to be isolated. The original image is $\mathbf{a}[\vec{x}]$, and the ROI (region of interest) is defined by a mask which is shown in Figure 3.3. The operator to construct the mask needs the frame size $\vec{w}$, the location of the center of the circle $\vec{v}$, and the radius r. The isolation of the ROI is then

$$\mathbf{c}[\vec{x}] = \mathbf{a}[\vec{x}] \times o_{\vec{w};\vec{v},r} \tag{3.40}$$

The image shown in Figure 3.3 is 256×256, which is too large to examine the method of creating the mask. So a much smaller example is used starting with Code 3.68. This created matrices of the size 4×6. The values in the first matrix are the distances to the top of the frame, and the values

Code 3.68 Example of the **indices** function

```
1  >>> ndcs = np.indices( (4,6) )
2  >>> ndcs[0]
3  array([[0, 0, 0, 0, 0, 0],
4         [1, 1, 1, 1, 1, 1],
5         [2, 2, 2, 2, 2, 2],
6         [3, 3, 3, 3, 3, 3]])
7  >>> ndcs[1]
8  array([[0, 1, 2, 3, 4, 5],
9         [0, 1, 2, 3, 4, 5],
10        [0, 1, 2, 3, 4, 5],
11        [0, 1, 2, 3, 4, 5]])
```

Figure 3.3 A circular mask.

Code 3.69 Creating a solid circle

```
1  >>> ndcs[0] -= 2
2  >>> ndcs[1] -= 3
3  >>> ndcs[0]
4  array([[-2, -2, -2, -2, -2, -2],
5         [-1, -1, -1, -1, -1, -1],
6         [ 0,  0,  0,  0,  0,  0],
7         [ 1,  1,  1,  1,  1,  1]])
8  >>> ndcs[1]
9  array([[-3, -2, -1,  0,  1,  2],
10        [-3, -2, -1,  0,  1,  2],
11        [-3, -2, -1,  0,  1,  2],
12        [-3, -2, -1,  0,  1,  2]])
13 >>> dist = np.sqrt( ndcs[0]**2 + ndcs[1]**2 )
14 >>> dist
15 array([[ 3.606,  2.828,  2.236,  2.   ,  2.236,  2.828],
16        [ 3.162,  2.236,  1.414,  1.   ,  1.414,  2.236],
17        [ 3.   ,  2.   ,  1.   ,  0.   ,  1.   ,  2.   ],
18        [ 3.162,  2.236,  1.414,  1.   ,  1.414,  2.236]])
```

Code 3.70 The **Circle** function

```
# mgcreate.py
def Circle( size, loc,rad):
    b1,b2 = np.indices( size )
    b1,b2 = b1-loc[0], b2-loc[1]
    mask = b1*b1 + b2*b2
    mask = ( mask <= rad*rad ).astype(int)
    return mask
```

in the second matrix are the distances to the left wall of the frame. In order to create a circle at a location (v,h), the values in these matrices are shifted. In this example, the desired location of the circle is at (2,3). So, the values of the two matrices are shifted by these coordinates as seen in the first two lines in Code 3.69. The values of the first matrix are now the vertical distances to the row 2, and the values in the second matrix are the horizontal distances to the column 3.

Line 13 uses these distances to compute the Euclidean distance for all pixels from the location (2,3). These are shown in lines 15 through 18. The same process can be applied to a much larger array. Lines 1 and 2 in Code 3.69 would then be the location of the circle in the new array. Line 13 still computes the Euclidean distances. The circle is then defined as any pixel with a distance of less than r.

The operator for the creation of a circle is $o_{\vec{w};\vec{v},r}$ where $\vec{w}$ is the frame size and $\vec{v}$ is the location of the center of the circle and r is the radius. This creates a binary-valued image with a solid circle. The Python function **Circle** (from the *mgcreate* module) creates image with the given input parameters.

3.9 SCIPY

The NumPy package offers definitions and functions for creating and manipulating arrays. The SciPy package offers a large suite of scientific modules, each containing several functions. Four of these packages are used in this text. These are:

- *fftpack*: contains functions for Fourier transforms.
- *ndimage*: contains image processing functions.
- *signal*: contains signal processing functions.

Several other modules do exist in SciPy and readers are encouraged to scan the SciPy documentation to see what is available. A good rule of thumb is that: probably the desired function has already been written. Programmers should check with NumPy, SciPy, and other packages before venturing into writing a complicated function.

3.9.1 LOADING AND SAVING IMAGES

The Python Image Library (PIL) contains functions to read and write images of many different formats. However, these functions treat images as an image object and not as an array. The functions in *imageio* uses the PIL functions, but the variables used in these functions are arrays. Since arrays are more amenable to the function in this text, the functions from *imageio* will be used.

Reading an image is expressed as

$$\mathbf{a}[\vec{x}] = Y(\texttt{FileName}). \tag{3.41}$$

The Python script is shown in Code 3.71. Line 1 imports the *imageio* module and renames it. Line 2 reads in an image. Line 4 shows that this image has 519 pixels in the vertical direction, 653 pixels in the horizontal direction, and 3 color channels. These are discussed in Chapter 5. The elements in the matrix are unsigned integers. Line 7 loads the image with the `as_gray=True` switch. This reads the image as a grayscale image, and the math operations that convert the color image to gray scale use decimal values, so the output is now an array of floats.

If the preference is to have the color image in the $N \times V \times H$ arrangement, then the **transpose** function is used as shown in Code 3.72. The argument in line 1 defines the new order of the dimensions. So, axis 2 (the color channel in the original data) is moved to the first position.

Saving an image to disk is just as easy as reading an image. However, the filename must have an extension that defines the image compression format. An example is shown in Code 3.73. This

Code 3.71 Loading an image

```
1  >>> import imageio
2  >>> amg = imageio.imread( 'data/bird.jpg')
3  >>> amg.shape
4  (519, 653, 3)
5  >>> amg.dtype
6  dtype('uint8')
7  >>> cmg = imageio.imread( 'data/bird.jpg', as_gray=True)
8  >>> cmg.shape
9  (519, 653)
10 >>> cmg.dtype
11 dtype('float32')
```

Code 3.72 Rearranging the color channels

```
1  >>> bmg = amg.transpose((2,0,1))
2  >>> bmg.shape
3  (3, 519, 653)
```

Code 3.73 Saving an image

```
>>> imageio.imsave( 'myimage.jpg', amg )
```

saves the amg array as a JPG file. The array to be saved must be in the form of $V \times H \times N$, and thus bmg cannot be saved in this manner.

3.9.2 EXAMPLES FROM NDIMAGE

The *scipy.ndimage* module provides a suite of more than 80 image functions. It is not possible, or even necessary, to review all of these functions in this text. This section will demonstrate a couple of popular functions, and others that are used in subsequent sections will be discussed in their first presentation.

3.9.2.1 Rotation and Shift

Two common functions are **rotate** and **shift** . The Rotation operator is R_θ and the Shift operator is $D_{\vec{v}}$. Given an input image $\mathbf{a}[\vec{x}]$, the operation of rotating and shifting the image is

$$\mathbf{c}[\vec{x}] = D_{\vec{v}} R_\theta \mathbf{a}[\vec{x}], \tag{3.42}$$

where $\vec{v}$ is the amount of shift and θ is the amount of rotation. The Rotation operator is applied first since it is just in front of the image.

Code 3.74 shows the complete process in individual steps. Lines 1 and 2 load the necessary modules, and line 3 loads the color image. Line 4 applies the **rotate** function. There are four inputs to this function. The first is the input image, and the second is the angle (in degrees) of rotation. The next two arguments are optional. The **rotate** function will resize the frame to accommodate all pixels in the rotated image. The `reshape=False` throws a switch that forces the output frame to be the same size as the input frame. Thus, corners of the original image will be lost since they are outside the output frame. The `cval==255` sets the pixels that are not defined in the output to white. The original input and rotated image are shown in Figure 3.4(a) and (b).

The Shift operator is applied in line 6. The second argument is the tuple (25,10,0). The color image has three dimensions: vertical, horizontal, and color. Thus, there are three shift values. This image is shifted 25 pixels in the vertical direction, 10 pixels in the horizontal direction, and it is not shifted in the color dimension. The result is shown in Figure 3.4(c). The black pixels are those not defined by the shift and since no `cval` was set, the function uses the default value of 0, which is black.

Code 3.74 An example of melding the operators and functions from *ndimage*

```
1 >>> import imageio
2 >>> import scipy.ndimage as nd
3 >>> amg = imageio.imread('data/bird.jpg' )
4 >>> bmg = nd.rotate(amg,20,reshape=False,cval=255)
5 >>> imageio.imsave('bird1.jpg', bmg )
6 >>> cmg = nd.shift(bmg, (25,10,0))
7 >>> imageio.imsave('bird2.jpg', cmg )
```

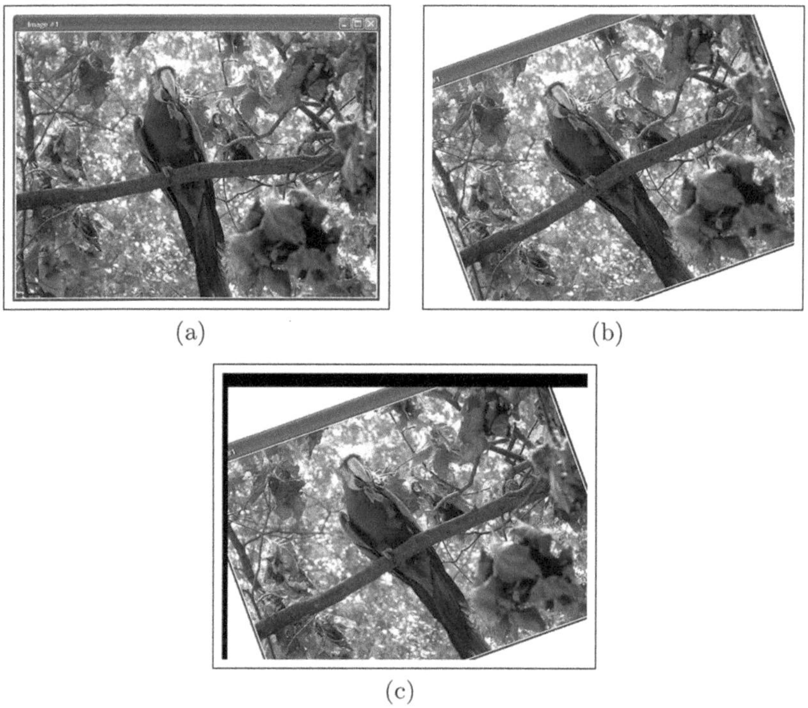

Figure 3.4 (a) The original image, (b) the image after rotation, and (c) the image after the subsequent shift.

Code 3.75 Finding the center of mass

```
1 >>> amg = imageio.imread('data/myth/horse1.bmp') < 128
2 >>> nd.center_of_mass( amg )
3 (295.35284937228892, 295.27580271281357)
```

3.9.2.2 Center of Mass

Another useful function is the center of mass computation, which returns the location of the center of mass of an input image. Consider the image shown in Figure 3.5. The horse has pixel values of 0 and the background has pixel values of 255. The task is to compute the center of mass of the horse. The operator notation is

$$\vec{v} = \boxtimes \mathbf{a}[\vec{x}]. \tag{3.43}$$

The *ndimage* module has the function **center_of_mass** which will perform the computation, but it is necessary to convert the image first so that the horse pixels each have a value of 1 and all other pixels have a value of 0. Thus, the correct procedure is

$$\vec{v} = \boxtimes \Gamma_{<128} \mathbf{a}[\vec{x}]. \tag{3.44}$$

Code 3.75 shows the two steps necessary to perform the computation. Line loads the image and converts the pixel values so the horse pixels each have a value of 1. Line 2 calls on the **center_of_mass** function to perform the final computation. The output shows the location of the center of mass.

Figure 3.5 An image of a black horse [3].

3.10 SUMMARY

This chapter reviewed some of the key concepts in Python and NumPy. It also presented just a few examples of functions from the SciPy package. These are the foundation tools for the following chapters. Programmers are encouraged to read documentation that list the functions that are available through these packages.

PROBLEMS

1. Create a vector with ten elements and all of the elements are integers with a value of 0.
2. Create a vector with ten elements and the values are (0,1,2,3,4...,9).
3. Create a vector of ten randomly valued elements.
4. Create a vector of ten randomly valued elements in which the values range between -1 and $+1$.
5. Create a vector of 1000 randomly valued elements (with an equal probability distribution) and confirm that the average value of the vector is 0.5.
6. Create a 4×3 matrix in which all elements are 1.
7. Create a 4×3 matrix of random elements and compute the sum of the entire matrix. Use a random seed of 6435.
8. Create a 4×3 matrix of random elements and compute the sum of the rows. Use a random seed of 6435.
9. Create a ten element vector of random values and determine the largest value. Use a random seed of 6435.
10. Create a ten element vector of random values and determine the location of the largest value. Use a random seed of 6435.
11. Create two ten element vectors and add them together.
12. Create a vector of 100 random elements using the random seed (6435). In a single Python line count the number of elements greater than 0.7.
13. Start with a random seed of 6435. Create two vectors $\vec{a}$ and $\vec{b}$, each with 100 elements. In a single Python line perform $\sum(\vec{a}-\vec{b})$.
14. Using the bird image, write a single Python line that performs $\sum_{\vec{x}} \Gamma_{>250} \mathcal{L}_L \mathbf{a}[\vec{x}]$.
15. Load the gray version of the bird image as $\mathbf{a}[\vec{x}]$. Construct the image in Figure 3.6.
16. Write the operator notation for the previous problem.

Figure 3.6 A new image consisting of the concatenation of the grayscale version of the original image and its mirror image.

Figure 3.7 The horse image with holes.

17. Load the horse image from Figure 3.7 [3] so that the horse pixels have a value of 1 and all other pixels have a value of 0. Use the **ndimage.binary_fill_holes** function to remove the line through the middle of the horse.
18. Load the bird image and use the **ndimage.shift** function to change the color of the image. The information in the first color channel should be placed in the second channel. The information of the second channel should be placed in the third, and the information of the third should be placed in the first.

4 Digital Images

In the world of modern computers and smart phones, images are generally collected as digital information. There are a discrete number of pixels, and each pixel has a finite range of values. Today's common cameras can capture an image with a tremendous resolution with a few thousand pixels in both the vertical and horizontal dimensions.

Each pixel contains a single value that represents a level of intensity or three values, which represent a color. An image of this size would contain 12,000,000 pixels, and each pixel can have more than 16,000,000 different combinations values to create a color. This amount of data is huge and so it is necessary to compress the information when storing in on a disk. The use of the incorrect compression technique can have devastating effects on the image. Therefore, it is prudent to understand the basics of image resolution and the most popular image compression methods.

4.1 IMAGES IN PYTHON

There are two methods by which Python can load an image. The first uses the Python Image Library to load an image data type and then convert this to a matrix or tensor. Line 3 in Code 4.1 loads the image and the variable `mg` is an image type. To manipulate this data, it is usually necessary to convert it to a matrix or tensor, which is shown in line 4.

The second method uses the *imageio* library commands, which is shown in Code 4.2. The array `data` in line 2 is a matrix or tensor depending on the type of input. Both the arrays of data will be presented in the form of $V \times H \times N$, where V and H are the vertical and horizontal dimensions. The value N is the number of channels. Often it is desired to rearrange the data in the form of $N \times V \times H$. This is accomplished in line 3 using the `transpose` command.

4.2 RESOLUTION

There are two types of resolution in an image. *Intensity resolution* represents the range of values that each pixel can have. *Spatial resolution* is the size of the image in terms of the number of pixels.

4.2.1 INTENSITY RESOLUTION

The intensity resolution of an image refers to the number of gray levels that each pixel can realize. Quite commonly a pixel in a grayscale image will have 256 different intensity levels. This is a

Code 4.1 Loading the image using Python Image Library

```
1 >>> import Image
2 >>> import numpy as np
3 >>> mg = Image.open( fname )
4 >>> data = np.array( mg )
```

Code 4.2 Loading the image using commands from *imageio*

```
1 >>> import imageio
2 >>> data = imageio.imread( fname )
3 >>> nvh = data.transpose( (2,0,1) )
```

convenient number since it extends beyond the discrimination ability of most humans, and it is also the intensity range of most computer monitors. In other words, if an image had 1,000 gray levels, a human would not be able to see the difference between consecutive gray levels, and a common computer display cannot display the full resolution.

The bird image (see Figure 3.4(a)) is a color image and loading it as a grayscale image from a file is represented by

$$\mathbf{b}[\vec{x}] = \mathcal{L}_L Y(''\texttt{data/bird.jpg}''), \tag{4.1}$$

where the subscript L represents the grayscale model. Figure 4.1(a) shows the image $\mathbf{b}[\vec{x}]$.

The image is loaded by line 3 in Code 4.3. Using the `as_gray=True` returns, the image as floating point values. These are converted to unsigned 8-bit integers so that each pixel represented by 8 binary values or 256 levels of intensity. Line 4 creates a new image in which the four highest bits are kept and the lowest four bits are all set to 0. This reduces the spatial resolution from 256 gray levels to only 16 gray levels. The process is described as

$$\mathbf{c}[\vec{x}] = \triangle_{F0}\mathbf{b}[\vec{x}], \tag{4.2}$$

which is equivalent of multiplying each pixel value by the hexadecimal value $F0$. This result is shown in Figure 4.1(b), and as seen there is very little difference in two images. The bird's body which has a smoothly changing intensity does show the effects of the reduced intensity resolution. However, the leaves tend to be very similar in both images. Lines 6 and 7 reduce the intensity resolution to 4 levels and then 2 levels, and these are shown in Figure 4.1(c) and (d). Clearly, as the levels of gray are reduced, the ability to distinguish different features is also reduced.

A second example starts with an image from an early version of the gene expression arrays. These images were to reveal circular regions of interactions of DNA samples. In this case, the detector was

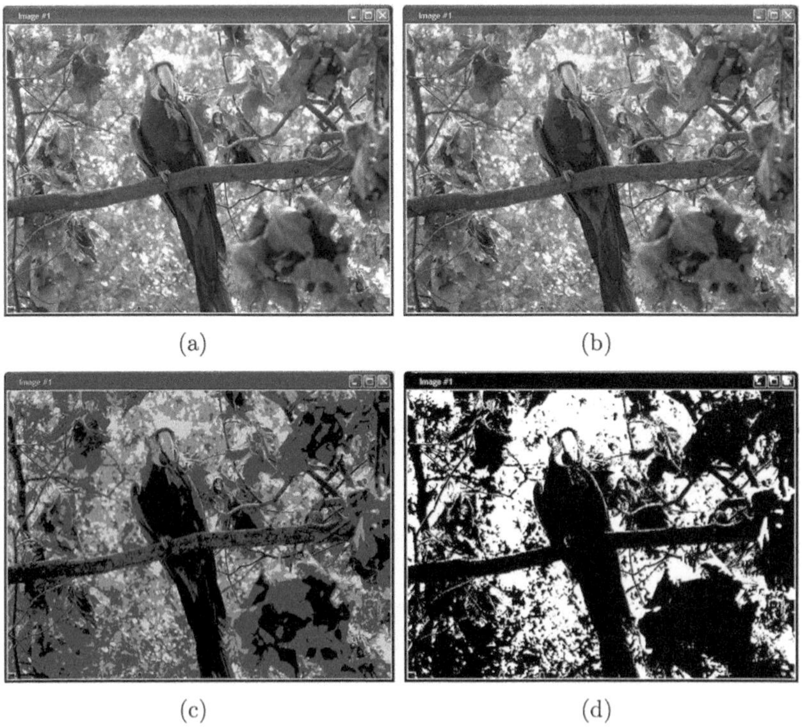

Figure 4.1 (a) The original grayscale image, (b) the image reduced to 6 bits or 64 different gray levels, (c) the image reduced to 4 bits or 16 gray levels, and (d) the image reduced to 1 bit or 2 gray levels.

Code 4.3 Reducing the intensity resolution

```
>>> import numpy as np
>>> import imageio
>>> mgdata = imageio.imread('data/bird.jpg', as_gray=True).astype(np.uint8)
>>> c1 = mgdata & 0xF0
>>> imageio.imsave('birdF0.png', c1 )
>>> c2 = mgdata & 0xc0
>>> c3 = mgdata & 0x80
>>> mgdata.size
(653, 519)
```

a 16-bit camera capable of detecting up to 16,384 different gray levels. However, when first viewed on the computer screen, the image from the detector appeared blank except for a few pixels. The reason why this image was blank was the computer monitor that only had the ability to display 256 gray levels, and thus, the conversion from 16-bit data to 8-bit data consisted of keeping the 8 highest bits of information from the original image. The computer monitor still shows 256 levels of gray, but the pixel intensities associated with those levels are: 0, 256, 512, ... , 16384. Any pixel with a value of less than 256 will appear as a black pixel on the screen. The histogram of the data shown on a log scale in image 4.2(a) shows that a large majority of the information was in the lower values well below the threshold. Thus, the intensity levels that were displayed on the computer screen had very little information. Selection of the correct intensity levels revealed the array spots which are shown in Figure 4.2(b). The data had a spatial resolution of 256 levels of resolution.

4.2.2 SPATIAL RESOLUTION

Spatial resolution is often equated to the number of pixels in an image, but this is not exactly correct. It is possible to have an image in which every 2×2 block of pixels have the same intensity. While the number of pixels may be $V \times H$, where V is the number of pixels in the vertical dimension and H is the number of pixels in the horizontal dimension, the spatial resolution would only be $V/2 \times H/2$.

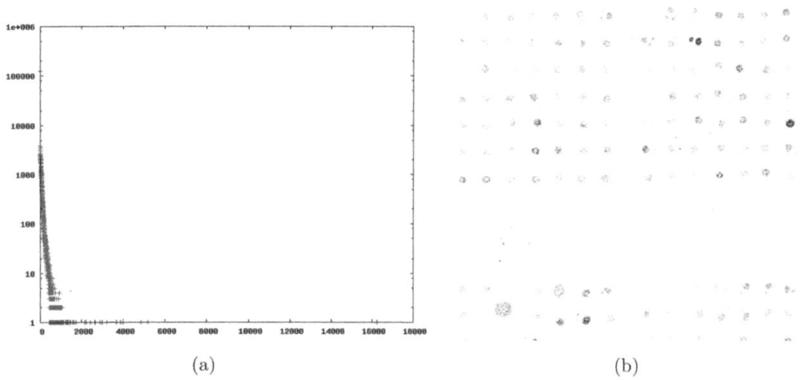

(a) (b)

Figure 4.2 (a) The log histogram of the pixel values and (b) selection of the middle 8 bits reveal the image with the most clarity. The *y*-axis is on a log scale from 1 up to 1,000,000. The *x*-axis is linear with each tic mark representing a value of 2,000.

Line 9 in Code 4.3 displays the spatial resolution as 653 horizontal pixels and 519 vertical pixels. The custom is that images provide the spatial dimensions with the horizontal value first. Custom also dictates that the opposite is true for matrices, and the vertical dimension is shown first.

The spatial resolution of an image can be reduced using the Downsample operator $\Downarrow_n$. The subscript n indicates the formula for sampling and can take many different forms. The operator $\Downarrow_2$ would extract every second row and every second column from the data, thus creating a new image that was half of the size of the original in both dimensions. The operator $\Downarrow_{m,n}$ would extract every m-th row and n-th column. More complicated extractions could be written in the form $\Downarrow_{\vec{w},\vec{v}}$ where $\vec{w}$ dictates which rows are extracted and $\vec{v}$ dictates which columns are extracted. Each of these would be integer vectors. The subscript can also be defined by the user as needed.

The operation

$$\mathbf{c}[\vec{x}] = \Downarrow_2 \mathbf{b}[\vec{x}], \tag{4.3}$$

would sample even numbered rows and columns creating a new image $\mathbf{c}[\vec{x}]$ that is half of the size of the original image $\mathbf{b}[\vec{x}]$ in every dimension.

The result is shown in Figure 4.3(a). To display the image, it is expanded to twice its new size so that it displays as the same size on the page. Most of the information is there but the title bar clearly shows that some of the finer details are ruined. Reducing the spatial resolution by another factor of 2 produces the result shown in Figure 4.3(b). Again, this image is visually expanded so that it has the same size on the display page. The details are destroyed. Figure 4.3(c) and (d) shows two more reductions, each of another factor of 4.

To reiterate the distinction between number of pixels and spatial resolution, it should be noted that all of these images consume the same amount of real estate on the page. Thus, they all have the same number of pixels. However, they clearly have different spatial resolutions.

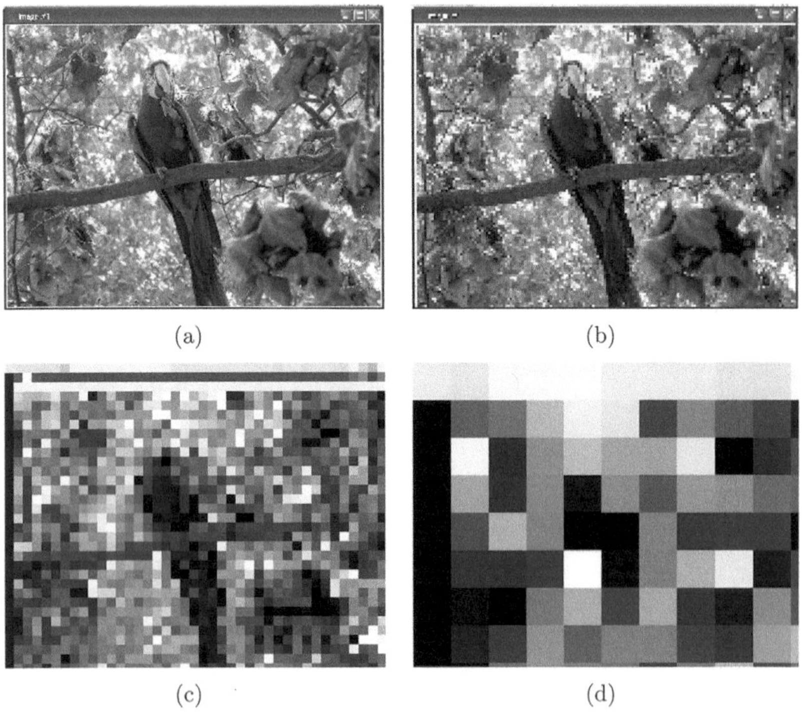

Figure 4.3 (a) The image with a spatial resolution of 128×128. (b) The image with a spatial resolution of 64×64. (c) The image with a spatial resolution of 16×16. (d) The image with a spatial resolution of 4×4.

4.3 DIGITAL FORMATS

Storing an image digitally can be a rather simple concept in which each pixel is stored in three bytes (one each for color channel). However, this is extremely inefficient and can produce very large files. For example, a common smart phone can capture an image that is $4,160 \times 2,340$, which is over 9.7 million pixels. Furthermore, this image is a color image which means that it would take over 29 megabytes to store the image. If images were stored in this fashion, the Internet would grind to a halt. Images are therefore usually compressed into one of several formats. The compression drastically reduces the size of the data file. Greater compressions can be achieved if the user is willing to sacrifice image quality. In some cases, this sacrifice is hardly noticeable.

The improper use of image compression, however, can seriously damage an image. So, this section provides a quick review of some of the compression algorithms and a discussion on when it is appropriate and inappropriate to use certain compressions.

4.3.1 BITMAPS

A bitmap image stores that image with one byte per pixel for grayscale images and three bytes per pixel for color images. Each of the formats has a few bytes at the beginning of the file to store information such as the dimensions of the image. Thus, a bitmap image is not compressed. Such file formats have filename extensions such as `.bmp`, `.tga`, `.pgm`, and `.ppm`. These are usually used only for very small images such as icons. In some rare cases, they are used for large images to ensure that there is no loss of information through compression.

4.3.2 JPEG

The JPEG (which is an acronym for Joint Photographic Experts Group) format is commonly used for photographs. This format sacrifices clarity of sharp edges for compression efficiency. Since most photographs do not contain exceedingly sharp edges, this sacrifice is often imperceptible. In fact, most digital cameras store images in the JPEG format.

The JPEG compression converts the image from RGB to YCbCr (see Sections 5.1 and 5.3), and the two chroma channels (Cb and Cr) are reduced to half of their original size. These are the channels that store color information, and since humans have a lower spatial resolution for color, these reductions are not noticeable. Each channel is then divided into small squares and the discrete cosine transform (DCT) is computed for each square which stores frequency information. The DCT coefficients are then quantized and many of the values are set to 0. The compression occurs in two places. The first is the reduction of the Cb and Cr channels in which the total number of pixels is reduced by a factor of 4. The second is the quantization of the DCT coefficients, as data with consecutive zeros compresses well. Compression by a factor of 10 is typical without harming the image in any significant manner.

However, there are certain images for which JPEG is the incorrect compression method to apply. Figure 4.4(a) shows an image with four letters in it. The image was stored through JPEG compression, and the results are shown in Figure 4.4(b), which has been digitally enhanced to exaggerate the intensity of the transients. In this image, the view has been zoomed in near the bottom of the "J" to demonstrate the errors that occurs near the edges within the image. There are spurious pixels both inside and outside the original image. This is the ringing that occurs through JPEG compression. While JPEG is good for photographs, it should be noted that it is not good for storing cartoons or drawings because they contain razor sharp edges.

4.3.3 GIF

The GIF format uses a palette to store color information. A look-up table is created that can store 256 different colors (RGB values), but this is far too few colors to use on photographic images. The

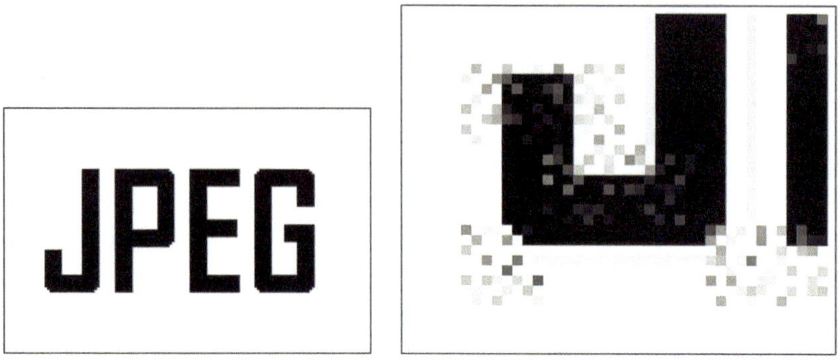

(a) (b)

Figure 4.4 (a) The original image and (b) a portion of the image after JPEG compression with the intensity of the ringing effects enhanced for viewing purposes.

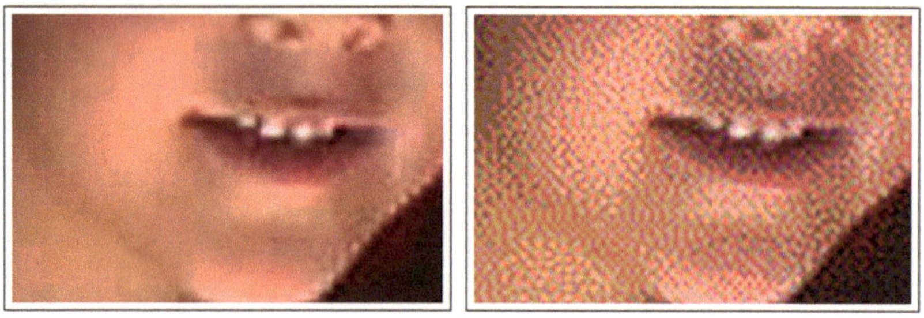

(a) (b)

Figure 4.5 (a) A portion of the original image and (b) the same portion saved in the GIF format.

result is that GIF compression will estimate which 256 colors best represents the image. A portion of an image is shown in Figure 4.5(a), which shows the color image stored in the PNG format. The same portion is saved in Figure 4.5(b), and the damage to the image is apparent. This image is shown zoomed in by a factor of 4 to display the dithering effect of GIF storage. Since only 256 colors can be used, the compression algorithm dithers these colors so that the image appears to contain more colors.

While GIF has problems with color images it does compress grayscale images very well without loss of information. The reason is that a grayscale image technically contains only 256 different colors (all of which are different shades of gray). So, there is no estimate necessary. GIF compression works well for grayscale images or color images with very few colors (such as banners or cartoons). GIF is also widely used for small "movies" on web sites. Basically, a GIF file can contain multiple images and the viewer will cycle through the images to create a small movie.

4.3.4 TIFF

The TIFF image format can store data with or without compression. If the latter is chosen, then TIFF files can be larger than bitmap files. TIFF images do have the ability to store 16-bit pixels which makes them a candidate for images being obtained from scanning equipment.

As stated before, some gene expression array scanners used the TIFF format to store their 16-bit images. Quite often, the scanner software will not compress a TIFF image which allows the user to view the raw data in the file. This is actually useful in some companies which will modify TIFF images by inserting new data fields in the file. Standard TIFF readers will not recognize these companies specific fields and therefore may not be able to display the images. For uncompressed data, the user can write small programs to read the TIFF information and create a viewable image.

4.3.5 PNG

The PNG format was designed for transmitting images over the Internet, thus gaining its acronym for PNG (portable network graphics). It is a lossless compression and so usually creates files that are bigger than JPEG. However, PNG images store any type of RGB image data without damaging the information. PNG compression algorithms tend to be slower than others and less efficient. However, they are commonly accepted by web browsers and do not generate transients.

One of the advantages of the PNG compression is that it is nonpatentable and thereby not causing usage issues. This is a successful strategy, and in early 2013, PNG surpassed GIF in popularity on the web [33].

4.3.6 OTHER COMPRESSIONS

There are many different image compression formats in use. Some are application specific. For example, some manufacturers of DNA sequencing machines stored their image information in a modified TIFF format by adding manufacturer information at the end of the image. DICOM formats are used by the medical industry as this format is designed to store a variety of images, patient information, and information about the collection of the images. Standard Python installations read neither of these, but third party software is freely available that users can install.

4.4 SUMMARY

Digital images have two different resolutions. One is the intensity resolution which is the number of intensity levels that any pixel can realize. The second is spatial resolution which is the number of distinct pixels in each dimension. There can be a difference in the number of pixels and the number of distinct pixels as demonstrated. If, for example, an image was doubled in size then the number of pixels in each dimension doubles, but the information in the image contained therein does not.

A few of the many image storage formats were reviewed. The most popular formats lightly described with an emphasis on the most important feature being the proper application of the compression format for the application. The rest of the chapters will use only the formats that are described in this chapter.

PROBLEMS

1. In this problem, the rotations should be performed without changing the frame size. Use a binary image such as the one shown in Figure 3.5. Perform the following operation,

$$\mathbf{b}[\vec{x}] = \mathcal{R}_{-\theta}\mathcal{R}_{\theta}\left[\mathbf{a}[\vec{x}] + 1\right],$$

   where $\theta = 25°$. If the frame is not expanded, this operation will lose information at the corners of the image. Since a value of 1 was added to $\mathbf{a}[\vec{x}]$, the only pixels with a value of 0 are those that are not defined after the rotation. Determine the number of pixels that are lost due to the rotations.
2. In the previous problem, determining the number of pixels that are not defined required the user to count the number of the pixels that have a value of 0. Write the operator for this function.
3. Create a plot of compression efficiency versus number of pixels, where the compression ratio is the number of pixels in the filed divided by (the number of pixels times the number of color channels). To create this plot, collect several images of different sizes and content

Figure 4.6 Created image.

and for each image plot of point, where the x-axis location is the number of pixels and the y-axis location is the compression ratio. Is there a trend?

4. Write a script that loads a GIF image and counts the unique color levels therein.
5. Compute the average pixel error for an image compressed with JPEG. Start with an image that is a bitmap (or PNG), $\mathbf{a}[\vec{x}]$, and save it as a JPEG, $\mathbf{b}[\vec{x}]$. Then compute,

$$f = \sum_{\mathcal{L}} \frac{\sum_{\vec{x}} |\mathbf{a}[\vec{x}] - \mathbf{b}[\vec{x}]|}{N\mathbf{a}[\vec{x}]}$$

6. To create the image in Figure 4.6, the following protocol was employed. Write a script to recreate this image using the given protocol.

$$\mathbf{r}[\vec{x}] = \Gamma_{p,>1} \frac{\begin{Bmatrix} 1 \\ \varnothing \\ \varnothing \end{Bmatrix} \mathbf{a}[\vec{x}]}{\begin{Bmatrix} \varnothing \\ 1 \\ \varnothing \end{Bmatrix} \mathbf{a}[\vec{x}] + 0.1}$$

$$\mathbf{b}[\vec{x}] = \triangle_{C0} \mathcal{L} \mathbf{a}[\vec{x}]$$

$$\mathbf{m}[\vec{x}] = \mathbf{b}[\vec{x}] > 0$$

$$\mathbf{c}[\vec{x}] = (1 - \mathbf{m}[\vec{x}]) \times \mathbf{b}[\vec{x}] + \begin{Bmatrix} \mathbf{m}[\vec{x}] \times \mathbf{r}[\vec{x}] \\ 0 \\ 0 \end{Bmatrix}$$

7. The fingerprint image `data/fing.png` has the dimensions 703×745. The reduction of the spatial resolution is defined by

$$\mathbf{b}[\vec{x}] = \Downarrow_n \mathbf{a}[\vec{x}].$$

What is the largest value of n that does not destroy ridge information (i.e., neighboring ridges are still separated by a valley)?

5 Color

In order to represent color, each pixel in an image, such as the one shown in Figure 5.1, is required to have three values. However, there are many different ways in which three values can represent color. This chapter will review some popular methods and how the correct application of a model can lower the complexity of an image analysis task.

5.1 THE RGB COLOR MODEL

The best-known format is RGB which encodes the color as three channels (red, green, and blue). The RGB channels for Figure 5.1 are shown in Figure 5.2. The first image corresponds to the red channel, and the pixels that are bright in this image contain significant red contributions. So obviously, the bird's body is brighter in this channel than it is in the others because it is red. Those same pixels are darker in the other two channels. The blue border is bright in the blue channel, and white pixels are bright in all channels.

The *Channel operator* allows the user the ability to manipulate individual channels of a color image. There are several ways that it can be used. For example, a single pixel can be set by representing the three color values (such as RGB) inside curly braces. Another application is to multiply each channel by individual values. For example, if the task is to multiply the red, green, and blue channels by the scalars r, g, and b, respectively, then the operation is written as

$$\mathbf{b}[\vec{x}] = \begin{Bmatrix} r \\ g \\ b \end{Bmatrix} \mathbf{a}[\vec{x}]. \tag{5.1}$$

The symbol $\varnothing$ is used to block a channel, and so the notation for isolating the red channel is

$$\mathbf{r}[\vec{x}] = \begin{Bmatrix} 1 \\ \varnothing \\ \varnothing \end{Bmatrix} \mathbf{a}[\vec{x}]. \tag{5.2}$$

The image in Figure 5.3 shows suppressed pixel values from the foliage. It was created by

$$\mathbf{a}[\vec{x}] = Y(\texttt{'data/bird.jpg'}), \tag{5.3}$$

$$\mathbf{b}[\vec{x}] = \sum_{\mathcal{L}} \begin{Bmatrix} \varnothing \\ 0.5 \\ 0.5 \end{Bmatrix} \mathbf{a}[\vec{x}], \tag{5.4}$$

$$\mathbf{c}[\vec{x}] = \Gamma_{>140} \mathcal{L}_L \mathbf{a}[\vec{x}], \tag{5.5}$$

and

$$\mathbf{d}[\vec{x}] = \begin{Bmatrix} \mathbf{b}[\vec{x}] \times \mathbf{c}[\vec{x}] + (1 - \mathbf{c}[\vec{x}]) \begin{Bmatrix} 1 \\ \varnothing \\ \varnothing \end{Bmatrix} \mathbf{a}[\vec{x}] \\ \mathbf{b}[\vec{x}] \\ \mathbf{b}[\vec{x}] \end{Bmatrix}. \tag{5.6}$$

Code 5.1 shows the associated script. Line 1 corresponds to Equation (5.3), line 2 corresponds to Equation (5.4), and line 3 corresponds to Equation (5.5). Lines 4 through 7 correspond to Equation (5.6).

Figure 5.1 An original color image.

(a) (b) (c)

Figure 5.2 The three color channels: (a) red, (b) green, and (c) blue.

Figure 5.3 Suppressing the color of the non-red pixels.

Code 5.1 Creating an image that suppresses the background

```
>>> adata = imageio.imread( 'data/bird.jpg')
>>> bdata = (adata[:,:,1] + adata[:,:,2] )/2
>>> cdata = imageio.imread( 'data/bird.jpg',as_gray=True) > 140
>>> ddata = np.zeros( adata.shape )
>>> ddata[:,:,0] = cdata*(bdata) + (1-cdata)*adata[:,:,0]
>>> ddata[:,:,1] = bdata + 0
>>> ddata[:,:,2] = bdata + 0
```

5.2 THE HSV COLOR MODEL

While RGB is an easy method of representing color, it is not a useful format for the purposes of analysis and recognition. In the RGB format, the intensity of a pixel is mixed in with its hue, and this creates a difficulty in many analysis techniques. There are several other formats, which separate the intensity from the hue that tends to work better. The first model considered is HSV.

The HSV color model represents data in three channels known as hue, saturation, and value. The V channel represents the intensity information. The H channel represents the hue, which is the ratio of the two largest values from the three RGB values for each pixel. Figure 5.4 displays a polar map in which red is at $\theta = 0$. If a pixel had G = B = 0, then the hue would be purely red or $\theta = 0$. If the two largest values are R and G, then the angle would be between 0° and 120°. Thus, the hue is measured in angles between 0° and 360°. The S represents the saturation which is a measure of the color brilliance which is ratio of the difference between the smallest of the RGB values to the largest value.

The *colorsys* provides a few functions to convert from one color model to another. The function **rgb_to_hsv** converts a single RGB pixel to its HSV values. Likewise, the **hsv_to_rgb** function converts HSV values to RGB values. A few examples are shown in Code 5.2. Line 2 converts a pure, bright red pixel to its HSV values. The H value is 0 because pure red is at an angle of 0 in Figure 5.4. The S value is 1.0 because this is a pure color. The V value of 255 indicates that this pixel has the highest possible intensity. Line 4 performs the same operation on a very dim red pixel, and as seen, the H and S values are the same. This is still a pure red pixel. The intensity, though, is small.

Line 6 converts a green pixel and line 8 converts a blue pixel. The H value for green is 1/3 because pure green is one-third of the total angle in Figure 5.4. Likewise, the H value for blue is 2/3. As the angle in Figure 5.4 heads toward the angle of 360° the color becomes red again. As seen in lines 10 and 11, an H value of 1 is the same as an H value of 0.

These functions convert the values of a single pixel, but quite often this conversion needs to be applied to the entire image. The **numpy.vectorize** function will create a new function that can be applied to an array. Line 1 creates a new function which is given the name **rgb2hsv**, and line 2 applies this new function to all of the pixels in an image `amg`.

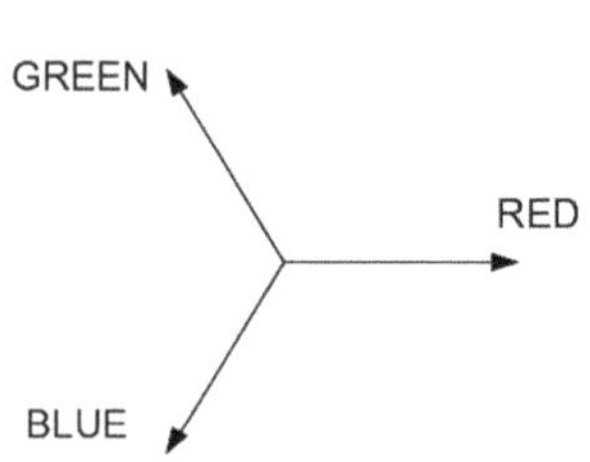

Figure 5.4 HSV Coordinates [28].

Code 5.2 Converting between HSV and RGB values

```
>>> import colorsys
>>> colorsys.rgb_to_hsv( 255,0,0)
(0.0, 1.0, 255)
>>> colorsys.rgb_to_hsv( 1,0,0)
(0.0, 1.0, 1)
>>> colorsys.rgb_to_hsv( 0,255,0)
(0.3333333333333333, 1.0, 255)
>>> colorsys.rgb_to_hsv( 0,0,255)
(0.6666666666666666, 1.0, 255)
>>> colorsys.hsv_to_rgb(1,1,255)
(255, 0.0, 0.0)
```

Code 5.3 The **vectorize** function applies the operation to all pixels

```
>>> rgb2hsv = np.vectorize( colorsys.rgb_to_hsv)
>>> h,s,v = rgb2hsv( amg[:,:,0], amg[:,:,1], amg[:,:,2] )
```

The HSV channels for the bird image are shown in Figure 5.5. The H channel follows Figure 5.4; in that, the darkest and brightest pixels are associated with the red hue. The values near one-third of the maximum are associated with green, and the values near two-thirds of the maximum are associated with blue. The bright pixels in the S channel are associated with the purest colors. So, the bird and particularly the blue border have high intensities in this channel. The V channel reflects the intensities of the pixel in the original image.

The $\mathcal{L}_m$ operator converts an image from one color model to the m model. So, for this conversion, the notation is

$$\mathbf{b}[\vec{x}] = \mathcal{L}_{\text{HSV}}\mathbf{a}[\vec{x}]. \tag{5.7}$$

Consider an example in which the task is to modify the image by replacing all of the H channel pixels with the square of their original values. The protocol is:

1. Convert the RGB image to HSV.
2. Compute the squares in the H channel.
3. Create a new image using these new H values along with the original S and V values.
4. Convert this image back to RGB for viewing.

(a) (b) (c)

Figure 5.5 The three HSV channels. (a) The H channel, (b) the S channel, and (c) the V channel.

Given the original RGB image, $\mathbf{a}[\vec{x}]$, the Color Transformation operator to HSV notion is

$$\mathbf{b}[\vec{x}] = \mathcal{L}_{\mathrm{HSV}}\mathbf{a}[\vec{x}], \tag{5.8}$$

and the modifications of the channels is expressed as

$$\mathbf{c}[\vec{x}] = \left\{ \begin{matrix} \varnothing \\ 1 \\ 1 \end{matrix} \right\} \mathbf{b}[\vec{x}] + \left(\left\{ \begin{matrix} 1 \\ \varnothing \\ \varnothing \end{matrix} \right\} \mathbf{b}[\vec{x}] \right)^2 \tag{5.9}$$

where the first term passes the S and V channels without alteration and the second term squares only the H values. The image $\mathbf{c}[\vec{x}]$ is still in the HSV format and so the final step is convert it back to RGB space:

$$\mathbf{d}[\vec{x}] = \mathcal{L}_{\mathrm{RGB}}\mathbf{c}[\vec{x}]. \tag{5.10}$$

The Python script for performing this tasks is shown in Code 5.4. Line 6 converts the array of RGB values to HSV values. Line 7 squares the H channel, and line 9 converts the values back to RGB values. However, line 9 creates a tuple, and this is converted to the properly designed array in line 10 so that the data can be saved as an image (Figure 5.6).

Code 5.4 Modifying the hue channel

```
1  >>> import imageio
2  >>> import numpy as np
3  >>> import colorsys
4  >>> rgb = imageio.imread( 'data/bird.jpg')
5  >>> rgb2hsv = np.vectorize( colorsys.rgb_to_hsv)
6  >>> h,s,v = rgb2hsv( rgb[:,:,0], rgb[:,:,1], rgb[:,:,2] )
7  >>> h *= h
8  >>> hsv2rgb = np.vectorize( colorsys.hsv_to_rgb)
9  >>> rgb2 = hsv2rgb( h,s,v )
10 >>> rgb2 = np.array( rgb2 ).transpose( (1,2,0) )
```

Figure 5.6 The result after modifying the hue.

Figure 5.7 (a) An image with a gray drive and (b) the hue channel depicting widely varying values in the gray regions.

The HSV format does well for some applications as long as the pixels are colorful. This format, however, does have problems with an image with gray pixels such as the driveway shown in Figure 5.7(a). When this image is converted to HSV, the hue channel (shown in Figure 5.7(b)) varies widely in this range. Consider a pixel that has RGB values of (128,128,128). A slight increase in the red value will create an H value of 0. However, that same slight increase in the green channel will create an H value of 120°. For gray pixels, HSV is extremely sensitive to small fluctuations as seen in Figure 5.7(b).

5.3 THE YUV FAMILY

The YUV family of color models has several similar variants, and for many applications, the selection of which variant to use offers very little advantage. The conversion to the YUV representation is a linear transformation, which for a single pixel is described by

$$\begin{bmatrix} Y \\ U \\ V \end{bmatrix} = \begin{bmatrix} 0.299 & 0.587 & 0.114 \\ -0.147 & -0.289 & 0.436 \\ 0.615 & -0.515 & -0.100 \end{bmatrix} \begin{bmatrix} R \\ G \\ B \end{bmatrix}. \tag{5.11}$$

Each of the channels is a weighted linear combination of the original RGB values.

The YIQ color model is similar except that some of the matrix values are slightly different:

$$\begin{bmatrix} Y \\ I \\ Q \end{bmatrix} = \begin{bmatrix} 0.299 & 0.587 & 0.114 \\ 0.596 & -0.275 & -0.321 \\ 0.212 & -0.528 & 0.311 \end{bmatrix} \begin{bmatrix} R \\ G \\ B \end{bmatrix}. \tag{5.12}$$

YIQ is the only conversion from this family that is provided in the *colorsys* module. The Python script is similar to the RGB to HSV conversion and is shown in Code 5.5. The Y channel is the intensity. The I and Q channels are associated with hue and are shown in Figure 5.8. The coat is actually one color of blue, but the creases create different intensities. The I and Q channels do not have intensity information, and thus, in these channels, the coat appears quite uniform. Furthermore, the gray drive does not have the problem as experience in the HSV conversion.

Code 5.5 The RGB to YIQ conversion

```
>>> rgb2yiq = np.vectorize( colorsys.rgb_to_yiq)
>>> yiq = rgb2yiq( amg[:,:,0], amg[:,:,1], amg[:,:,2] )
```

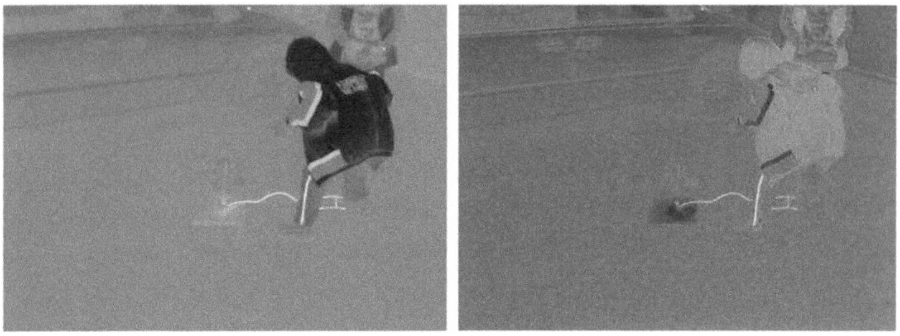

(a) (b)

Figure 5.8 The (a) I and (b) Q channels for Figure 5.7(a).

Code 5.6 Getting the Cb and Cr channels from the rocket image

```
1 >>> fname = 'data/rocket.jpg'
2 >>> mg = Image.open( fname)
3 >>> ycbcr = mg.convert('YCbCr')
4 >>> y,cb,cr = ycbcr.split()
5 >>> cb.show()
6 >>> cr.show()
```

A similar transformation is YCbCr, and it has similar matrix values but also adds a bias:

$$\begin{bmatrix} Y \\ Cb \\ Cr \end{bmatrix} = \begin{bmatrix} 0.2573 & 0.5013 & 0.0969 \\ -0.147 & -0.2907 & 0.4378 \\ 0.4378 & -0.3676 & -0.0712 \end{bmatrix} \begin{bmatrix} R \\ G \\ B \end{bmatrix} + \begin{bmatrix} 16 \\ 128 \\ 128 \end{bmatrix}. \tag{5.13}$$

The YIQ transformation is available through the *colorsys* module. The YCbCr conversion is available through the PIL (Python Image Library) Line 3 in Code 5.6 shows the conversion from an RGB image to an YCbCr image. Since PIL is being used, the variable `mg` is still an image, and any further manipulations to the data may require that they be converted to matrices.

The YUV family of transformations does provide some advantages over the HSV and RGB color models. However, that does not imply that it is a better model. Each application has its unique properties, and there is no single color model that addresses all of these properties. It can be prudent to view the images from a given problem in all of the color models before deciding which one best served the goal of the project.

5.4 CIE L*A*B*

The final conversion model to be reviewed here is CIE L*a*b*, which was designed to emulate human perception as it is based on human responses of different color frequencies. It is an involved conversion and only the protocol is reviewed here.

To compute the L*a*b* values, the RGB values are first converted to XYZ space through a linear conversion. This is performed through three functions shown in Code C.1, which is placed in Code C.1 in Appendix C. The first receives the RGB values and computes the linear transformation to XYZ space similar to `RGB2YCbCr`. The `XYZ2LAB` function performs the second conversion, and the results are shown in Figure 5.9. Once again, the intensity is displayed in one channel (the L channel), and the hue is represented in two other channels.

(a) (b) (c)

Figure 5.9 L*a*b* channels of the bird image.

5.5 IMPROVEMENTS IN RECOGNITION

Consider the case of isolating the boy's coat in Figure 5.7(a). In the RGB model, this would be somewhat difficult since the coat values have large ranges in the each of the three-color channels. In the YIQ model, the pixels on the coat have a small range of intensities, and thus, isolation by threshold would capture the coat and reject most of the other pixels.

Certainly, there are several models which can be used, and there is no magical transformation. Often, a task begins with the conversion of the image to several color formats and through visual inspection determines which model best isolates the pertinent information for the assigned task. Once the desired color model is selected, then the rest of the protocol can be implemented.

The task of isolating the boy's coat from the rest of the image can be accomplished by the following protocol:

1. Select the correct color model and convert the image,
2. Manually select a region of interest (ROI),
3. Collect statistics over the ROI,
4. Decide on threshold values,
5. Apply the thresholds, and
6. Display the result.

Step 1 is to determine the best color model and to convert the image to that model. After visually examining the results from RGB, HSV, and YIQ, the decision was that the best model to use was YIQ. The main criterion was the coat needed to be presented with a small range of pixel values that are unique to other regions of the image. Given the original RGB image $\mathbf{a}[\vec{x}]$, the conversion is

$$\mathbf{b}[\vec{x}] = \mathcal{L}_{\mathrm{YIQ}}\mathbf{a}[\vec{x}]. \tag{5.14}$$

Step 5 will require the application of threshold values, but before this can occur, it is necessary to determine those values. Step 2 is to determine the on-target ROI (region of interest). This is a region that contains only pixels from the coat that will then be used for the next step. In this case, it is easy to use a rectangular region which makes the capture of this region easy. The Window operator represents the isolation of the ROI and is presented in Section 6.1. Since it is rectangular, the ROI is defined by opposing corners $\vec{v}_1$ and $\vec{v}_2$, where $v_1 \in \mathbf{X}$ and $v_2 \in \mathbf{X}$. To keep the notation clean, the shorthand notation is employed,

$$\square_1 \equiv \square_{\vec{v}_1,\vec{v}_2}.$$

Step 3 applies this operator over a region and collects information such as the maximum value. This is applied to both the I and Q channels with the application to the I channel described as

$$\mathbf{c}[\vec{x}] = \begin{Bmatrix} \varnothing \\ 1 \\ \varnothing \end{Bmatrix} \mathbf{b}[\vec{x}], \tag{5.15}$$

$$m_1 = \bigvee \Box_1 \mathbf{c}[\vec{x}], \tag{5.16}$$

where $\bigvee$ is the Max operator. The minimum value is likewise obtained by

$$n_1 = \bigwedge \Box_1 \mathbf{c}[\vec{x}]. \tag{5.17}$$

The average and standard deviation over this region are determined by

$$\mu_1 = \mathcal{M}\mathbf{c}[\vec{x}], \tag{5.18}$$

and

$$\sigma_1 = \mathcal{T}\mathbf{c}[\vec{x}]. \tag{5.19}$$

In this example, the threshold values chosen were $\mu_1 \pm \sigma_1$. A similar process is followed for the Cr channel which repeats the above equations to determine m_2, n_2, μ_2, and σ_2. Again, the same criteria of $\mu_2 \pm \sigma_2$ is applied to determine the threshold values completing step 4.

Step 5 applies these thresholds, and in this case,

$$\mathbf{d}[\vec{x}] = \left(165.6 < \begin{Bmatrix} \varnothing \\ 1 \\ \varnothing \end{Bmatrix} \mathbf{b}[\vec{x}] < 185.4 \right) \times \left(78.7 < \begin{Bmatrix} \varnothing \\ \varnothing \\ 1 \end{Bmatrix} \mathbf{b}[\vec{x}] < 98.5 \right). \tag{5.20}$$

The output $\mathbf{d}[\vec{x}]$ is a binary-valued image, and the goal is that the only pixels that are set to 1 are on-target. The negative result, $(1 - \mathbf{d}[\vec{x}])$, of this process is shown in Figure 5.10.

This may seem like an easy problem since the coat is blue and there is little in the image that is of the same color. However, in the RGB color space, it is still troublesome to directly isolate the coat by color. Figure 5.10 shows the plot of the pixel values in RGB space. The dots denoted by green are from the ROI, and the red dots are from the rest of the image. The 3D perspective shown on a flat page does not show all of the relationship between the ROI pixels and the rest of the image; thus, the other two images in Figure 5.11 show 2D perspectives of the same data. Figure 5.11(b) shows the plot of R (x-axis) vs G (y-axis) pixel values, and Figure 5.11(c) shows R vs B. As seen in the images, the ROI pixels are separate from the rest of the image since there are no other blue objects. However, the extent of the ROI pixels along any axis significantly overlaps that of non-ROI pixels. There is no single value in these channels that can act as a threshold to isolate the on-target pixels. These channels are coupled, and that complicates the isolation of the target.

The same style of plot is created for the data in IQ space as shown in Figure 5.12. Since there are only two channels representing color, just a single plot is shown. As seen, the ROI pixels are isolated from the rest of the image and the boundary between ROI and non-ROI pixels is distinct. It is possible to isolate the coat using thresholds in the Cb and Cr spaces that are not coupled. Thus, isolation of the target is much easier in YCbCr space.

Figure 5.10 The target is isolated and shown as black pixels in the image.

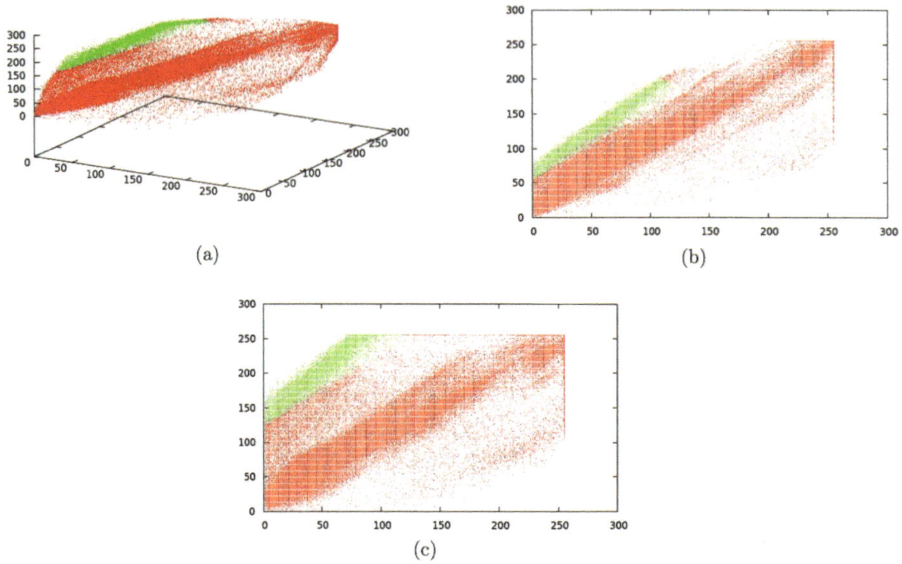

Figure 5.11 (a) contains the 3D plot of the pixels in RGB space. The pixels marked in green are from the ROI Figures (b) and (c) show 2D plots of the same data for red-green and red-blue, respectively.

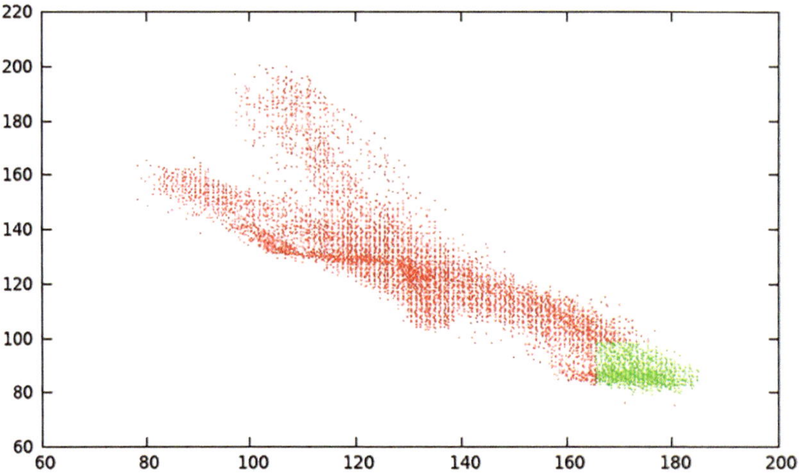

Figure 5.12 Pixels in I vs Q space with the dots in green denoting ROI pixels.

While this example depicts the usefulness of YIQ, it does not actually answer the question as to why it is better. Consider again the transformation from RGB to YIQ in Equation (5.13). This is a linear transformation and thus is basically a coordinate transformation. The Y channel is multiplied by three positive constants with the R, G, and B channels. Therefore, this axis ventures out into the first quadrant of the 3D RGB space. It is somewhat parallel with the long axis of the data in Figure 5.11(a). The other two axes (I and Q) are perpendicular and point into other quadrants. In combination, they separate the data from the perspective of looking down the long axis. The ROI data shown in Figure 5.11(a) also runs somewhat parallel along the long axis. A view along the long axis of the data is shown in Figure 5.13, and clearly the ROI pixels are isolated from the others.

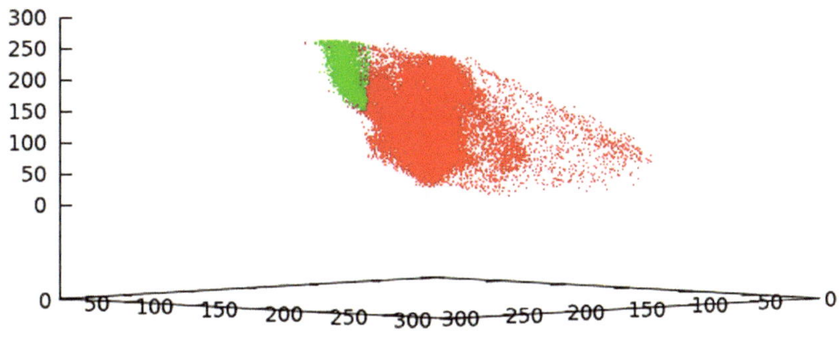

Figure 5.13 A view along the long axis of the data from Figure 5.11(a).

5.6 SUMMARY

Color can be represented by several different models. The most popular of these are RGB, models from the YIQ family, models from the HSV family, and the CIE L*a*b* model. The models other than RGB attempt to isolate the intensity information from the hue information. Thus, these models are more relevant to image processing applications. However, the choice of model is application dependent and there can even be cases in which RGB is the best selection.

PROBLEMS

For these problems, use a color image that has regions with brilliant reds, greens, and blues. In the following problems, this is called the test image.

1. Split the test image into its three components and save each as a grayscale image. Write the operator notation for this process.
2. Load the test image, exchange the RGB components such that the green information is now in the red channel, the blue information is now in the green channel, and the red information is now in the blue channel. Save this as a color image. Write the operator notation for this process.
3. Convert the test image to the HSV components and save each channel as a grayscale image. Write the operator notation for this process.
4. Convert the test image to YIQ and save each channel as a grayscale image. Write the operator notation for this process.
5. Write a script that will load the test image, convert it to HSV, multiply the H values by 2/3 and create a new RGB image using the modified H values. Write the operator notation for this protocol.
6. If color conversions were lossless, then the conversion from one space and then back to the original should produce exactly the same image as the original. Demonstrate that the color conversion process does lose some information by employing the following operations on the test image, $\mathbf{a}[\vec{x}]$,

$$\mathbf{b}[\vec{x}] = \left| (\mathcal{L}_{\text{RGB}} \mathcal{L}_{\text{YIQ}} \mathbf{a}[\vec{x}]) - \mathbf{a}[\vec{x}] \right|.$$

7. Perform the following operations using the test image as $\mathbf{a}[\vec{x}]$ and save the answer as an image.

$$\mathbf{b}[\vec{x}] = \mathcal{L}_{\mathrm{HSV}}\mathbf{a}[\vec{x}]$$

$$\mathbf{c}[\vec{x}] = \left\{\begin{matrix}\varnothing \\ 1 \\ 1\end{matrix}\right\}\mathbf{b}[\vec{x}] + \left(\left\{\begin{matrix}1 \\ \varnothing \\ \varnothing\end{matrix}\right\}\mathbf{b}[\vec{x}]\right)^{1/2}$$

$$\mathbf{d}[\vec{x}] = \mathcal{L}_{\mathrm{RGB}}\mathbf{c}[\vec{x}]$$

8. Starting with Figure 5.7(a), determine which color model would be best for isolating the green van. The choices to consider are RGB, YIQ, and HSV. Justify your selection.
9. Write the operator notation for the method selected in the previous problem.

Part II

Image Space Manipulations

6 Geometric Transformations

A geometric transformation converts an image by rearranging the location of the pixels. A couple of simple examples are a linear shift or a rotation. This section discusses the theory of geometric transformation functions and methods by which they can be achieved using Python functions. Some of the functions are provided by *scipy.ndimage*, while others are constructed from basic principles.

6.1 SELECTIONS

The most basic geometric operations are to extract a subimage from a larger image or to create a larger image from a set of smaller images. The Window operator extracts a subimage from the original. Usually, this will be a rectangular region defined by opposing corners $\vec{v}_1$ and $\vec{v}_2$. However, other applications may require the user to define the shape of the window. The Window operator is denoted by $\square_{\vec{v}_1,\vec{v}_2}$ and returns an image that is smaller in size than the original.

The opposite of the Window operator is the Plop operator, which places a smaller image inside a larger frame. The size of this frame is defined by the vector $\vec{w}$. The Plop operator is denoted by $U_{\vec{w}}$, and places the smaller image in the center of a larger frame.

Consider a small example which starts with the image in Figure 6.1(a). The task is simply to cut out a rectangular region around the clock and paste that into a new image that is the same size as the original. Given a color image $\mathbf{a}[\vec{x}]$, the notation to apply a Window and then a Plop is

$$\mathbf{b}[\vec{x}] = U_{\vec{w}} \square_{\vec{v}_1,\vec{v}_2} \mathcal{L}_L \mathbf{a}[\vec{x}], \tag{6.1}$$

where $\vec{v}_1$ and $\vec{v}_2$ define the opposing corners of the cutout region and $\vec{w}$ defines the output frame size. Code 6.1 shows the Python steps where line 4 cuts out the clock image and line 5 places it in a larger frame. This line uses the **mgcreate.Plop** function which is lengthy, and therefore, it is shown in Code C.3 in the Appendix.

A related operator is the Concatenate operator, which creates a larger image by placing two smaller images side by side. For two-dimensional images, the concatenation can either place the images next to each other in the horizontal dimension or the vertical dimension. Therefore, the operator is denoted with a subscript that indicates the direction of the concatenation. For a two-dimensional case, the vertical concatenation is denoted by $\mathcal{C}_V$ and the horizontal concatenation by $\mathcal{C}_H$. For higher dimensions, the user can define the meaning of the subscripts that are used. This operator receives a list of input images and produces a single output image.

The Downsample operator extracts rows and/or columns according to a user formulation. For instance, this operator can be used to extract every second row or every third column. Actually, the user can define any pattern and just denote this pattern with a user-defined subscript to the operator $\Downarrow$.

Consider a very simple example starting with the image in Figure 6.2(a). This image is created by adding 1 to every pixel in each even numbered row and again adding 1 to each even numbered column. Thus, any pixel that is located on an even numbered row and column has a value of 2, any pixel on an odd row and column has a value of 0, and all other pixels have a value of 1. The code for generating these images is shown in the first three lines in Code 6.2.

The goal is to reorganize the pixels shown in Figure 6.2(a) to the arrangement shown in Figure 6.2(b). This is accomplished by creating a new image from downsampling and concatenation processes. Four smaller images are created from downsampling the input and then these are arranged through concatenation. The first image is created by

$$\mathbf{b}[\vec{x}] = \Downarrow_{w_1} \mathcal{L}_L \mathbf{a}[\vec{x}]. \tag{6.2}$$

(a) (b)

Figure 6.1 (a) Original image and (b) portion of the image that has been cut out of the original.

Code 6.1 Using the Window and Plop operators

```
>>> import imageio
>>> import mgcreate as mgc
>>> amg = imageio.imread( 'data/bananas.png', as_gray=True )
>>> bmg = amg[102:317,387:544]
>>> cmg = mgc.Plop(bmg, (640,585), 255 )
```

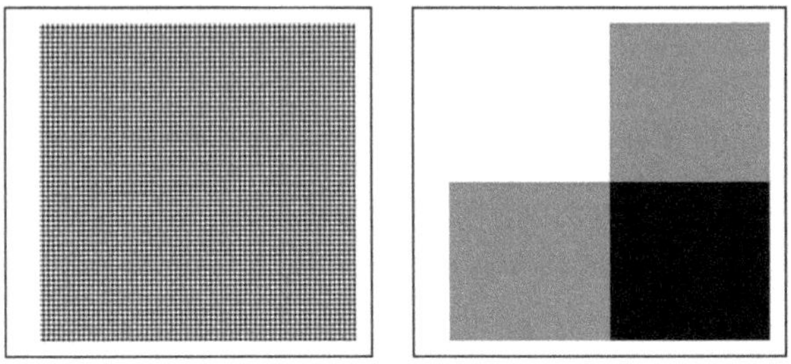
(a) (b)

Figure 6.2 (a) The original image and (b) an image created from four downsample operations and three concatenations.

The $\mathcal{L}_L$ operator converts the image to gray scale and the $\Downarrow_{w_1}$ function downsamples the image according to the prescription defined as w_1. In this case, w_1 is defined as only those pixels that are one even numbered rows and even numbered columns. Basically, it is extracting data from every other row and every other column.

The other images are obtained by,

$$\mathbf{c}[\vec{x}] = \Downarrow_{w_2} \mathcal{L}_L \mathbf{a}[\vec{x}] \tag{6.3}$$

$$\mathbf{d}[\vec{x}] = \Downarrow_{w_3} \mathcal{L}_L \mathbf{a}[\vec{x}] \tag{6.4}$$

$$\mathbf{f}[\vec{x}] = \Downarrow_{w_4} \mathcal{L}_L \mathbf{a}[\vec{x}]. \tag{6.5}$$

Code 6.2 Demonstrating the Downsample and Concatenate operators

```
>>> data = np.zeros( (128,128) )
>>> data[0:128:2] = 1
>>> data[:,:128:2] += 1
>>> b = data[0:V:2,0:H:2]
>>> c = data[1:V:2,0:H:2]
>>> d = data[0:V:2,1:H:2]
>>> f = data[1:V:2,1:H:2]
>>> g = np.concatenate( (b,c) )
>>> h = np.concatenate( (d,f) )
>>> m = np.concatenate( (g,h),1 )
```

The prescription w_2 extracts pixels from the odd numbered rows and the even numbered columns. The w_3 extracts pixels from the even numbered rows and the odd numbered columns. Finally, the w_4 prescribes the extraction of pixels from the odd numbered rows and the odd numbered columns.

The next step is to use the Concatenate operator to create a larger image from these newly created smaller images. The first step is to create a new image from stacking $\mathbf{b}[\vec{x}]$ and $\mathbf{c}[\vec{x}]$ vertically as in

$$\mathbf{g}[\vec{x}] = \mathcal{C}_V\{\mathbf{b}[\tilde{\mathbf{x}}], \mathbf{c}[\tilde{\mathbf{x}}]\}. \tag{6.6}$$

This is repeated to create another image that stacks $\mathbf{d}[\vec{x}]$ and $\mathbf{f}[\vec{x}]$:

$$\mathbf{h}[\vec{x}] = \mathcal{C}_V\{\mathbf{d}[\tilde{\mathbf{x}}], \mathbf{f}[\tilde{\mathbf{x}}]\}. \tag{6.7}$$

Finally, these two images are concatenated in the horizontal direction to form the final image:

$$\mathbf{m}[\vec{x}] = \mathcal{C}_H\{\mathbf{g}[\tilde{\mathbf{x}}], \mathbf{h}[\tilde{\mathbf{x}}]\}. \tag{6.8}$$

The result is shown in Figure 6.2(b), which shows the four subimages combined to create a new image that is the same size as the original.

6.2 LINEAR TRANSLATION

One of the simplest interpolations is a linear shift in which all of the pixels are moved in the vertical and/or horizontal directions. Problems that arise from shifting are that some pixels will become undefined, and in the cases where the shift includes fractions of pixel widths, an estimation of pixel intensity values may be required.

6.2.1 SIMPLE SHIFTING

The shift vector, $\vec{v}$, moves the information en masse to a new location. When an image is shifted, then part of the information moves out of the frame and pixels on the other side of the frame are undefined. For example, if an image is shifted to the right then the rightmost columns will leave the frame, while the leftmost columns are not defined. For now, those undefined pixels are set to 0. The Shift operator is formally defined as

$$\mathbf{b}[\vec{x}] = D_{\vec{v}}\mathbf{a}[\vec{x}] = \mathbf{a}[\vec{x} - \vec{v}]. \tag{6.9}$$

The *scipy* package provides the function **ndimage.shift** which translates the information within an image as shown in Code 6.3. Line 2 reads in the original image that is shown in Figure 6.3(a).

Code 6.3 Shifting an image

```
1 >>> fname = 'data/bird.jpg'
2 >>> adata = imageio.imread( fname, as_gray=True )
3 >>> bdata = nd.shift( adata, (10,25))
4 >>> cdata = nd.shift( adata, (10,25), cval=255)
```

(a) (b)

Figure 6.3 (a) The original image and (b) the shifted grayscale version.

Line 3 performs the shift which moves the image 10 pixels in the vertical direction and 25 pixels in the horizontal direction. The result is shown in Figure 6.3(b), and the operator notation is

$$\mathbf{b}[\vec{x}] = D_{\vec{v}} \mathcal{L}_L \mathbf{a}[\vec{x}], \tag{6.10}$$

where $\vec{v} = (10, 25)$.

The undefined pixels values are set to 0, but this can be controlled via the parameter *cval* as shown in line 4 which sets the undefined pixels to a specified value. The operator includes an optional term that defines the values for undefined pixels or the wrap-around. It is the b term in

$$\mathbf{c}[\vec{x}] = D_{\vec{v},b} \mathbf{a}[\vec{x}]. \tag{6.11}$$

6.2.2 NONINTEGER SHIFTS

The previous section shifted images by integer amounts, whereas shifting an image by a noninteger amount requires a bit more care. Pixels are not points in space but rather are rectangular regions $v_p \times h_p$. If the image is shifted by an amount nv_p, mh_p, where m and n are integers, then the rectangular pixels align with the rectangular grid of the original image. The grid shown in Figure 6.4 is the original location of the pixels, and the image shown is the new location after a shift in which m and n are non-integers. As seen, the pixels no longer align with the grid.

Figure 6.5(a) shows a few of the pixels and the original grid. In order to create the new image, each pixel as defined by the grid can only be represented a single value. One approach is to consider the pixel values as located at the center of the newly shifted pixels which are represented as dots in Figure 6.5(b). While this approach works for shifts, it will not work so well for other operations such as scaling as shown in Figure 6.5(c) in which it is possible for some rectangular grid regions to have zero or multiple pixel centers.

A better method is to use smoothing or spline fits to estimate the value for each rectangular grid. The function **ndimage.shift** does this estimation and the order of the spline is controlled by the value `order`, which has a default value of 3. To determine the effect of this order, two copies of an input image $\mathbf{a}[\vec{x}]$ were shifted by the same amount but with a different order value. These were

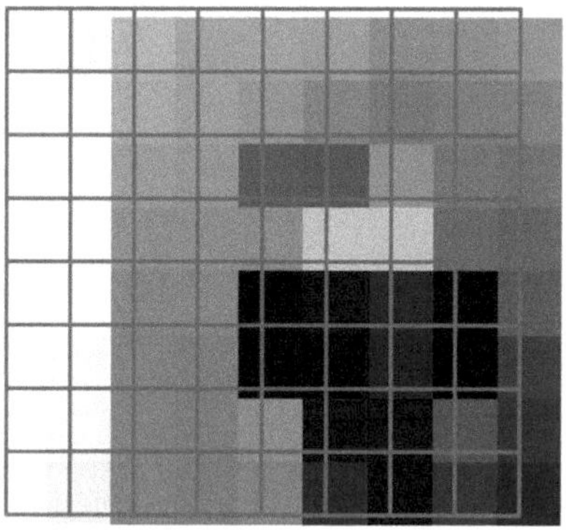

Figure 6.4 Shifting by a non-integer amount.

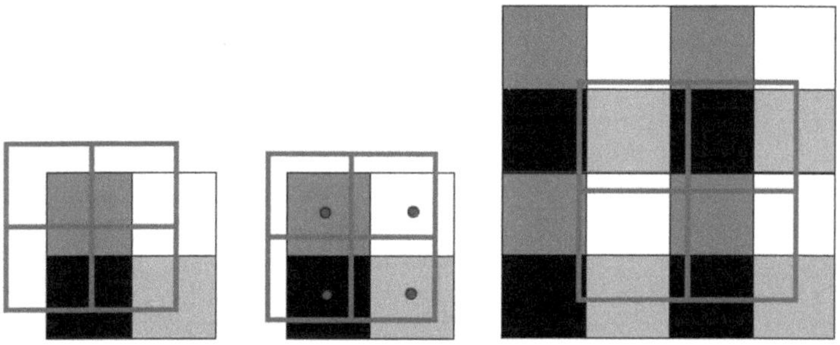

(a) (b) (c)

Figure 6.5 (a) The pixels and the shifting grid are not aligned. (b) The selection of the value for a pixel (box) is the gray value of the old pixel at the center of the square. (c) In cases where the image is also scaled, it is seen that several pixels may contribute to the value of the new pixel.

then subtracted and the average of the magnitude of the resultant pixels was computed. The nonzero result indicates that there is a difference between the two values, but the magnitude of the result is small indicating that the difference is not very significant in this example.

This test was performed by

$$\chi = \mathcal{M}\left| D_1 \mathbf{a}[\vec{x}] - D_2 \mathbf{a}[\vec{x}] \right|, \tag{6.12}$$

where D_1 represents a shift with `order` = 1 and D_2 represents the same shift with `order` = 3. The average of the absolute value of the subtraction is represented by $\mathcal{M}$. The answer χ is a scalar that is the average difference between the pixels from the two shifted images. A nonzero value indicates that there are differences between the two shifted images due to the order of their estimations.

The process is implemented in Code 6.4. The shift distance is 20.5 pixels in the vertical direction. Lines 1 and 2 create the two shifted images and line 3 computes the average of the magnitude difference between them. As seen, the result is a very small number, which means that this difference between the two shifts is not significant.

6.3 SCALING

Scaling increases or decreases the spatial extent of the image. If, for example, the image were scaled by a factor of 2, then it would use twice as many pixels in each dimension to contain the information. The Scaling operation is defined as

Code 6.4 Noninteger shifts

```
>>> m2 = shift( a, (20.5,0), order=1)
>>> m1 = shift( a, (20.5,0), order=3)
>>> abs(m1-m2).mean()
0.0072713632558151081
```

$$\mathbf{c}[\vec{x}] = S_m\mathbf{a}[\vec{x}], \tag{6.13}$$

where S_m is the Scaling operator and m is the magnitude of the scaling. If m is a scalar, then the scaling is the same in all dimensions. Of course, it is possible to scale each dimension differently by placing different scaling factors along the diagonal of a matrix $\mathbf{D}$ and employing

$$\mathbf{c}[\vec{x}] = \mathbf{a}[\mathbf{D}\vec{x}]. \tag{6.14}$$

The function **scipy.ndimage.zoom** performs the scaling functions. Examples are shown in Code 6.5 in which lines 1 through 4 are used to read in the image. As seen in line 6, this image is color and measures 653×519. The **zoom** function is applied in line 7, but as seen in line 9, this doubles the dimension in all three axes including the color axes. In this case, the zoom should be applied to only the first two axes and the method of doing this is shown in line 10 where the zoom factor for each axis is specified. This function can also be used to shrink an image using zoom factors that are less than 1.0 as shown in line 13. The result is shown in Figure 6.6.

Code 6.5 Scaling the image

```
1  >>> import imageio
2  >>> import scipy.ndimage as nd
3  >>> fname = 'data/bird.jpg'
4  >>> data = imageio.imread( fname )
5  >>> data.shape
6  (519, 653, 3)
7  >>> data2 = nd.zoom( data, 2 )
8  >>> data2.shape
9  (1038, 1306, 6)
10 >>> data2 = nd.zoom( data, (2,2,1))
11 >>> data2.shape
12 (1038, 1306, 3)
13 >>> data2 = nd.zoom( data, (0.5,0.9,1))
```

Figure 6.6 Scaling the image differently for each dimension.

6.4 ROTATION

The Rotation operator also moves pixels to a new location. Formally, the rotation of vector $\vec{x}$ is described by

$$\vec{y} = \mathbf{R}\vec{x}, \tag{6.15}$$

where $\mathbf{R}$ is the rotation matrix,

$$\mathbf{R} = \begin{bmatrix} \cos\theta & -\sin\theta \\ \sin\theta & \cos\theta \end{bmatrix} \tag{6.16}$$

and θ is the angle of rotation. To rotate an image, the coordinate of each pixel is considered as an input to Equation (6.15).

In operator notation, it is represented by

$$\mathbf{b}[\vec{x}] = \mathcal{R}_{\theta;\vec{v}}\mathbf{a}[\vec{x}], \tag{6.17}$$

where the $\mathcal{R}$ is the Rotation operator, θ is the rotation angle (where CCW is positive), and $\vec{v}$ is an optional argument that places the center of rotation someplace other than the center of the frame.

Code 6.6 uses the function **ndimage.rotate** to rotate the image about the center of the frame. The second argument in Line 1 is the angle of rotation in degrees and a positive value will rotate the image counterclockwise. The parameter *reshape* has a default value of `True` and controls the frame size which can be either the original frame size or enlarged to include all pixels from the original image. The result from line 1 is shown in Figure 6.7(a). In this case, the frame is resized so that all of the image is included in the result. When `reshape` is `False`, the frame size remains the same as the original and the corners of the original image are off the frame as shown in Figure 6.7(b).

Rotations are not perfect in that a small amount of information is lost during the estimation of the pixel values. Consider the following process,

$$\mathbf{b}[\vec{x}] = \square_{\vec{v}_1,\vec{v}_2}\mathcal{R}_{-\theta}\mathcal{R}_{+\theta}\mathbf{a}[\vec{x}], \tag{6.18}$$

where θ is a user-defined angle and $\vec{v}_1$ and $\vec{v}_2$ define the a rectangular region in $\mathbf{a}[\vec{x}]$ that is centered in the frame. Since the $\mathcal{R}$ changes the size of the frame, the $\square_{\vec{v}_1,\vec{v}_2}$ operator is used to extract a

Code 6.6 Rotation using scipy.ndimage

```
1 >>> data2 = nd.rotate(data, 10 )
2 >>> data2 = nd.rotate(data, 10, reshape=False )
```

(a) (b)

Figure 6.7 (a) The image with `reshape=True` and (b) the image with `reshape=False`.

Figure 6.8 Difference from the original and a result after many canceling rotations.

Code 6.7 Multiple rotations

```
1 >>> m2 = data + 0
2 >>> for i in range( 25 ):
3         m2 = nd.rotate(m2, 1)
4 >>> for i in range( 25 ):
5         m2 = nd.rotate(m2, -1)
6 >>> cut = m2[401:401+479, 355:355+638]
7 >>> answ = abs(cut-mat)
```

sub-image that is the same size as the original image. If there are no errors in the rotation algorithm then the $\mathbf{b}[\vec{x}] = \mathbf{a}[\vec{x}]$ should be true.

Now consider repetitive rotations of a small angle as in

$$\mathbf{b}[\vec{x}] = \Box_{\vec{v}_1,\vec{v}_2} \mathcal{R}^n_{-\theta} \mathcal{R}^n_{+\theta} \mathbf{a}[\vec{x}], \tag{6.19}$$

where the superscript n indicates that the process is repeated n times. In this example, $\theta = 1^\circ$ and $n = 25$. Again, if there were no errors in the rotation process then $\mathbf{b}[\vec{x}] = \mathbf{a}[\vec{x}]$ would be true. The image $|\mathbf{b}[\vec{x}] - \mathbf{a}[\vec{x}]|$ is shown in Figure 6.8 with the intensity inverted so that the dark pixels represent the largest values. Obviously, errors do exist. The images are pixelated and there is not a one-to-one mapping of the pixels in the rotation process.

Code 6.7 shows the steps that were used to obtain the image. Lines 2 and 3 rotate the image 1° 25 times, and lines 4 and 5 perform the same process but with the rotation in the opposite direction. Line 6 performs the window operation, and line 7 creates the final image.

6.5 DILATION AND EROSION

The Dilation and Erosion operators are often used to fill in holes and remove single pixel noise. The Dilation operator replaces pixel $a_{i,j}$ with the maximum value of its immediate neighbors,

$$b_{i,j} = \max_{k,l}(a_{k,l}; k = i-1, i, i+1; l = j-1, l, l+1).$$

Likewise, the Erosion operator replaces pixel $a_{i,j}$ with the minimum value of its immediate neighbors,

$$b_{i,j} = \min_{k,l}(a_{k,l}; k = i-1, i, i+1; l = j-1, l, l+1).$$

Code 6.8 Dilation operations

```
>>> data2 = nd.binary_dilation( data )
>>> data2 = nd.binary_dilation( data, iterations=3 )
>>> data2 = nd.grey_dilation( data )
>>> data2 = nd.grey_dilation( data, iterations=3 )
```

The operator notation for dilation is $\lhd_k$ where the optional subscript k indicates the number of iterations. The operator notation for an erosion is $\rhd_k$ again with an optional iteration value.

The *scipy.ndimage* module offers Dilation and Erosion operators for binary valued images and for gray-valued images. Example calls for dilation are shown in Code 6.8 showing the application to binary arrays, iterations, and gray arrays.

Consider the binary valued image shown in Figure 6.9(a), where, again, the image is shown with inverted intensities so that the black pixels correspond a value of 1 and the white pixels correspond a value of 0. Figure 6.9(b) is created by repeated uses of the Dilation operator.

Since the target is solid and binary valued, the Dilation operator will expand the perimeter one pixel in every direction. The operator $\lhd_3$ performs three iterations of the Dilation operator and thus expands the perimeter 3 pixels in every direction. Then, $\lhd_3\mathbf{a}[\vec{x}] - \mathbf{a}[\vec{x}]$ defines only the expansion and not the body of the original image. It is just the 3-pixel wide perimeter. The image in Figure 6.9(b) was created by

$$\mathbf{b}[\vec{x}] = \lhd_{12}\mathbf{a}[\vec{x}] - \lhd_9\mathbf{a}[\vec{x}], \tag{6.20}$$

$$\mathbf{c}[\vec{x}] = \lhd_6\mathbf{a}[\vec{x}] - \lhd_3\mathbf{a}[\vec{x}], \tag{6.21}$$

and

$$\mathbf{d}[\vec{x}] = \mathbf{a}[\vec{x}] + \mathbf{b}[\vec{x}] + \mathbf{c}[\vec{x}]. \tag{6.22}$$

The Python script for this process is shown in Code 6.9. This particular image is stored such that the background has the highest value. So, the loading function in line 3 sets to 1 all of the pixels that originally have a value less than 100. These are the target pixels. Likewise, the background pixels are set to 0. Lines 4 through 7 perform the dilation operations and subtractions. The final result is computed in line 8.

(a) (b)

Figure 6.9 (a) The original image and (b) the image with the two perimeters.

Code 6.9 The perimeters are created by computing the difference between two dilations

```
>>> import imageio
>>> import scipy.ndimage as nd
>>> amg = imageio.imread( 'data/alien73.png',as_gray=True) < 100
>>> bmg = nd.binary_dilation( amg, iterations=12 ) -
 nd.binary_dilation( amg, iterations=9 )
>>> cmg = nd.binary_dilation( amg, iterations=6 ) -
 nd.binary_dilation( amg, iterations=3 )
>>> dmg = amg + bmg + cmg
```

6.6 COORDINATE MAPPING

While scipy does have functions for shifting, rotation, and affine transformations, the creation of other types of transformations relies on two functions. The first is a function, perhaps user defined, that creates a coordinate map. The second is **scipy.ndimage.map_coordinates** which receives input image data and an array containing new pixel locations. This moves the pixels in the input array to new locations defined by the mapping array.

The idea is that each pixel in the input space is moved to a new location in the output space. In practice, though, this would leave some pixels in the output undefined. The function actually works in the reversed logic, where each pixel in the output space is mapped to the input space. Then, the value of the pixel in the input is placed in the output space. In this manner, every pixel in the output space is defined as long as it mapped to a pixel that is within the frame of the input space.

The Mapping operator is $C_{\mathbf{M}}$, where $\mathbf{M}$ is the array of new coordinates; thus, an operation is described as

$$\mathbf{b}[\vec{x}] = C_{\mathbf{M}}\mathbf{a}[\vec{x}]. \tag{6.23}$$

Consider a case in which M is of the size $2 \times V \times H$, which is basically two matrices of size $V \times H$. The value of the first matrix corresponds to the row index. In other words, the values in the first row are zeros, the values in the second row are ones, and so forth. The second matrix has the same behavior along its columns. Thus, for any location (v,h), the values of the two matrices are v and h. The Python function that creates these matrices is **numpy.indices**. This version of M will not alter the position of any pixel. Thus, Equation (6.23) would produce an image $\mathbf{b}[\vec{x}]$ that is the same as the input image $\mathbf{a}[\vec{x}]$.

Now, consider the task of creating an image such as the one shown in Figure 6.10. The pixel located at (v,h) in the output obtains its value from a pixel located within a random distance d from (v,h). Using $\vec{w}$ to represent the size of the image and the random number operator $\mathbf{q}$, the generation of a new mapping matrix is

$$\mathbf{M}' = \mathbf{M} + 2d\mathbf{q}_{\vec{w}} - d. \tag{6.24}$$

The creation of the new image is then,

$$\mathbf{b}[\vec{x}] = C_{\mathbf{M}'}\mathbf{a}[\vec{x}]. \tag{6.25}$$

Line 4 in Code 6.10 loads the image as a gray scale, and line 6 creates the identity mapping. Line 7 establishes the maximum travel distance. Lines 8 and 9 define $\mathbf{M}'$. The output image is created in line 10 and shown in Figure 6.10.

6.7 POLAR TRANSFORMATIONS

Arrays in Python contain information as rows and columns; however, there are some objects in images that are circular and angular in nature. Processing this information is easier in polar coordinates, but programming languages prefer rectilinear coordinates. Thus, in these cases, it is useful

Figure 6.10 The results from Code 6.10.

Code 6.10 Creation of the image in Figure 6.10

```
>>> import imageio
>>> import numpy as np
>>> import scipy.ndimage as nd
>>> adata = imageio.imread( 'data/bird.jpg', as_gray=1)
>>> V,H = adata.shape
>>> M = np.indices( (V,H) )
>>> d = 5
>>> q = 2*d*np.random.ranf( M.shape ) - d
>>> Mp = (M + q).astype(int)
>>> bdata = nd.map_coordinates(adata, Mp)
```

to have functions that transform information between rectilinear and polar coordinates. Three such functions are presented here. The first is **RPolar** which maps a rectilinear image into a polar space, and the second is **IRPolar** which performs the reverse transformation. The third is **LogPolar**, which converts image data into a log-polar mapping space. This section will review the theory of these transformations, present the operator notation, present the Python scripts, and finally, incorporate one transformation into a simple transformation.

6.7.1 THEORY

Figure 6.11 shows a single point with both the rectilinear coordinates (i, j) and the polar coordinates (r, θ). The conversion from the rectilinear coordinates to the polar coordinates is,

$$r = \sqrt{i^2 + j^2} \tag{6.26a}$$

$$\theta = \arctan(j/i), \tag{6.26b}$$

and the inverse transformation is,

$$x = r\cos(\theta) \tag{6.27a}$$

$$y = r\sin(\theta). \tag{6.27b}$$

Figure 6.11 shows that the center of the coordinate space is in the interior of the image. The location of this center is defined by the user and is often placed at the center of the object that is begin

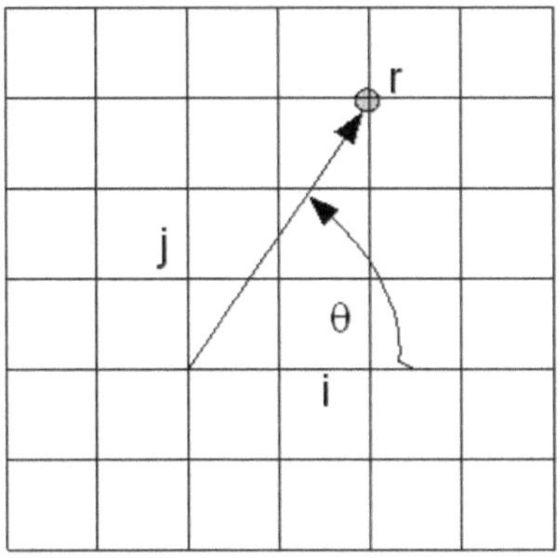

Figure 6.11 Conversion of rectilinear coordinates to polar coordinates.

analyzed. Therefore, the operator notation also includes the location of the center as a vector $\vec{v}$, if it is not located at the center of the frame. The **RPolar** operator transforms image $\mathbf{a}[\vec{x}]$ by

$$\mathbf{b}[\vec{x}] = \mathcal{P}_{\vec{v}}\mathbf{a}[\vec{x}]. \tag{6.28}$$

The Reverse operator is the **IRPolar** operator which is represented by $\mathcal{P}^{-1}$ as in

$$\mathbf{c}[\vec{x}] = \mathcal{P}_{\vec{v}}^{-1}\mathbf{b}[\vec{x}]. \tag{6.29}$$

In theory, $\mathbf{c}[\vec{x}] = \mathbf{a}[\vec{x}]$, but in practice, there will be small errors since the images are pixelated. The final operator in this family is the **LogPolar** operator that converts the rectilinear image into a log-polar image. This is used to also zoom in on the details near the center of the transformation at the sacrifice of the spatial resolution for points far away from the center. It is represented as

$$\mathbf{b}[\vec{x}] = \mathcal{P}_{L;\vec{w}}\mathbf{a}[\vec{x}]. \tag{6.30}$$

6.7.2 PYTHON IMPLEMENTATION

A Python implementation for **RPolar** is shown in Code 6.11, which receives the input grayscale image data and the center of the transformation. In line 3, the **indices** function is called to create the mapping array named `ndx`. This mapping array is considered to be in (r, θ) space, and therefore, it is necessary to find the locations of the pixels back in rectilinear space. The matrix `ndx[0]` is the radial component, and `ndx[1]` is the angle which is converted to a scale of 0 to 2π in line 6. The conversion occurs in lines 7 and 8 which are adjusted to the location so the user-defined center of rotation, `pxy`, in lines 9 and 10. The pixel mapping occurs in line 11, and the process is finished.

Code 6.12 shows the **IRPolar** function, which is the reverse process of **RPolar**. This function is quite similar to its predecessor except that the original coordinate starts in rectilinear space and maps back to polar coordinates. Thus, Equations (6.27) are used as seen in lines 7 and 8. The **LogPolar** function is similar to the **RPolar** function except for the computation of the radial component. The function is shown in Code 6.13 and the alteration is in lines 7 through 9.

Consider the image shown in Figure 6.12(a). The clock face is nearly circular, and the center is at $\vec{v} = (146, 125)$. The image in Figure 6.12(b) is created by

$$\mathbf{b}[\vec{x}] = \mathcal{P}_{\vec{v}}\mathcal{L}_L\mathbf{a}[\vec{x}]. \tag{6.31}$$

The trim around the dial appears to be almost a straight, horizontal line in $\mathbf{b}[\vec{x}]$. In the original image, the dial is nearly circular and the center of the transformation is located at the center of this circle.

Code 6.11 The RPolar function

```
# rpolar.py
def RPolar( data, pxy ):
    ndx = np.indices( data.shape )
    v,h = data.shape
    a = ndx[1].astype( float)
    a = a / h * 2 * np.pi
    y = ndx[0] * np.cos(a)
    x = ndx[0] * np.sin(a)
    ndx[0] = x.astype(int) + pxy[0]
    ndx[1] = y.astype(int) + pxy[1]
    answ = nd.map_coordinates( data, ndx )
    return answ
```

Code 6.12 The IRPolar function

```
# rpolar.py
def IRPolar( rpdata, pxy ):
    ndx = np.indices( rpdata.shape )
    ndx[0] -= pxy[0]
    ndx[1] -= pxy[1]
    v,h = rpdata.shape
    r = np.sqrt( ndx[0]**2 + ndx[1]**2 )
    theta = np.arctan2( -ndx[0], -ndx[1] )/2/np.pi*h
    ndx[0] = r.astype(int)
    ndx[1] = theta.astype(int) +h/2
    answ = nd.map_coordinates( rpdata, ndx )
    answ[pxy[0],pxy[1]:] = answ[pxy[0]-1,pxy[1]:]
    return answ
```

Code 6.13 The LogPolar function

```
# rpolar.py
def LogPolar( data, pxy ):
    ndx = np.indices( data.shape )
    v,h = data.shape
    a = ndx[1].astype( float)
    a = a / h * 2 * np.pi
    r = np.exp( ndx[0]/v * np.log(v/2))-1.0
    y = r * np.cos(a)
    x = r * np.sin(a)
    ndx[0] = x.astype(int) + pxy[0]
    ndx[1] = y.astype(int) + pxy[1]
    answ = nd.map_coordinates( data, ndx )
    return answ
```

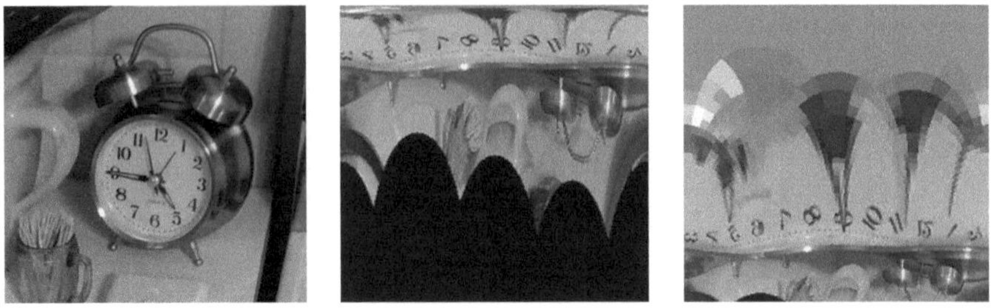

Figure 6.12 (a) The original image, (b) the radial-polar transformation of the image, and (c) the log-polar transformation.

Thus, the radial distance from the center to the trim is almost the same for any angle. In $\mathbf{b}[\vec{x}]$, the vertical dimension is the radial distance in $\mathbf{a}[\vec{x}]$ and so the trim is a horizontal line.

Figure 6.12(c) displays the log-polar transformation of the image, which is created by

$$\mathbf{c}[\vec{x}] = \mathcal{P}_{L;\vec{v}} \mathcal{L}_L \mathbf{a}[\vec{x}]. \tag{6.32}$$

The main difference is that more of the pixels are dedicated to the smaller radii in the original image. The trim, though, is still a horizontal line in $\mathbf{c}[\vec{x}]$.

6.7.3 EXAMPLE

An example is to extract the radius of the cytoplasm from the red cell in Figure 6.13(a). The proposed process is

$$\vec{z} = \mathcal{N}_{1V} \mathcal{P}_{\vec{v}} \Gamma_{<83} \left(\frac{\left\{ \begin{matrix} \varnothing \\ \varnothing \\ 1 \end{matrix} \right\} \mathbf{a}[\vec{x}]}{\left\{ \begin{matrix} 1 \\ \varnothing \\ \varnothing \end{matrix} \right\} \mathbf{a}[\vec{x}] + 40} \right), \tag{6.33}$$

where $\vec{v} = (178, 288)$ is the center of the target cell. The numerator extracts the blue channel, and the denominator extracts the red channel and adds a bias to prevent divide by zero errors. The ratio will be larger in the blue part of the image and smaller in red portions. Figure 6.13(b) shows the ratio values for each pixel, and the pixels that were blue are now the brightest in the image, making them easy to isolate.

The Γ operator applies a binary threshold and the result is shown in Figure 6.13(c), but this display is inverted so that the black pixels represent a value of 1 and the white pixels represent a value of 0. The $\mathcal{P}_{\vec{v}}$ applies the radial-polar transformation, and this result is shown in Figure 6.13(d). The radii from the center of the cell to the edge of cytoplasm are the distances from the top of this image to the first 0 value (white pixel) in each column. The $\mathcal{N}_{1V}$ operator will extract this pixels location for each column, and it is related to the **nonzero** function. The result is a vector $\vec{w}$ that contains the radii of the cytoplasm for each angle.

Code 6.14 shows the Python script for this entire process. Line 3 loads the image, and Equation (6.33) is computed in line 4. Line 5 performs the inequality with $\gamma = 0.83$ and converts the image to the radial polar space. The `for` loop in lines 8 and 9 grab the first element that is nonzero in every column.

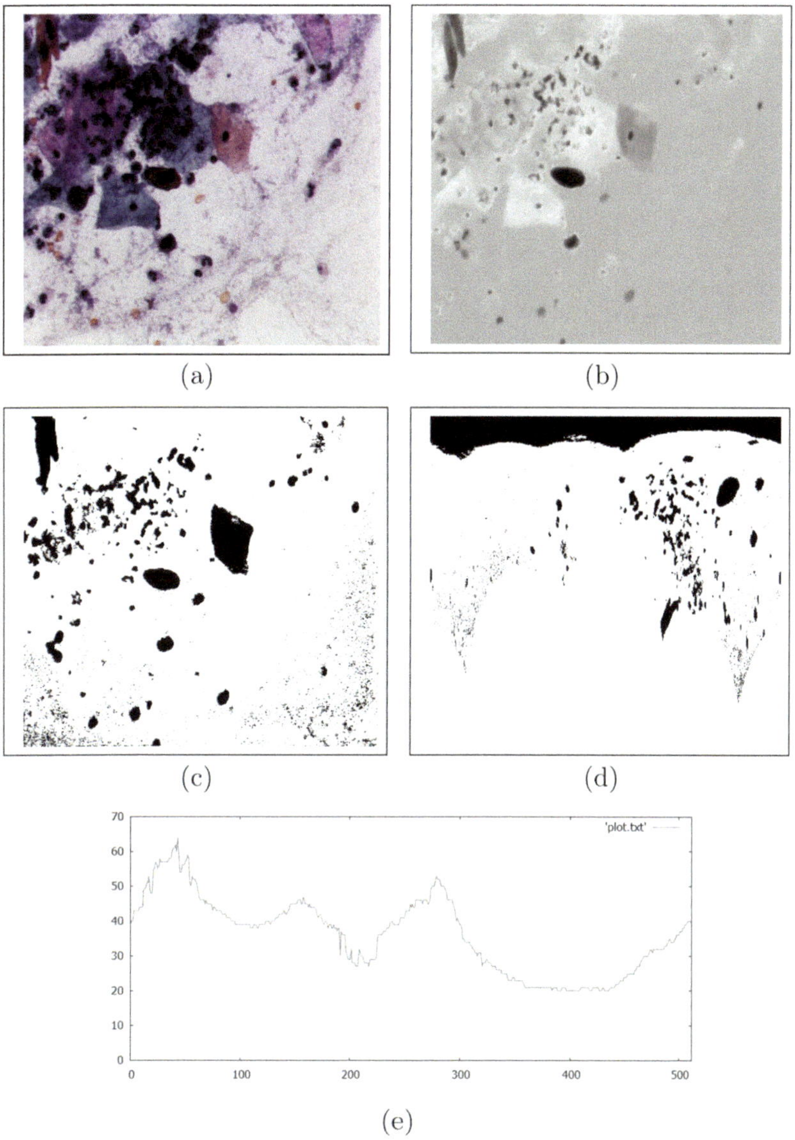

Figure 6.13 (a) The original image. (b) The image after computing the ratio of the blue to the red channel. (c) Isolation of the target via a threshold. (d) The same image in radial polar space. (e) The radial distance of the perimeter of the target.

6.8 PINCUSHION AND BARREL TRANSFORMATIONS

Pincushion and barrel transformations alter the radial distance to points in a nonlinear fashion. The two transformations are shown in Figure 6.14, where Figure 6.14(a) is the original image, Figure 6.14(b) is the pincushion transformation, and Figure 6.14(c) is the barrel transformation.

These two transformations are actually the same operator, but with a different bending parameter. The bending process for a single pixel is described by

$$r' = \frac{r^{\beta}}{V^{\beta-1}}, \tag{6.34}$$

Code 6.14 Finding the perimeter of the cell

```
>>> import imageio
>>> import rpolar
>>> amg = imageio.imread('data/00003312.png')
>>> a = amg[:,:,2]/(amg[:,:,0]+40.)
>>> rp = rpolar.RPolar((a<0.83)+0.0,(178,288))
>>> V,H = rp.shape
>>> vec = np.zeros(H)
>>> for i in range(H):
        vec[i] = (rp[:,i]<0.5).nonzero()[0][0]
```

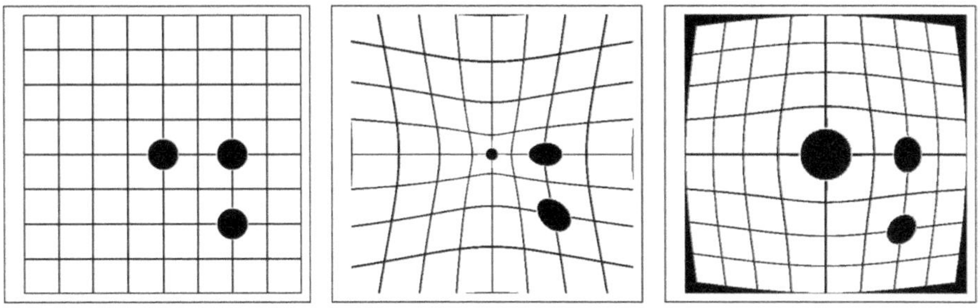

(a) (b) (c)

Figure 6.14 (a) The original image, (b) the pincushion transformation, and (c) the barrel transformation.

where β is the bend parameter, r is the radial distance, and V is the vertical dimension of the image.

The barrel transform appears to enlarge the pixels in the center of the image at the sacrifice of the resolution at the outer portions of the image to accomplish this $\beta > 1$. The pincushion transform shrinks the center of the image while enlarging the outer portions. This is accomplished with $0 < \beta < 1$.

The notation is rather simple. Given a grayscale input $\mathbf{a}[\vec{x}]$ the notation is

$$\mathbf{b}[\vec{x}] = B_{\beta,\vec{v}}\mathbf{a}[\vec{x}], \tag{6.35}$$

where β is the bending factor and $\vec{v}$ is the center of transformation.

The **Bend** function shown in Code 6.15 receives a matrix, the bend factor, the center of the transformation, and returns the new matrix. Basically, it transforms the coordinates to polar coordinates in lines 7 and 8 and performs the bending in line 9. The following lines return the points to rectilinear coordinates and in lines 12 and 13 the original image is transformed. The last two lines create the images shown in Figure 6.14(b) and (c).

6.9 OTHER TRANSFORMATIONS

The radial-polar and bend transformations are a subset of the Geometric operators, which map pixels to new locations. There are a wide range of functions that fall into this category. This section reviews methods by which the user defines the mapping function.

Code 6.15 The **Barrel** function

```
# geomops.py
def Bend( data, bend, cvh ):
    cv,ch = cvh
    a = np.indices(data.shape)
    V,H = data.shape
    a[0],a[1] = cv-a[0],a[1]-ch
    r = np.sqrt( a[0]*a[0] + a[1]*a[1] )
    t = np.arctan2( a[0], a[1] )
    r = (r ** bend)/(cv**(bend-1))
    x = r* np.cos( t );      y = r* np.sin( t )
    x = x + ch;      y = cv - y
    coords = np.array([y.astype(int),x.astype(int)])
    z = nd.map_coordinates( data, coords )
    return z

>>> b = geomops.Bend( a, 0.7, (256,256) )
>>> b = geomops.Bend( a, 1.3, (256,256) )
```

6.9.1 GENERIC TRANSFORMATIONS

Of course, it is not possible to conceive of every possible geometric transformation and assign a unique operator symbol to it. So, the Generic operator, F, is available for used defined functions as in

$$\mathbf{b}[\vec{x}] = F\mathbf{a}[\vec{x}]. \tag{6.36}$$

The *scipy.ndimage* module also offers a generic transform function named **geometric_transform** which maps pixels from one space to another according to a user-defined function.

Consider the altered bird image shown in Figure 6.15. In this example, the pixels are moved according to cosine functions in both the vertical and horizontal dimensions. The function **GeoFun** shown in Code 6.16 receives a single (v,h) coordinate and computes the new coordinate that will map back to the input space. Line 8 uses the **geometric_transform** function to determine the mapping for all pixels and returns the resultant image.

Figure 6.15 The results from Code 6.16.

Code 6.16 An example using **scipy.ndimage.geometric_transform**

```
# geomops.py
def GeoFun( outcoord ):
    a = 10*np.cos( outcoord[0]/10. )+outcoord[0]
    b = 10*np.cos( outcoord[1]/10. )+outcoord[1]
    return a,b

>>> import scipy.ndimage as nd
>>> amg = imageio.imread('data/bird.jpg',as_gray=True)
>>> bmg = nd.geometric_transform( amg, geomops.GeoFun )
```

Figure 6.16 The results from Code 6.17.

6.9.2 AFFINE TRANSFORMATION

The Rotation operator (see Section 6.4) multiplied the vector representing the coordinate by a matrix (see Equation (6.16)). The affine transformation is quite similar but user defines the matrix. The example in Code 6.17 creates a matrix `mat`, which is altered from the rotation matrix, and the `affine_transform` function is then called that maps the coordinates according to this prescription. The result is shown in Figure 6.16.

The user can define any 2×2 matrix that suits their purpose. The application of the matrix would follow that in Code 6.17.

Code 6.17 An example using `scipy.ndimage.affine_transform`

```
# geomops.py
def AffineExample( data ):
    theta = 11 * np.pi/180 # 11 degreess in radians
    mat = np.array( [ [np.cos(theta), -np.sin(theta)], \
        [np.sin(theta/4), np.cos(theta)]] )
    data2 = nd.affine_transform( data, mat )
    return data2

>>> mg2 = AffineExample( 'data/bird.jpg')
```

Figure 6.17 An image after the shift with the wraparound option.

6.10 SUMMARY

The family of Geometric operators is dedicated to moving pixels to new locations within the image rather than the alteration of intensities. Simple operations including shifting or rotating the image. Erosion and Dilation operators change the perimeter of shapes within an image. More complicated operations require a re-mapping of the image pixels. These include transformations to radial-polar space, bending, and other morphological functions.

PROBLEMS

1. Write a script to shift an image 10 pixels to the right.
2. Write a script to perform the following

$$\mathbf{b}[\vec{x}] = D_{(5,-3)}\mathbf{a}[\vec{x}]$$

3. Compute the error in a noninteger shift by performing the following:

$$f = \frac{\sum_{\vec{x}} \left| D_{\vec{w}}\, D_{-\vec{w}}\, \mathbf{a}[\vec{x}] \right|}{N\mathbf{a}[\vec{x}]}$$

   where $\vec{w} = (4.5, -3.4)$.
4. Write the notation for placing an image $\mathbf{a}[\vec{x}]$ inside a larger frame of size $\vec{w}$ such that the upper left corner of the image is located in the upper left corner of the frame. The size of $\mathbf{a}[\vec{x}]$ is $\vec{w}$.
5. Create an image from, $\mathbf{b}[\vec{x}] = \lhd_3\mathbf{a}[\vec{x}] - \rhd_3\mathbf{a}[\vec{x}]$, where $\mathbf{a}[\vec{x}] = Y('\texttt{data/myth/horse1.bmp}')$.
6. Consider the image shown in Figure 6.17 which shows that the original image is shifted to the right and that the portion that was out of the frame after the shift is placed on the left side of the image. Write an operator notation for this horizontal wraparound shift.
7. Write the operator notation that will rotate an image about an angle θ with the center at location $\vec{v}$.
8. Write a script that will rotate the color image of the bird 20° CCW about the center of the beak.

9. Write a script that will:
   a. Convert an image to Radial-Polar space
   b. Shift to the right each row by n pixels were n is the row number. This shift should use the wraparound effect such that pixels that are shifted out of the frame are placed on the left end of the row. (See Problem 4.)
   c. Convert back to the original space.
10. Using the knowledge of the previous problem write a script that creates the Figure 6.18 from the original bird image. (Figure 6.18).
11. Write the operator notation that will create Figure 6.19 from the original bird image. (Note, a black background is also acceptable.)
12. Perform the following operations on $\mathbf{a}[\vec{x}] = \mathcal{L}_L Y(\texttt{'data/bananas.jpg'})$, using $\vec{w} = (640, 585)$, $\vec{v} = (245, 458)$, $r = 45$, $\theta = 180^\circ$, and $\vec{z} = (-75, 166)$.

$$\mathbf{c}[\vec{x}] = \mathbf{o}_{\vec{w};\vec{v},r}[\vec{x}]$$
$$\mathbf{d}[\vec{x}] = \mathbf{a}[\vec{x}] \times (1 - \mathbf{c}[\vec{x}])$$
$$\mathbf{b}[\vec{x}] = R_{\vec{v},\theta}\left(\mathbf{c}[\vec{x}] \times \mathbf{a}[\vec{x}]\right)$$
$$\mathbf{f}[\vec{x}] = \mathbf{b}[\vec{x}] + \mathbf{d}[\vec{x}]$$

Figure 6.18 A transformed image.

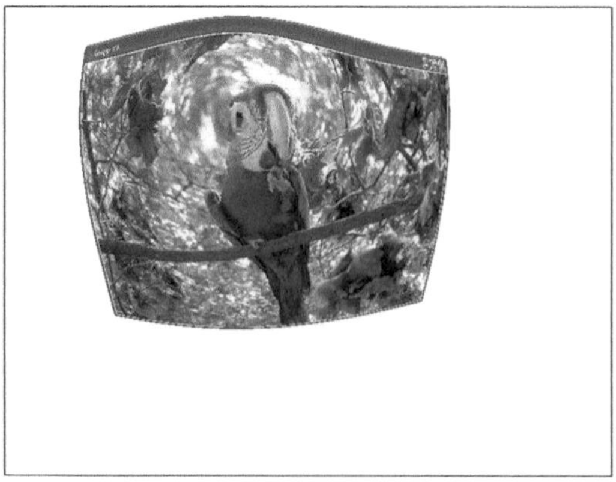

Figure 6.19 Zooming in on the bird's eye.

7 Image Morphing

Image morphing is the process of gradually converting one image into another. This process has been used in the past two decades in the entertainment industry, but it is also very useful in image classification algorithms. This chapter will demonstrate the basic methods and practice, but it should be noted by the reader that there are more complicated methods available with advanced theory and expensive computations. For the purpose of image recognition algorithms, the methods reviewed here are sufficient.

There are two processes used here in the morphing. The first is an image *warp* which smoothly transitions one image to a specified grid, and the second is the *morph* which uses the warp to gradually transition one image into another. The warp process requires that the user mark several anchor points (fiducial points) in the image. Generally, these are located at the edges and corners of major features. A large number of fiducial points will create a better warp result but also increase the computational cost.

7.1 WARP

There are a few steps required to warp an image. First, fiducial points in the image are identified. The process will require a grid for the image and a second grid which is the target for warping. Basically, points in the image will move from locations in the first grid to locations in the second grid. The grid defines a lattice structure which is created by Delaunay tessellation. Finally, all of the components are in place, and the warping process can be applied. These components are discussed in the following subsections.

7.1.1 MARKING FIDUCIAL POINTS

Fiducial points are the anchor points with in an image that guide the warp. Commonly, the fiducial points are located at important features such as corners and edges. These fiducial points are the nodes in the lattice structure. In face recognition algorithms, fiducial points are commonly located about the perimeter of the eyes, nose, and mouth. An example of a set of fiducial points for a face is shown in Figure 7.1. In this case, there are 45 points and to warp this face it is necessary to define the location of these 45 points in the new space.

Methods are being developed for automated detection of the fiducial points [35], but they are not employed here. For small-scale jobs, it is possible to manually mark the fiducial points. This requires the use of an image viewing program that displays the location of the mouse. This method is slow and prone to making errors.

7.1.2 IMAGE DANCER

Many programs provide the ability to place a mouse at a location on an image and then display the mouse coordinates. However, this can be a tedious process to mark several fiducial points in an image. A module named *Dancer* was created for just this purpose for earlier versions of Python. However, the Python Image Library for Python 3.x does not yet have all of the tools necessary to run this program in all operating systems. The contents of this chapter work on version Python 2.7 and earlier. For now, Python 3.x users will need to use other software to report the location of the mouse on the screen. One suggestion is the GIMP program which is freely available.

The *dancer* module will work for Python 2.7 and two files are required. The first is the *dancer.py* file, and the second is an image named "init.png." Any image can be used as the initialization just as

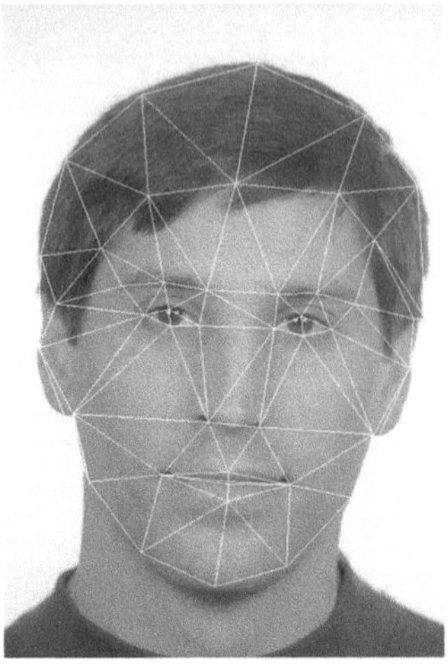

Figure 7.1 Fiducial points and grid.

long as it is named "init.png" and is the current working directory. The *dancer* module uses Tkinter which has a contradiction with most editing environments for Python. Thus, in Windows, the *dancer* is more interactive if it is run in a command window rather than a Python environment such as IDLE or Spyder.

Code 7.1 shows the Python commands that will start the Dancer program. Line 1 imports the *dancer* module which does require that the directory that contains this module be in the list `sys.path`. Line 2 creates an instance of the Dancer class where the variable `mydir` is the directory where the image *init.png* is stored. This line will create a new window with that image in it. Line 3 is used to change that image by replacing `myImage` with the name of the face image to be loaded. This will replace the image showing in the new window. The final step is to right click on locations according the list below or a user-defined list of fiducial points. The right mouse click will report the location and RGB values of the pixel to the console. 6 of the 45 points are shown. This list can then be copied and pasted into a text editor and then stored for future use.

Code 7.1 Starting Dancer

```
1 >>> import dancer
2 >>> q = dancer.Dancer(mydir)
3 >>> q.Load( myImage )
4 >>> (239, 171) 193 140 96
5 (37, 273) 167 144 162
6 (37, 273) 167 144 162
7 (67, 442) 140 80 113
8 (67, 442) 140 80 113
9 (67, 442) 140 80 113
```

The user can define any fiducial grid, but commonly the points are at the edges and corners of key features. It is important that the pixels in the image are selected in the same order for all images. So, only one pattern of points should be used for a project. The grid used here is:

- 1,2: left and right corners of left eye.
- 3,4: upper and lower part of left eye.

- 5,6,7,8: left, right corners, and upper, lower parts of right eye.
- 9,10,11: left, middle, right points on left eyebrow.
- 12,13,14: left, middle, right points on right eyebrow.
- 15,16: left and right nostril.
- 17,18: left and right corners of the mouth.
- 19,20: upper and lower points on the middle of the upper lip.
- 21,22: upper and lower points on the middle of the lower lip.
- 23–29: points around the chin from bottom of left ear to bottom of right ear.
- 30,31: top and middle of left ear.
- 32,33: top and middle of right ear.
- 34–38: five points (from left to right) across the hairline.
- 39–44: seven points across the top of the hair outline.

The output from the Dancer interaction is saved into a text file. The warp process requires three files. The first two are the image and the fiducial grid for that image. The last file is the fiducial grid to the destination for the warping process. The function **ReadDancer** reads in the text file that contains the fiducial points which assumes the format in lines 4 through 9 of Code 7.1.

Code 7.2 The **DelaunayWarp** function

```
# delaunaywarp.py
def ReadDancer( fname ):
    grid = []
    rows = open( fname ).read().split('\n')
    for r in rows:
        if len(r) > 10:
            a = r.split()
            h = int(a[0][1:-1])
            v = int(a[1][:-1])
            grid.append( np.array( (v,h) ))
    grid = np.array( grid )
    return grid
```

It is also possible that the fiducial points may be obtained through a different program. In this case, the format of the data will be different from that which is obtained through Dancer. The *pandas* package has many data analysis tools, but that suite also includes programs to read Excel spreadsheets or text data. The example in Code 7.3 is designed to read in data that is stored in two columns in a text file. In this case, a space separates the two numbers on each row (as opposed to a tab character). Line 2 reads in this data as integers, and line 3 converts the data to a matrix. This is another method in which a fiducial file can be easily read into a Python array.

Code 7.3 Reading a CSV file

```
1 >>> import pandas as pd
2 >>> a = pd.read_csv('data/003.txt',dtype=int,sep=' ')
3 >>> fida = a.get_values()
```

7.1.3 DELAUNAY TESSELLATION

Delaunay tessellation produces a set of triangles much like that in Figure 7.1. The intersections of the triangles are the fiducial points, and the triangles have the following conditions:

- Edges of the triangles do not cross other edges,
- A circle that connects three points of a triangle will not contain other points from other triangles, and
- Acute angles are avoided.

The *scipy.spatial* module offers the Delaunay object and some of its features are listed in Code 7.4. Line 2 reads in a fiducial grid file, and line 4 performs the Delaunay tessellation. Each triangle is defined by the three corners. The listing of these is obtained from line 5, but only a small part of the response is printed here. The first triangle is defined by three points referenced in line 6. Lines 12 and beyond print out the location of these three points, thus defining the first triangle.

The warping process will consider each point in the image. It will be necessary to determine which simplex (or triangle) that each pixel is in. This is accomplished by the **find_simplex** as shown in Code 7.5. The pixel located at (100,200) is in simplex 27. This triangle uses the points listed in line 6. Now, each point has a defined simplex and the final step is to move this pixel to a new location.

Code 7.4 Extracting information from the tessellation

```
1  >>> import delaunaywarp
2  >>> fid = delaunaywarp.ReadDancer( fname )
3  >>> from scipy.spatial import Delaunay
4  >>> dela = Delaunay( fid )
5  >>> dela.simplices
6  array([[39, 34, 40],
7         [40, 35, 41],
8         [34, 35, 40],
9  ...
10 >>> dela.vertices[0]
11 array([39, 34, 40], dtype=int32)
12 >>> dela.points[39]
13 array([ 111.,   57.])
14 >>> dela.points[34]
15 array([ 129.,   76.])
16 >>> dela.points[40]
17 array([ 67.,  83.])
```

Code 7.5 Finding a simplex

```
1  >>> import numpy as np
2  >>> pt = np.array( (100,200))
3  >>> dela.find_simplex( pt )
4  array(27, dtype=int32)
5  >>> dela.simplices[27]
6  array([42, 36, 43], dtype=int32)
```

7.1.4 APPLYING THE WARP

The Warp operator moves pixels to new locations according the prescription defined by a fiducial grid G, as in

$$\mathbf{b}[\vec{x}] = W_G \mathbf{a}[\vec{x}]. \tag{7.1}$$

The tessellation creates the triangles and can even identify the triangles in which a point is in. The fiducial points are the corners of the triangles and moving them from input space to output space is quite easy, since the points are well defined in both spaces as shown in Figure 7.2. The points that are interior to the triangle are not well defined in the output space. The method used in the **Warp** function computes the distance from a point to the three corners (see Code C.4 in the Appendix). The ratios of these distances are used to compute the new points. The corners of a triangle are defined as: $\vec{p}_1$, $\vec{p}_2$, and $\vec{p}_3$. Any point, $\vec{p}_4$, within a triangle is defined as a linear combination of the the corners as in

$$\vec{p}_4 = \lambda_1 \vec{p}_1 + \lambda_2 \vec{p}_2 + \lambda_3 \vec{p}_3, \tag{7.2}$$

where

$$\lambda_1 + \lambda_2 + \lambda_3 = 1. \tag{7.3}$$

Determining the location of the point in the new space T is performed through a linear transformation:

$$\begin{bmatrix} \lambda_1 \\ \lambda_2 \end{bmatrix} = \mathbf{T}^{-1}(\vec{p}_4 - \vec{p}_3), \tag{7.4}$$

where

$$T = \begin{bmatrix} x_1 - x_3 & x_2 - x_3 \\ y_1 - y_3 & y_2 - y_3 \end{bmatrix}, \tag{7.5}$$

where λ_3 is defined using Equation (7.3),

$$\lambda_3 = 1 - \lambda_1 - \lambda_2. \tag{7.6}$$

This process is repeated for all points using the appropriate triangles. Transformation of points that are outside of the face boundary is accomplished by adding four fiducial points for the corners of the image frame. Thus, each point in the input image is within a triangle.

The **Warp** function is lengthy and so is printed in Code C.4. It receives three inputs. The first is the item return in from the **Delaunay** function (line 4 from Code 7.4). The second is the target fiducial grid, and the third is the matrix that contains the grayscale image data.

Code 7.6 shows the steps to warp an image. Lines 3 and 4 read in the image fiducial grid and the target fiducial grid. Line 5 reads in the image, and line 6 performs the tessellation. Line 7 performs the warping process.

There are three files required for the warping process. These are the original image, the fiducial grid for the original image (not shown here), and the target fiducial grid shown in Figure 7.3(b). Figure 7.3(c) shows the grid overlain on the original image, and as seen, the fiducial features of the image and the grid are not aligned. The warping process moves the pixels in the original image so that they do align with this grid. The results are shown in Figure 7.3(d) with the target grid and again in Figure 7.3(e) as a final result.

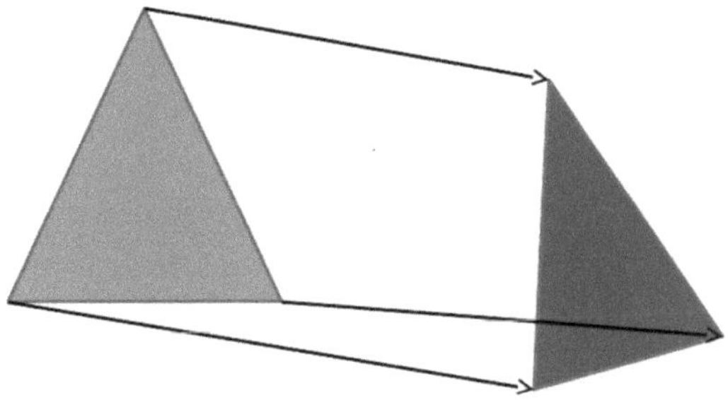

Figure 7.2 The points in one triangle are moved to form points in a new triangle.

Code 7.6 Commands to warp an image

```
>>> import imageio
>>> import delaunaywarp
>>> fid1 = delaunaywarp.ReadDancer( fname1 )
>>> fid2 = delaunaywarp.ReadDancer( fname2 )
>>> mg = imageio.imread( mgname, as_gray=True)
>>> dela = delaunaywarp.Delaunay( fid1 )
>>> mat3 = delaunaywarp.Warp( dela, fid2, mg )
```

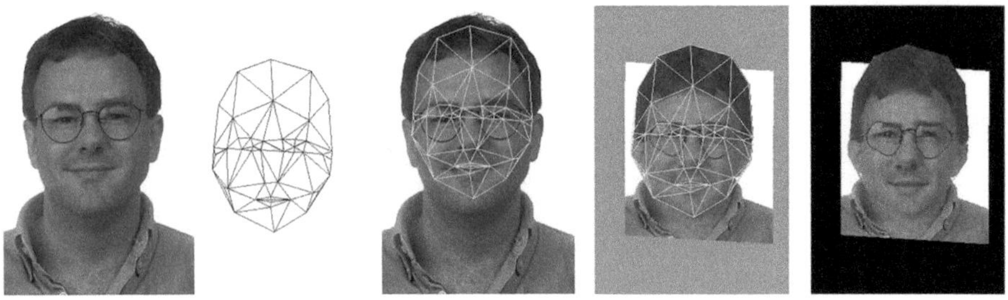

(a) (b) (c) (d) (e)

Figure 7.3 (a) The original image, (b) the target grid, (c) the target grid overlain on the original image, (d) the image after warping with the grid, and (e) the image after warping.

7.2 AVERAGE FACE

The average face is a composite image created from the facial images of many people. The phrase "average" implies that the entities are summed and then the result is divided by the number of entities. Given a set of vectors $\{\vec{x}_i; i = 1, \ldots, N\}$, the average is usually computed by $\frac{1}{N}\sum_{i=1}^{N} \vec{x}_i$. This, however, does not work for face images as depicted in Figure 7.4. This was created from a few images and all of the features appear blurry. The reason is that the features in the original images were not aligned. For example, the left eye was in different locations in the different images.

In order to create a realistic average face, it is necessary to align significant features such as the eyes, mouths, noses, etc. This is accomplished by aligning all of the faces to a single fiducial grid, G, before computing the average. Formally, the average face is

Figure 7.4 Simple averaging of face images.

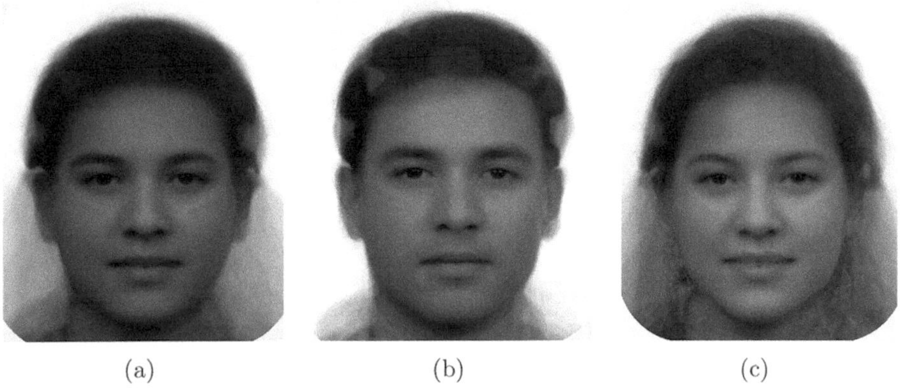

Figure 7.5 (a) The average face using both genders and all races. (b) The average face from males of multiple races. (c) The average face for females of multiple races.

$$\mathbf{b}[\vec{x}] = \frac{1}{N}\sum_i W_G \mathbf{a}_i[\vec{x}] = \frac{1}{N}\sum\{W_G \mathbf{a}[\vec{x}]\}, \quad (7.7)$$

where $\mathbf{a}_i[\vec{x}]$ is the i-th face image and W_G is the warping operator that maps the image to an average fiducial grid G. Figure 7.5(a) shows the average of several faces including multiple ethnicities and both genders. Figure 7.5(b) and (c) shows the averages for the male faces and female faces.

It is interesting to note that humans identify faces based on deviations from averages. We describe people with terms like large eyebrows, small ears, eyes that are too far apart, etc. These are all relative terms that depict the differences of a single individual from the average. Faces that are close to average are difficult to describe. This is the case with Figure 7.5(b) and (c). While the images have definite gender features, the face is still difficult to describe. Figure 7.5(a) was created with images from both genders and it is even difficult to describe gender features in this face. This is the property of the average face.

7.3 IMAGE MORPHING

Image morphing is the gradual transformation from one image into another. The process is fairly simple once the **Warp** function has been established. Consider Figure 7.6 which shows two images at either end of the line, and the goal is to create a set of images that gradually change from one image into another. The marker α can slide anywhere along this line and its position indicates which percentage of each input image will contribute to the output. Basically, if $\alpha = 0.6$, then the image on the left will contribute to 60% of the resultant image, and the image of the right will be responsible for 40% of the resultant image.

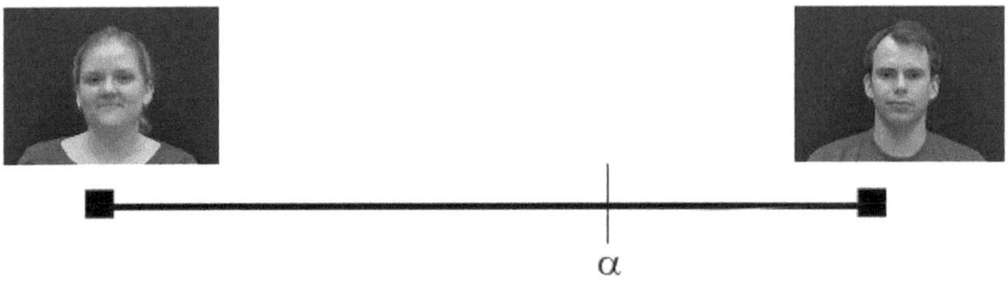

Figure 7.6 The process of morphing two images is to project their warps into a location α.

Code 7.7 Morphing two images

```
>>> fida = # read the fiducial grid
>>> mga = imageio.imread( 'data/003.png', as_gray=True)
...
>>> fidc = (1-alpha)*fida + alpha*fidb
>>> dela = delaunaywarp.Delaunay( fida )
>>> mga1 = Warp( dela, fida, mga )
>>> dela = Delaunay( fidb )
>>> mgb1 = Warp( dela, fidb, mgb )
>>> mgc = (1-alpha)*mga1 + alpha*mgb1
```

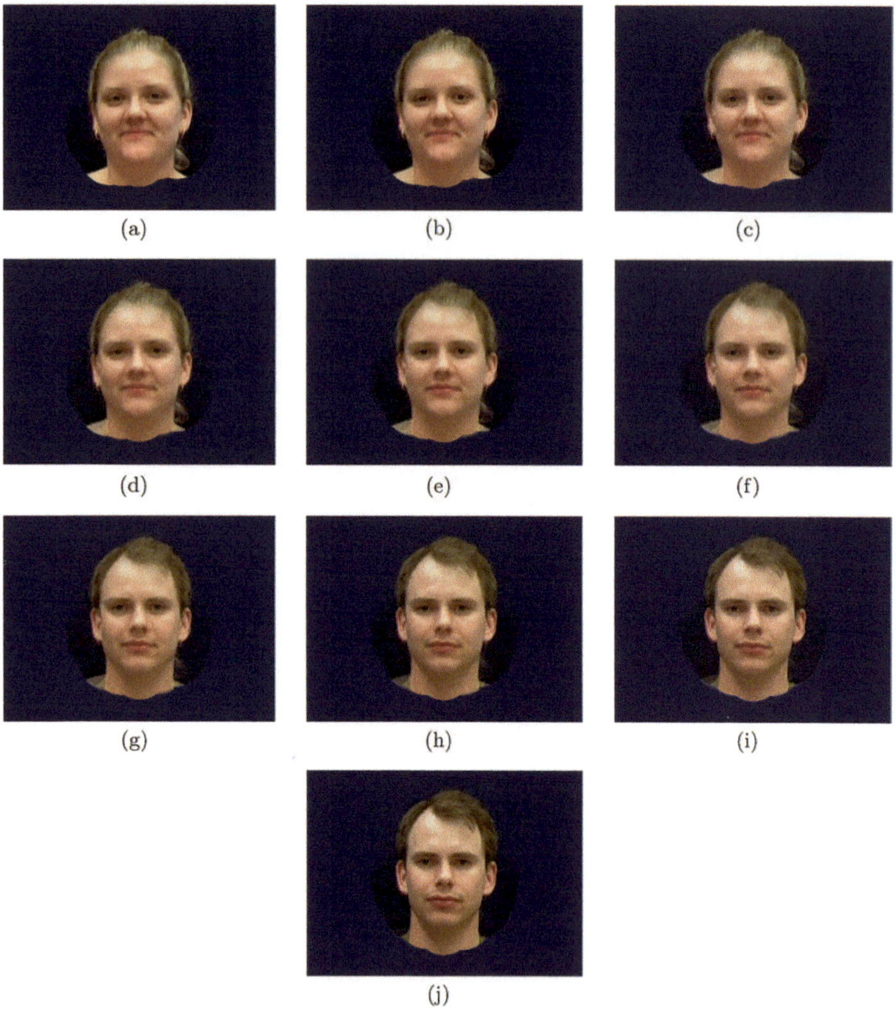

Figure 7.7 The series of images produced using all three RGB channels.

The two input images are designated at $\mathbf{a}[\vec{x}]$ and $\mathbf{b}[\vec{x}]$, and their respective fiducial grids are G_a and G_b. For a given value of α between 0 and 1, the target grid is

$$G_c = (1-\alpha)G_a + \alpha G_b. \tag{7.8}$$

Both images are warped to the G_c grid, and the intensity contributions are determined by α. So, the new image is obtained by

$$\mathbf{c}[\vec{x}] = (1-\alpha)W_{G_c}\mathbf{a}[\vec{x}] + \alpha W_{G_c}\mathbf{b}[\vec{x}]. \tag{7.9}$$

The morphing process generates a series of images for different values of α.

Code 7.7 shows the process of creating a morph image for a single α value. Line 1 is used to read in a fiducial file by a method shown in Section 7.1.2 or by a means selected by the user. Line 2 reads in the image and converts to gray scale. These two lines load the information for the first image, and so they need to be repeated for the second image. Line 4 computes the target grid. Lines 5 and 6 warp the first image to the target grid, and lines 7 and 8 do the same for the second image. Line 9 creates the morphed image for α. To create a series of images, lines 4 through 9 are repeated for different values of α.

The series of images shown in Figure 7.7 used values $\alpha = (0, 0.1, 0.2, \ldots, 1.0)$. Morphing color images is achieved by applying the morphing process to each channel.

The operator notation for morphing images is

$$\mathbf{b}[\vec{x}] = M_\alpha \mathbf{a}[\vec{x}], \tag{7.10}$$

where α is the same α used in Equations (7.8) and (7.9). It is possible that the two α's in these equations are different values for an effect or that more than two images are used in creating the morph images. In these cases, the subscript α may be replaced with multiple parameters as defined by the user.

PROBLEMS

The following problems will require face images and their fiducial grids. The data set should consist of two different images of one person and single images of two other people. The images should be front views of the person's face, and the frame sizes should be the same. The author's website will have a few images that can be used, but these assignments will be more interesting if the reader uses images of their own. Unless otherwise noted, the grayscale versions of the images can be used.

In the following problems the image notations are:

- $\mathbf{a}[\vec{x}]$, G_a: First image and fiducial grid of person 1. In this image, the person should not be smiling
- $\mathbf{b}[\vec{x}]$, G_b: Second image and fiducial grid of person 1. In this image, the person should be smiling. Except for the smile, $\mathbf{b}[\vec{x}]$ should be as similar as possible to $\mathbf{a}[\vec{x}]$.
- $\mathbf{c}[\vec{x}]$, G_c: Image and fiducial grid of person 2.
- $\mathbf{d}[\vec{x}]$, G_d: Image and fiducial grid of person 3.

1. Warp image $\mathbf{a}[\vec{x}]$ to the grid G_c.
2. Create a new grid, $G_t = G_a$. Flip the new grid upside down by subtracting the vertical grid values from, V, the vertical dimension of $\mathbf{a}[\vec{x}]$. Warp $\mathbf{a}[\vec{x}]$ to G_t.
3. Create a new grid, $G_t = G_a$. Subtract 5 from all of the horizontal numbers in g_t. Thus, the grid no longer aligns with the fiducial points in $\mathbf{a}[\vec{x}]$. Using G_t instead of G_a, warp $\mathbf{a}[\vec{x}]$ do G_c. Did the fact that the starting grid was misaligned with the original image produce significant anomalies in the output?
4. Create a new grid, $G_t = G_a$. Modify the values in G_t so that the eyes are 20% larger. Warp $\mathbf{a}[\vec{x}]$ from G_a to G_t.
5. Find the changes in the fiducial point locations for when a person smiles. Create a new grid, $G_t = G_b$. Shift G_t so that the nose is in the same location as that of the nose in G_a. Define $G_{t;i}$ as the i-th fiducial point in G_t. Compute the vector differences $G_{t;i} - G_{a;i}$ for all i.

6. Create ten images that follow the morph of $\mathbf{a}[\vec{x}]$ to $\mathbf{c}[\vec{x}]$.
7. Create ten images that follow the morph of $\mathbf{b}[\vec{x}]$ to $\mathbf{c}[\vec{x}]$.
8. Compute a three-way morph. Using $\mathbf{a}[\vec{x}]$, $\mathbf{b}[\vec{x}]$, and $\mathbf{c}[\vec{x}]$. Compute the new grid,

$$G_d = \alpha G_a + \beta G_b + \gamma G_c,$$

where $\alpha = \beta = \gamma = 0.333$. Compute

$$\mathbf{d}[\vec{x}] = \alpha W_{G_d}\mathbf{a}[\vec{x}] + \beta W_{G_d}\mathbf{b}[\vec{x}] + \gamma W_{G_d}\mathbf{c}[\vec{x}].$$

9. Repeat the previous problem by using $\alpha = 0.15$, $\beta = 0.5$, and $\gamma = 0.35$.
10. Create a new image by morphing two images, $\mathbf{a}[\vec{x}]$ and $\mathbf{c}[\vec{x}]$, using to a grid defined by

$$G_d = \alpha G_a + (1-\alpha)G_c,$$

where $0 < \alpha < 1$, and

$$\mathbf{d}[\vec{x}] = (1-\alpha)\mathbf{a}[\vec{x}] + \alpha\mathbf{c}[\vec{x}].$$

The effect of this problem works well if the value of α is chosen to be near 0.25 or 0.75.
11. Websites such as `www.tanmonkey.com/fun/dog-looks-like-owner.php` contain images of people and dogs that have similar characteristics. Pick one pair of images and create the fiducial grids for each. Create a set of ten images that morph from one image to the other.
12. Use color images $\mathbf{a}[\vec{x}]$ and $\mathbf{c}[\vec{x}]$. Convert both color images to the Hue, Saturation, Value (HSV) color format (see Section 5.2). Create a morphed image using $\alpha = 0.25$. Apply Equation (7.9) to the V channel and apply $\mathbf{c}[\vec{x}] = \alpha W_G\mathbf{a}[\vec{x}] + (1-\alpha)W_g\mathbf{b}[\vec{x}]$ to the H and S channels. Convert the result back to the RGB color model.
13. Use color images $\mathbf{a}[\vec{x}]$ and $\mathbf{c}[\vec{x}]$. Convert both color images to the YIQ color format. Create a morphed image using $\alpha = 0.25$. Apply Equation (7.9) to the Y channel and apply $\mathbf{c}[\vec{x}] = \alpha W_G\mathbf{a}[\vec{x}] + (1-\alpha)W_g\mathbf{b}[\vec{x}]$ to the I and Q channels. Convert the result back to the RGB color model.

8 Principle Component Analysis

Data generated from experiments may contain several dimensions and be quite complicated. However, the dimensionality of the data may far exceed the complexity of the data. A reduction in dimensionality often allows simpler algorithms to effectively analyze the data. A common method of data reduction is principal component analysis (PCA).

8.1 THE PURPOSE OF PCA

PCA is an often used tool that reduces the dimensionality of a problem. Consider a set of vectors which lie in $\mathbb{R}^N$ space. It is possible that the data is not scattered about but contains inherent structure. When viewed in one view, the data appears scattered, but if viewed from a different orientation the data appears more organized. In this case, the data does not need to be represented in the full $\mathbb{R}^N$ space but can be viewed in a reduced dimensional space as represented by this view. PCA is then used to find the coordinate axes that best orient this data, thus possibly reducing the number of dimensions that are needed to describe the data.

Consider three vectors:

$$\begin{aligned} \vec{x}_1 &= \{2,1,2\} \\ \vec{x}_2 &= \{3,4,3\}. \\ \vec{x}_3 &= \{5,6,5\} \end{aligned} \tag{8.1}$$

The data is three dimensional but the first dimension and the data in the third dimension are exactly the same. Thus, one dimension is redundant and so the data can be viewed in two dimensions without any loss of information.

The example shown is quite simple and often the data redundancy is not so easily recognized. Every dimension contributes to the organization of the data, but some offer very small contributions. Thus, the process of reducing the dimensionality of the data will induce errors. Often these errors are tame and do not impeded on the knowledge gained through the reduction process.

The approach of PCA is equivalent to shifting and rotating the coordinates of the space. Figure 8.1(a) shows a case of a set of data points presented in two dimensions. Each data point requires an (x, y) coordinate to locate it in the space. Figure 8.1(b) shows that same data after rotating the coordinate system. In this case, all of the y coordinates of the data are the same, and therefore, this coordinate is no longer needed. The only necessary description of the data is the x coordinate as shown in Figure 8.1(c). Thus, this rotation can reduce the dimensions without inducing changes to the relationships between the data points.

The computation of the principal components uses the eigenvectors of the covariance matrix. It is necessary to review the covariance matrix and the importance of eigenvectors before proceeding to the determination of the principal components.

8.2 COVARIANCE MATRIX

The dimensionality can be reduced when the location of the data points along one coordinate is similar to another or a linear combination of others. This type of redundancy becomes evident in the covariance matrix which has the ability to indicate which dimensions are dependent on each other.

The covariance matrix is defined as

$$c_{i,j} = \sum_k \left(c_i^k - \mu_i\right) \cdot \left(c_j^k - \mu_j\right), \tag{8.2}$$

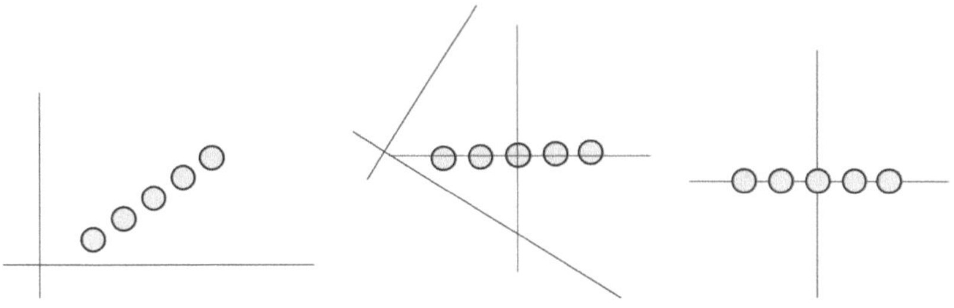

Figure 8.1 (a) The data in the original $\mathbb{R}^2$ space. (b) The data locations after rotation. (c) The data locations after a shift. Now the data is in a view in which the y-axis is no longer important.

where c_i^k is the i-th element of the k-th vector, and μ_i the i-th element of the average vector. The covariance matrix is symmetric and of size $D \times D$, where D is the number of elements in the vector. The diagonal elements are related to the variance of the elements in the vectors. Element c_{ii} is related to the variance of the i-th among all of the vectors. The $c_{i,j}$ element is related to the covariance between the i-th and j-th elements in the data. If $c_{i,j}$ is large positive value, then the i-th and j-th elements are somehow synchronized. A vector with a large i-th element will most likely have a large j-th element as well. A large negative value means that an increase in one element corresponds to a decrease in the other element. Values of $c_{i,j}$ near 0 mean that the data elements are not linked, at least in a first order sense. Data containing vectors with independent elements will produce a diagonal covariance matrix.

The PCA process finds a new view of the data that best diagonalizes the data. Such a perspective is the view in which the elements of the data are most independent, and therefore, this view is often the most informative view.

8.3 EIGENVECTORS

One method of finding the view with the optimally diagonal covariance matrix is through eigenvectors of the covariance matrix. The standard eigenvector-eigenvalue equation is

$$\mathbf{A}\vec{v}_i = \mu_i \vec{v}_i, \tag{8.3}$$

where $\mathbf{A}$ is a square, symmetric matrix, $\vec{v}_i$ is a set of eigenvectors, and μ_i is a set of eigenvalues where $i = 1, \ldots, N$ and the matrix $\mathbf{A}$ is $N \times N$.

In Python *numpy*, package provides an eigenvector solution engine. Code 8.1 creates a matrix $\mathbf{A}$ that is square and symmetric (which emulates the type of matrices that will be used in the PCA analysis). Line 9 calls the function to compute both the eigenvalues and eigenvectors. Since $\mathbf{A}$ is 3×3, there are three values and vectors. The eigenvectors are returned as columns in a matrix. Lines 18 and 19 show that Equation (8.3) holds for the first eigenvalue eigenvector pair and similar tests would reveal that it also holds for the other two pairs.

Another property of the eigenvectors of a symmetric matrix is that they are orthonormal, which means that they have a length of 1 and are perpendicular to each other. Code 8.2 demonstrates that the length of the first eigenvector is 1.0 (the self dot product is 1), and that the first two vectors are perpendicular (the dot product is 0). Since the eigenvectors are orthonormal, they define a coordinate system. Data viewed in this coordinate system produces the optimally diagonal covariance matrix.

Code 8.1 Testing the eigenvector engine in NumPy

```
>>> import numpy as np
>>> np.set_printoptions( precision =3)
>>> d = np.random.ranf( (3,3) )
>>> A = np.dot( d,d.transpose() )
>>> A
array([[ 0.796,  0.582,  0.622],
       [ 0.582,  0.456,  0.506],
       [ 0.622,  0.506,  0.588]])
>>> evl, evc = np.linalg.eig( A )
>>> evl
array([ 1.774,  0.062,  0.004])
>>> evc
array([[ 0.656,  0.698,  0.284],
       [ 0.505, -0.127, -0.853],
       [ 0.560, -0.704,  0.436]])
>>> np.dot( evc[:,0], A )
array([ 1.165,  0.896,  0.993])
>>> evl[0]*evc[:,0]
array([ 1.165,  0.896,  0.993])
```

Code 8.2 Proving that the eigenvectors are orthonormal

```
>>> np.dot( evc[:,0], evc[:,0] )
1.0
>>> np.dot( evc[:,0], evc[:,1] )
0.0
```

8.4 PCA

The process of PCA is to diagonalize the covariance matrix. In doing so, the elements of the data become more independent. If there are first order relationships within the data, then this new representation will often display these relationships more clearly than the original representation. Diagonalization of the covariance matrix is achieved through mapping the data through a new coordinate system.

These eigenvector also provide information about the distribution of the data. Consider the data set that consists of 1000 vectors in $\mathbb{R}^3$. The distribution of data is along a diagonal line passing through (0,0,0) and (1,1,1) with a Gaussian distribution about this line centered at (0.5, 0.5, 0.5) with standard deviations of (0.25, 0.05, 0.25) in the respective dimensions. Two views of the data are shown in Figure 8.2.

The first eigenvector is $(-0.501, -0.508, -0.700)$ which defines a line that follows the long axis of the data. This is shown in Figure 8.3. Removing this component from the data is equivalent of viewing the data along the axis as shown in Figure 8.4. This view is down the barrel of the first axis, as though the viewer is placing his eye at (0,0,0) and looking toward (1,1,1). The second and third eigenvectors define axes that are perpendicular to the first and to each other. The second eigenvector is along the axis of the data in this view as shown in Figure 8.4. The third eigenvector must be perpendicular to the other two, and so there is only one orientation that is possible.

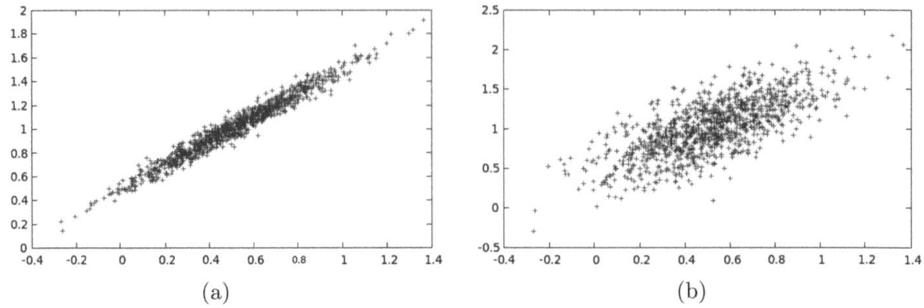

Figure 8.2 Two views of the data set. (a) y vs x. (b) z vs x.

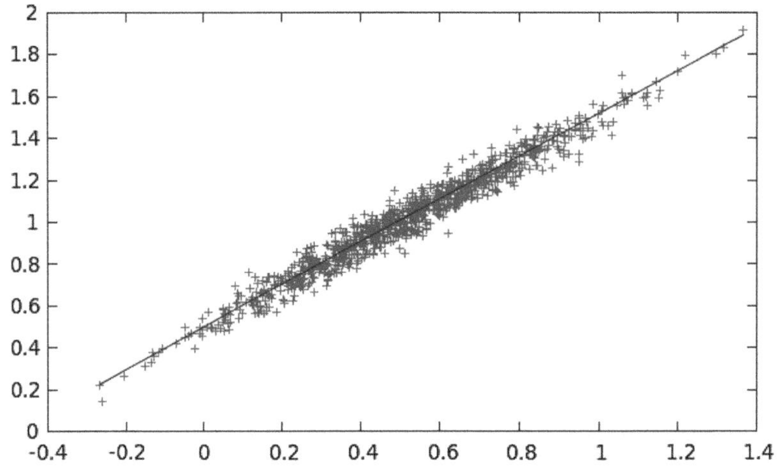

Figure 8.3 First principal component.

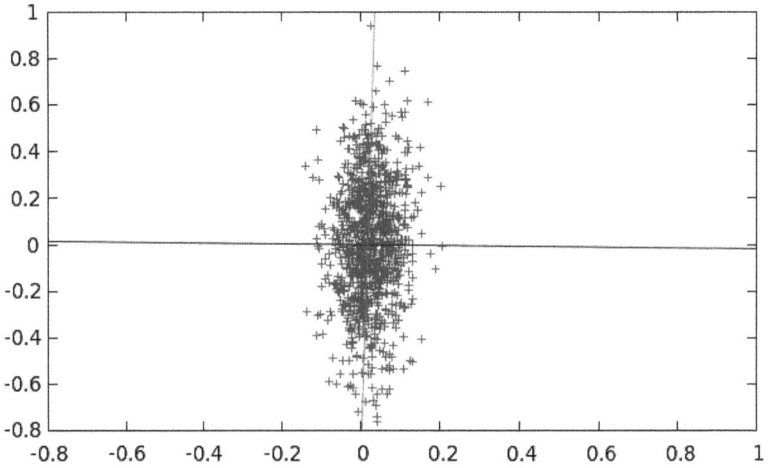

Figure 8.4 Second and third principal components.

PCA uses eigenvectors to find the axes along the data distributions and in doing so diagonalize the covariance matrix. It should be noted that these axes are dependent solely on first-order information. Higher order information is not detected which is discussed in Section 8.5.

The protocol for PCA is,

1. Compute the covariance of the data.
2. Compute the eigenvectors and eigenvalues of the covariance matrix.
3. Determine which eigenvectors to keep.
4. Project the data points into the new space defined by the eigenvectors.

Mathematical representation of data in matrices does tend to oppose the representation preferred in Python scripting. In theory, data is contained as columns in a matrix, whereas Python scripts tend to store data as rows in a matrix. The data vectors are stored in the columns of **X**. The transformation of this data into a new coordinate system is described as the multiplication of a transformation matrix **P** with the data. Thus, the data in the new space are stored as columns in **Y** where

$$\mathbf{Y} = \mathbf{P}\mathbf{X}. \tag{8.4}$$

In this case, the transformation **P** is the eigenvectors stored as rows. However, in Python, if the data is stored in rows then the eigenvectors are stored as columns (which is how the function returns the data). The previous equation could be applied if the transposes of the matrices were considered, which can also be represented by a matrix-matrix multiplication with the order of the matrices reversed. So, for Python the transformation is

$$\mathbf{Y} = \mathbf{X}\mathbf{P}, \tag{8.5}$$

where **X** stores the data in rows, **P** stores the eigenvectors as columns, and **Y** is the data in the new space stored in rows.

The steps to project the points into the new space are complete, and the entire process is accomplished in three lines of script as shown in Code 8.3. Line 1 computes the covariance matrix. Line 2 computes the eigenvectors and eigenvalues. Line 3 uses the **dot** function to perform the matrix-matrix multiplication. The output, is in fact, several dot products as each element in **Y** is,

$$y_{i,j} = \vec{x}_i \cdot \vec{v}_j,$$

where $\vec{x}_i$ is the i-th data vector and $\vec{v}_j$ is the j-th eigenvector.

The process in Code 8.3 maps vectors from N dimensional space to a new N dimensional space. This new space has an optimal covariance matrix, but it also has the same number of dimensions as the original space. If the data has first-order organization, then likely several of the dimensions in the new space provide very little information. The contribution of each eigenvector is offered by the magnitude of its associated eigenvalue. Eigenvectors with small eigenvalues are therefore deemed to be unimportant and can be removed.

The reduction of dimensions is performed by discarding the dimensions with small eigenvalues as seen in Code 8.4. Lines 1 and 2 compute the covariance matrix, the eigenvalues, and the eigenvectors. The **eig** function does not necessarily return the eigenvector/eigenvalue pairs in a sorted order. So, it is necessary to find the sort order as in line 3. The user defines N, the number of eigenvectors

Code 8.3 Projection of data into a new space

```
1 >>> cv = np.cov( data.T )
2 >>> evl, evc = np.linalg.eig( cv )
3 >>> ndata = np.dot( data, evc.T )
```

Code 8.4 Projection of data into a new space

```
1 >>> cv = np.cov( data.transpose() )
2 >>> evl, evc = np.linalg.eig( cv )
3 >>> ag = evl.argsort()[::-1]
4 >>> x = ag[:N]
5 >>> ndata = np.dot( data, evc[:,x] )
```

to keep. This can be a preset value or one determined after analysis of the eigenvalues. Line 4 keeps the indices of the N eigenvectors with the largest eigenvalues. Line 5 maps the data into this new, reduced dimension space.

One method of selecting N is to sort the eigenvalues from high to low as shown in Figure 8.8. Well-behaved data will create a plot that drops quickly and then abruptly flattens out as x increases. In this type of plot, there is a definable elbow demonstrating that there are a limited number of high eigenvalues and several remaining low values. The selection of N is to keep just those high values. The graph shown in Figure 8.8 does not have a definable elbow and so the selection of N is vague. Such is the reality of actual data.

The process in Code 8.3 maps the points to an output space with the same number of dimensions. Thus, there is no error in the projection. The distances between any two pair of points is the same in either space. The process of reducing the dimensions, as in Code 8.4, will induce some error. The distances between any two points will not be the same in the input and output spaces. The amount of error is dependent on the magnitude of the eigenvalues that correspond to the removed eigenvectors. Thus, if the curve of eigenvalues had a steep decline and then a flat region in the higher x values, then the error is small. The example shown in Figure 8.8 indicates that there will be a significant trade-off between error and the number of dimensions removed from consideration. However, as seen in the examples in Section 8.4.2, this error may not drastically alter the significant information garnered from this transition. In such a case, the errors are acceptable, and the reduction in space is beneficial.

8.4.1 DISTANCE TESTS

The projection of the points into a new space will not rearrange the points. The only change is that the viewer is looking at the data from a different angle. Thus, the distances between pairs of points will not change. In the PCA process, some of the dimensions will be eliminated thus inducing an error. The amount of error can be quantified by measuring the relative distances between points before the reduction compared to the relative distances after the reduction.

8.4.2 ORGANIZATION EXAMPLE

This example will create several vectors from the rows of an image. The order of the data vectors will be randomized, thus creating a vector set in which the data has similarity relationships, but the data is not received by the detector in any particular order. The dimensionality of the data will be reduced using PCA, and then there will be an attempt to reconstruct the original image from the relationships of the data in the reduced dimensional space.

The test image is shown in Figure 8.5, which comes from the Brodatz image set [6] that has been used as a library for texture recognition engines. The computation of the covariance matrix and eigenvectors are shown in Code 8.5. Line 4 projects the data into the new space using only the first two eigenvectors. The original image is 640×640, thus producing 640 vectors in a 640 dimensional space. The PCA process created the matrix `ndata`, which is a projection of that data into a two

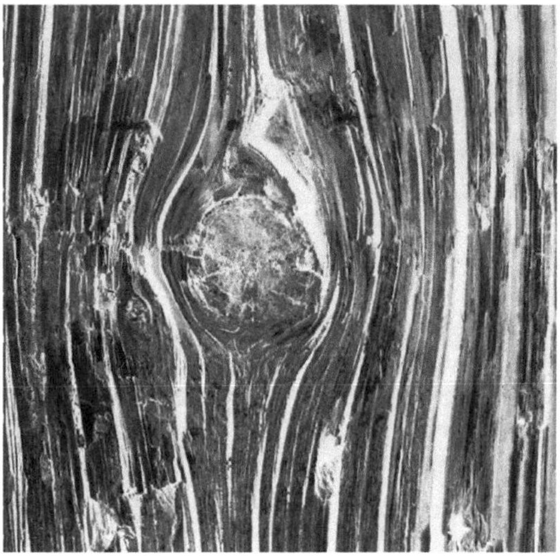

Figure 8.5 Image D72 from the Brodatz image set.

Code 8.5 The first two dimensions in PCA space

```
>>> data = imageio.imread( 'data/D72.png', as_gray=True)
>>> cv = np.cov( data.transpose() )
>>> evl, evc = np.linalg.eig( cv )
>>> ndata = np.dot( data, evc[:,:2] )
```

dimensional space. These points are plotted in Figure 8.6, where each point represents one of the rows in the original image. The top row is associated with the point at $(-584, -66)$ in the graph.

The original image had the quality that consecutive rows had considerable similarity. This is evident in the PCA plot as consecutive points are nearest neighbors. The line connecting the points shows the progression from the top row of the image to the bottom. The clump of points to the far

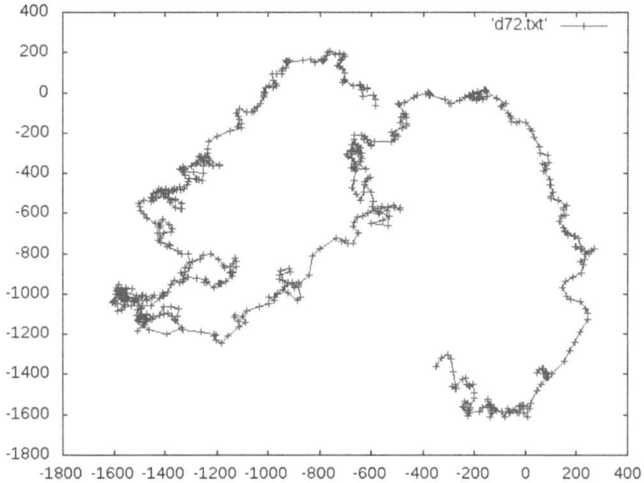

Figure 8.6 The points projected into $\mathbb{R}^2$ space.

left are associated with the bright knothole in the image. This feature of similarity demonstrates that most of the information is contained within the first few dimensions of the PCA space. If the first two dimensions were inadequate, then consecutive rows from the original data would not be close to each other in Figure 8.6.

The test starts with the random ordering of the original vectors. The eigenvectors are computed, and the points are projected into the new space. The data points will be the same as those in Figure 8.6, but the order is not known, so it is not possible to draw the connecting lines. The proposed system begins with the vector associated with the first row of the image and then finds the data point that is closest to it in the reduced space. The vector that is associated with this point is declared to be the second row in the image. The data point closest to the second point is found and the vector associated with it is declared to be the vector for the third row. The process repeats as distances between points in Figure 8.6 are used to determine how the vectors are ordered, and in turn recreate the original image. Success relies on two assumptions. The first is that the consecutive vectors in the original image are uniquely similar, and therefore, the distance between their representative points in PCA space should be small. The second assumption is that the PCA process does not destroy significant information.

The image rows are shuffled by the **ScrambleImage** function shown in Code 8.6. Line 5 scrambles the order of the rows. Lines 6 and 7 are used to locate the first row of data in the newly ordered data so that the process can start with the first row of the image. The function returns the scrambled rows and an integer that is the location of the first row of the original data within this newly ordered data. The scrambled image is shown in Figure 8.7.

Line 10 scrambles the rows, and line 11 projects this scrambled data into a PCA space. The data points in this PCA are in the same location as in Figure 8.6, but the lines cannot be drawn between the points.

Code 8.7 shows the function **Unscramble** , which performs the reconstruction of the image. The inputs are the scrambled data, `sdata`, the location of the first row of the image in the scrambled data, `seedrow`, and the projected data, `ndata`. Currently, all 640 dimensions are contained in `ndata` but these will be restricted in subsequent examples. The variable `udata` will become the unscrambled image and the first row is placed by line 5. The list `unused` maintains a list of rows that have not been placed in `udata`, so, the first row is removed in line 7. The variable `k` will track which row is selected to be placed into `udata`.

Line 11 computes the Euclidean distance from a specified row to all of the other unused rows. Thus, in the first iteration, `k = 0`, and so this computes the distance from the seed vector to all other rows using the projected data. Basically, it is finding the closest point in the PCA space shown in Figure 8.6. Line 12 finds the smallest distance and thus finds the vector that is closest to the `k` row.

Code 8.6 The **ScrambleImage** function

```
1  # pca.py
2  def ScrambleImage( fname ):
3      mgdata = imageio.imread( fname, as_gray=True)
4      sdata = mgdata + 0
5      np.random.shuffle( sdata )
6      dists = np.sqrt(((mgdata[0]-sdata).sum(1)))
7      seedrow = (dists==0).nonzero()[0]
8      return sdata, seedrow
9
10 >>> sdata, seedrow = pca.ScrambleImage( fname )
11 >>> ndata = pca.Project( sdata )
```

Figure 8.7 The scrambled image.

The corresponding row of data is then placed in the next available row in `udata`, and the vector in PCA space is removed from further consideration in line 15.

In the first example, all 640 dimensions of the projected space are used. Thus, there should be absolutely no loss of information. The call to the function is shown in line 1 of Code 8.8. The output `udata` is an exact replicate of the original image. Not all of the dimensions in the PCA space are required. Consider the plot of the first 20 eigenvalues shown in Figure 8.8. When data is organized, the eigenvalues fall rapidly thus indicating the importance of each eigenvector.

Line 2 in Code 8.8 reconstructs the image using only 7 of the 640 eigenvectors. The result, shown in Figure 8.9, is nearly perfect reconstruction with only a few rogue lines at the bottom of the image. This used the data points projected into an $\mathbb{R}^7$ space and then computed the Euclidean

Code 8.7 The **Unscramble** function

```
1  # pca.py
2  def Unscramble( sdata, seedrow, ndata ):
3      V,H = sdata.shape
4      udata = np.zeros((V,H))
5      udata[0] = sdata[seedrow] + 0
6      unused = list( range( V) )
7      unused.remove( seedrow )
8      nndata = ndata + 0
9      k = seedrow
10     for i in range( 1, V ):
11         dist = np.sqrt(((nndata[k]-nndata[unused])**2).sum(1))
12         ag = dist.argsort()
13         k = unused[ag[0]]
14         udata[i] = sdata[k]
15         unused.remove( k )
16     return udata
```

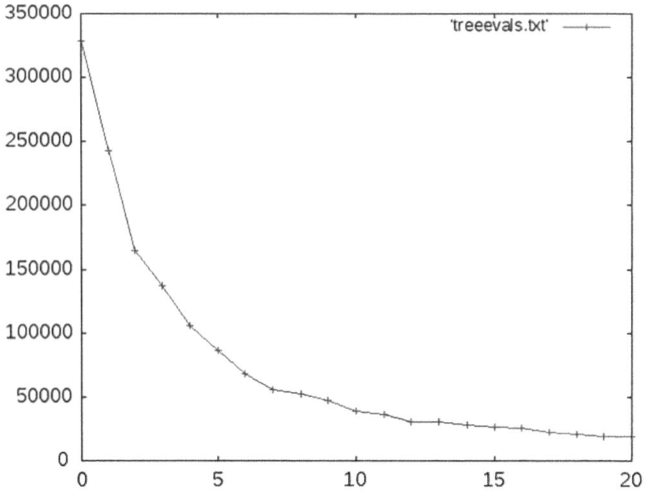

Figure 8.8 The first 20 eigenvalues.

Code 8.8 Various calls to the **Unscramble** function

```
1 >>> udata = pca.Unscramble(sdata, seedrow, ndata )
2 >>> udata = pca.Unscramble(sdata, seedrow, ndata[:,:7] )
3 >>> udata = pca.Unscramble(sdata, seedrow, ndata[:,:2] )
```

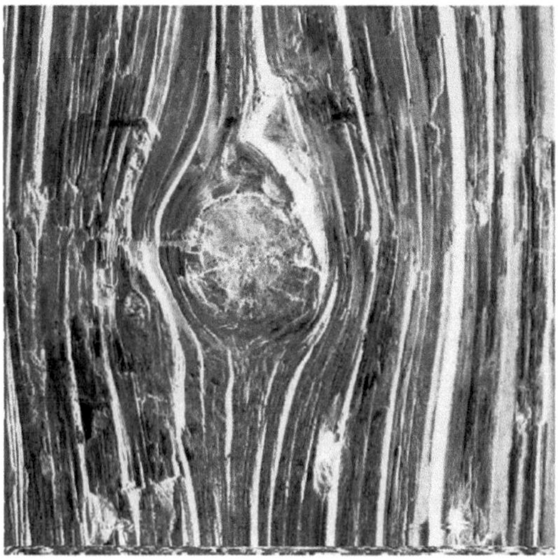

Figure 8.9 Reconstruction using only 7 dimensions.

distances between the projected points in that space. The few rows at the bottom were skipped during reconstruction.

Line 3 in Code 8.8 reconstructs the image using only 2 of the 640 eigenvectors. The result is shown in Figure 8.10. As seen, there are a few more errors in the reconstruction, but most of the reconstruction is intact. This is not a surprise since more than two eigenvalues had significant magnitude in Figure 8.8. However, even with this extreme reduction in dimensionality, most of the

Figure 8.10 Reconstruction using only 2 dimensions.

image could be reconstructed. This indicates that even in the reduction from 640 dimensions to 2 that there was still a significant amount of information that was preserved. In some applications of PCA, this loss of information is not significant in the analysis that is being performed.

8.4.3 RGB EXAMPLE

This second example uses PCA to help isolate pixels in an image according to the RGB colors. Data for this example starts with the image in Figure 8.11. This image is 480×640, and each pixel is represented by 3 values (RGB). The data set is thus 307,200 vectors of length 3. The task in this example is to isolate the blue pixels. This seemingly simple task is complicated by the shadows on the coat, which means that the pixel values on the coat have a wide range that also overlaps with other objects in the image.

Since there are no other blue objects in the image, an attempt was to made to isolate the blue pixels using

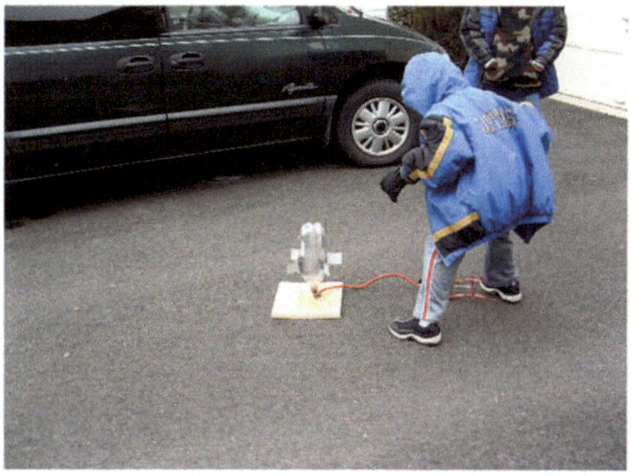

Figure 8.11 An input image.

$$\mathbf{b}[\vec{x}] = \left(\Gamma_{>1,5} \frac{\begin{Bmatrix} \varnothing \\ \varnothing \\ 1 \end{Bmatrix} \mathbf{a}[\vec{x}]}{\begin{Bmatrix} \varnothing \\ 1 \\ \varnothing \end{Bmatrix} \mathbf{a}[\vec{x}] + 1} \right) \times \left(\Gamma_{>1,5} \frac{\begin{Bmatrix} \varnothing \\ \varnothing \\ 1 \end{Bmatrix} \mathbf{a}[\vec{x}]}{\begin{Bmatrix} 1 \\ \varnothing \\ \varnothing \end{Bmatrix} \mathbf{a}[\vec{x}] + 1} \right). \tag{8.6}$$

Each of the two terms favors the blue channel over one of the others. The only pixels in $\mathbf{b}[\vec{x}]$ that are set to 1 survived both thresholds. Code 8.9 shows the **LoadRGBchannels** function, which separates the image into its three-color channels. The function **IsoBlue** performs the steps of Equation (8.6).

As seen in Figure 8.12(a), most of the pixels from the coat are turned on, but so also are some pixels from the other objects. Equation (8.6) works fairly well, but there is room for improvement. Figure 8.13(a) shows a plot in $\mathbb{R}^3$, where each dot represents a pixel from the image. The three axes are the R, G, and B value of the pixel. The pixels that survived Equation (8.6) are green, while the rest of them are red. Figure 8.13(b) shows the same data from a different perspective.

Improvements to the performance is possible by fine tuning the parameters in Equation (8.6). Adjustments can be made to the threshold values, the biases in the denominators, or it is possible to include multiplication factors in the numerator. In any case, it would be necessary for the user to determine the best values for these parameters. This equation creates a flat decision plane in the space shown in Figure 8.13(a), and the adjustments of the parameters changes the location and

Code 8.9 The **LoadImage** and **IsoBlue** functions

```
# pca.py
def LoadRGBchannels( fname ):
    data = imageio.imread( fname )
    r = data[:,:,0]
    g = data[:,:,1]
    b = data[:,:,2]
    return r,g,b

def IsoBlue( r,g,b ):
    ag = b/(g+1.0)>1.5
    ab = b/(r+1.0)>1.5
    isoblue = ag*ab
    return isoblue
```

(a) (b)

Figure 8.12 (a) Isolation of the pixels using Equation (8.6), and (b) isolation of the pixels using PCA.

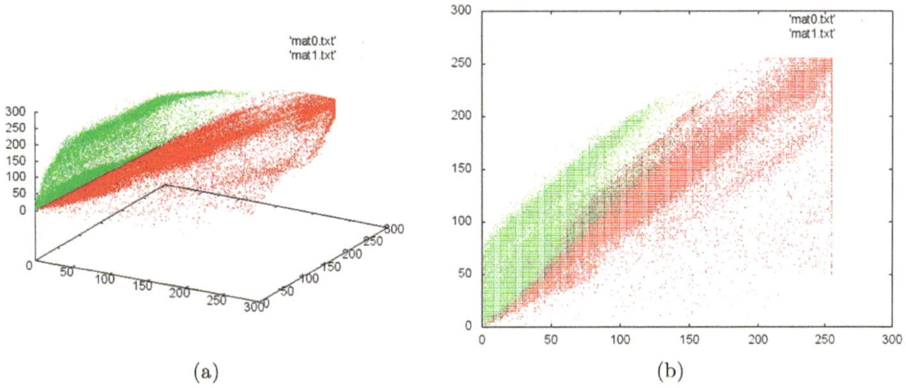

Figure 8.13 (a) The map of the RGB values in $\mathbb{R}^3$. The green points are those denoted in Figure 5.7(a). The map of the RGB values in $\mathbb{R}^2$.

orientation of this plane. Adjustments to the parameters do not change the shape of the decision surface. Even by looking at the images in Figure 8.13(a) and (b), it is not easy to determine which adjustments to the decision plane need to be made.

Now consider the application of PCA to the data. The covariance matrix is computed from the complete set of $\mathbb{R}^3$ vectors and from that the eigenvectors are computed. The eigenvector associated with the smallest eigenvalue is discarded, and the data is projected into this new $\mathbb{R}^2$ space. The result is shown in Figure 8.14. The color coding still relates to the previous solution, and it is noted that in this projection, the plane is not horizontal.

Before adjustments are made, information about the image needs to be reviewed. The blue coat is quite different in color the rest of the image. Therefore, we expect that the pixels from the coat should be grouped separately from the rest of the image in a proper view. The PCA projection does, in fact, display regions of higher density. If the decision surface were to be horizontal at about $x = 45$ (shown as a thick line), then it would separate regions based on these densities. This is a small adjustment from the decision surface of the previous approach, but it has a significant effect.

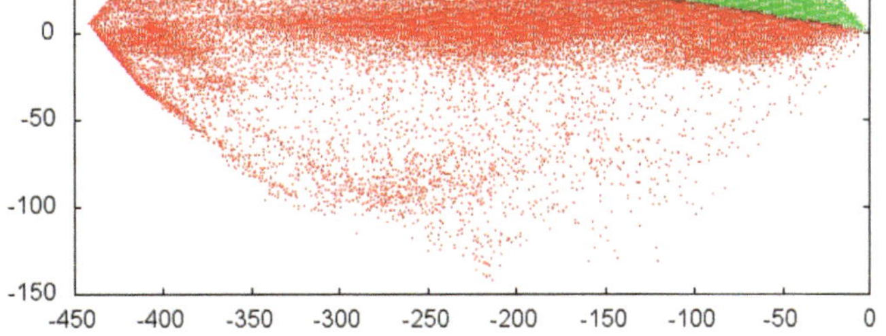

Figure 8.14 First two axes in PCA space.

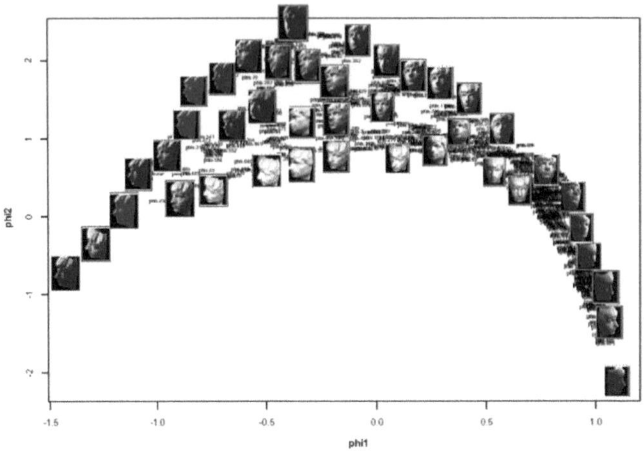

Figure 8.15 PCA map of face pose images.

The pixels above $y = 45$ are classified as blue and these are depicted in Figure 8.12(b). The isolation of the coat is much improved, which was aided by the presentation of the data in PCA space.

The PCA operation is a rotation of the viewing space. As seen in Section 6.4, a rotation is performed by multiplying the data vector by a matrix. Recall that the YIQ conversion was also a linear transformation performed by the multiplication of a matrix with vectors representing the RGB values. Both of these operations merely multiply the RGB vector of a pixel with a matrix so that the information appears in a different view. Examples have been given in which both methods demonstrate the isolation of objects by color.

The mapping to PCA space did not drastically change the data. It did, however, represent the data in a manner such that a simple threshold (only one axis) could isolate the desired color.

8.5 FIRST ORDER NATURE OF PCA

Consider a case in which the data consists of several images of a single face at different poses. In this case, the face is a computer generated face and so there are no factors such as deviations in expression, hair, glasses, etc. Figure 8.15 shows data mapped to a PCA space. In this figure, there are several data points in the PCA space, and a few of them have their original image placed next to the point. As seen, the pose of the face gradually changes from left to right. A possible conclusion is that the PCA was capable of mapping the pose of the face.

This would be an incorrect conclusion. PCA can only capture first order data. In other words, it can compare pixel (i, j) of the images to each other but not pixel (i, j) with pixel (k, l). The reason that the faces sorted as they did in this example is more of a function of the location of the bright pixels in the images. The idea of "pose" is not captured, and it is only that the face was illuminated from a single source that there was a correlation between the pose and the regions of illumination.

8.6 SUMMARY

The principal components are a new set of coordinates in which the data can be represented. These components are orthonormal vectors and are basically a rotation of the original coordinate system. However, the rotation minimizes the covariance matrix of the data and thus some of the coordinates may become unimportant. In this situation, these coordinates can be discarded and thus PCA space uses fewer dimensions to represent the data than the original coordinate system.

PROBLEMS

1. Consider the data shown in Figure 8.1(a). The coordinates are (2,1), (3,2), (4,3), (5,4), and (6,5). Compute the covariance matrix for this data. (The covariance matrix should be 2×2.)
2. Consider the data similar to Figure 8.1(c). Use coordinates (−2,0), (−1,0), (0,0), (1,0), and (2,0). Compute the covariance matrix for this data. (The covariance matrix should be 2×2.)
3. Use the data from Problem 1. Compute the covariance matrix, the eigenvectors, and the eigenvalues. Project the data into the new two-dimensional space. (Do not reduce the number of dimensions.) Compute the covariance matrix of this new data. Is the new matrix more diagonal than the original?
4. Given a set of N vectors. In this case, the eigenvalues of this set turn out to be 1,0,0,0, What does this imply?
5. Given a set of N vectors. In this case, the eigenvalues of this set turn out to be 1, 1, 1, 1 What does this imply?
6. Given a set of purely random vectors, describe what you expect the nature of the eigenvalues to be. Confirm your prediction.
7. Given a set of N random vectors of length D. Compute the covariance matrix. Why is this matrix not diagonal?
8. Repeat the process of Section 8.4.2 on the grayscale version of the bird image.
9. Plot the eigenvalues from the previous problem from high to low. How many dimensions are siginficant? Justify your answer.
10. Repeat the process of Section 8.4.2 on the grayscale version of the bird image using only two eigenvectors.

9 Eigenimages

The concept of average face was presented in Section 7.2, and it was noted that humans identify faces by their deviations from the average face. This chapter will explore one method of describing these differences within a population. This method has been used for face recognition systems but does not include all of the important features of a face. Just as in the computation of the average face, the face images will be warped to a single grid, which does remove shape information from the image. Shape is important, and so, necessarily the method shown here should be considered as only one tool in a suite available for face recognition.

The method of eigenimages is based on principal component analysis (PCA), and it creates a new space to represent a set of images. Advantages of this new space include a significantly reduced dimensionality and an optimal view of the first order data inherent in the data set. This chapter will review the foundations required of the eigenimage method as well as the generation and use of eigenimages coupled with examples.

9.1 EIGENIMAGES

Eigenimages are created in a process similar to PCA, but with a modification. The traditional PCA process is:

1. Warp images to a single grid.
2. Compute the covariance matrix of the images.
3. Compute the eigenimages from the covariance matrix.
4. Downselect the eigenimages to use.

Step 2, though, is an untenable step. The size of the covariance matrix is prohibitively large. If the images were 256×256, which is a small image, then the covariance matrix would be $65{,}536 \times 65{,}536$. Using single precision floats, this array would consume 16 GB. Thus, this method needs to be modified in order to calculate the eigenimages.

Formally, the process starts with a set of images, $\mathbf{a}_i[\vec{x}]; i = 1,\ldots,P$. Every image is warped to a single grid, G:

$$\{\mathbf{b}[\vec{x}]\} = \{W_G \mathbf{a}[\vec{x}]\}. \tag{9.1}$$

The average images is computed:

$$\bar{\mathbf{b}}[\vec{x}] = \frac{1}{N}\sum\{\mathbf{b}[\vec{x}]\}, \tag{9.2}$$

and subtracted from each image,

$$\{\mathbf{c}[\vec{x}]\} = \left\{\mathbf{b}[\vec{x}] - \bar{\mathbf{b}}[\vec{x}]\right\}. \tag{9.3}$$

The notation $\left\{\mathbf{b}[\vec{x}] - \bar{\mathbf{b}}[\vec{x}]\right\}$ is a set. There are several images in $\{\mathbf{b}[\vec{x}]\}$ but only a single image $\bar{\mathbf{b}}[\vec{x}]$. Since the latter resides inside the brackets, it is subtracted from each image. The output is a set of images, which are the original images with the average image subtracted.

In the original theory, the covariance matrix would be computed from these images. From this, the eigenimages are computed by

$$\{\mathbf{d}[\vec{x}]\} = \frac{1}{\sqrt{\lambda_j}}\left\{\begin{matrix}\varnothing \\ \Downarrow_{\lambda_j<\gamma}\end{matrix}\right\}\mathcal{E}V\{\mathbf{c}[\vec{x}]\} \tag{9.4}$$

but as stated, that matrix would be extremely large. Fortunately, the number of pixels far exceeds the number of images.

9.1.1 LARGE COVARIANCE MATRIX

The solution [32] is to consider an alternate but similar problem. The eigen-equation is

$$\mathbf{C}\vec{x}_i = \lambda_i \vec{x}_i \tag{9.5}$$

and that the covariance matrix can be expressed by $\mathbf{C} = \mathbf{X}\mathbf{X}^T$, where $\mathbf{X}$ is the matrix that contains the original data in columns. Therefore,

$$\mathbf{X}\mathbf{X}^T \vec{v}_i = \lambda_i \vec{v}_i. \tag{9.6}$$

For image data, this matrix is prohibitively large. Consider a new eigen-equation:

$$\mathbf{L}\vec{w}_i = \lambda_i \vec{w}_i, \tag{9.7}$$

where $\mathbf{L} = \mathbf{X}^T\mathbf{X}$. Left multiply Equation (9.6) by $\mathbf{X}^T$ to get

$$\mathbf{X}^T\mathbf{X}\mathbf{X}^T \vec{v}_i = \lambda_i \mathbf{X}^T \vec{v}_i, \tag{9.8}$$

and thus,

$$\mathbf{L}\mathbf{X}^T \vec{v}_i. = \lambda_i \mathbf{X}^T \vec{v}_i. \tag{9.9}$$

Comparing this to Equation (9.7) reveals $\mathbf{X}^T\vec{v}_i = \vec{w}_i$. The matrix $\mathbf{L}$ is $P \times P$ where P is the number of images and since $P << N$ the size of $\mathbf{L}$ is much smaller than the size of $\mathbf{C}$. The solution is to then solve for the eigenvectors and eigenvalue of $\mathbf{L}$ and then compute $\vec{v}$'s from $\vec{w}$'s.

9.1.2 PYTHON IMPLEMENTATION

The operator symbol for collecting eigenimages is T_m. The subscript m can have multiple meanings. If m is an integer, then it indicates how many eigenimages are used. If m is a float, then it indicates that the eigenimages that are used are those with eigenvalues greater than m. For the case of keeping, the first three eigenimages with the largest eigenvalues the operator is

$$\{\mathbf{d}[\vec{x}]\} = T_3 \{\mathbf{a}[\vec{x}]\}. \tag{9.10}$$

Code 9.1 shows a Python script that performs all of the necessary steps as the function **EigenImages**. It receives a list or array of two-dimensional matrices containing pixel intensities. The matrix $\mathbf{L}$ is initialized in line 5 and populated in lines 6 through 9. The eigenvectors are computed in line 10, and the rest of the function creates the eigenimages. The function returns a list of matrices (eigenimages) and the associated eigenvalues. At this time, there is no selection of eigenimages since there are different manners that a user may wish to use to make this selection, and therefore, the process is performed outside of **EigenImages**.

A simple example starts with three images shown in Figure 9.1. The images shown here have a white background for printing, but the original data have a black background to lessen the effect of the background on the computation.

The function **EigenImages** produces three eigenimages, each associated with eigenvalues. In this case, the eigenvalues were $(1.07 \times 10^8, -4.2 \times 10^{-6}, 2.11 \times 10^7)$. The second eigenvalue is many orders of magnitude less than the others, which indicates that the second eigenimage is not important. So, only the first and third are kept and these are shown in Figure 9.2. These define the new two-dimensional coordinate system. Clearly, evident is that the eigenimages are weighted linear combinations of the original images.

Projection of the data points onto the new space is performed by **ProjectEigen** which is shown in Code 9.2. The inputs are the selected eigenimages and a single input image (as a matrix). The process simply projects the image onto the eigenimages through dot products, which produces a

Code 9.1 The **EigenImages** function

```
# eigenimage.py
def EigenImages( d ):
    N,V,H = len( d )
    dd = d.astype(float) - d.astype(float).mean(0)
    L = np.zeros( (N,N) )
    for i in range( N ):
        L[i,i] = (dd[i] * dd[i]).sum()
        for j in range( i ):
            L[i,j] = L[j,i] =  (dd[i] * dd[j] ).sum()
    evls, evcs = np.linalg.eig( L )
    emgs = []
    for j in range( N ):
        a = np.zeros( (V,H) )
        for i in range( N ):
            a += evcs[i,j]*dd[i]
        emgs.append( a/np.sqrt(evls[j]) )
    return emgs, evls
```

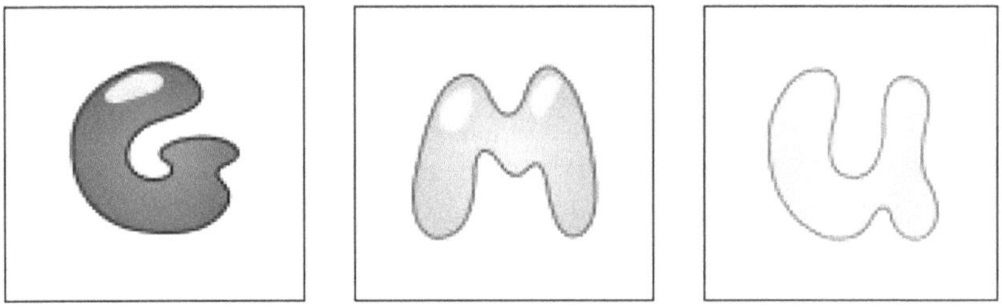

(a) (b) (c)

Figure 9.1 The three input images.

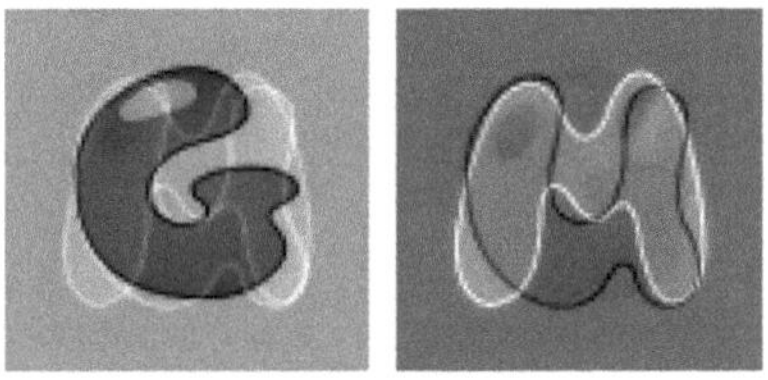

(a) (b)

Figure 9.2 Two eigenimages.

Code 9.2 The **ProjectEigen** function

```
# eigenimage.py
def ProjectEigen( emgs, indata ):
    vec = np.zeros( len( emgs ))
    for i in range( len( emgs )):
        vec[i] = (emgs[i] * indata).sum()
    return vec
```

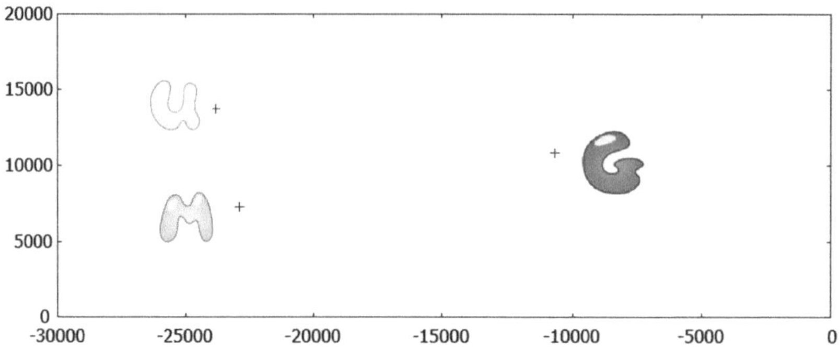

Figure 9.3 A plot of eigenvalues.

position vector in the new space. In this example, there are only two eigenimages, and therefore, the new space is two-dimensional.

The result is shown in Figure 9.3. There are three markers which show the position of the three images in the new space. Besides each one is a replication of the input image manually placed on the graph. The eigenimage in Figure 9.2(a) corresponds to the horizontal axis in Figure 9.3, and Figure 9.2(b) corresponds to the vertical axis. As seen, there the horizontal location of the "G" is distinct from the "M" or the "U." Obviously, this eigenimage was more sensitive to the "G" shape which is evident both through visual inspection of the eigenimage and the position in the graph. The eigenimage in Figure 9.2(b) shows the "M" and "U" but in opposing intensities. This shows that this image is sensitive to these two shapes but in an opposing manner, which is very useful in separating them. As seen in Figure 9.3 the "M" and "U" are separated vertically which is congruent with the idea that the vertical axis is corresponds to the second eigenimage.

9.1.3 FACE RECOGNITION EXAMPLE

Face recognition is a famous application for eigenimages, and this section will review an application on a small data set. The method of eigenfaces is first order, which means that when an image is compared to the eigenfaces that all features must align. Therefore, the training images and the testing images must all be warped to a single grid.

Each training face requires a fiducial file as well as the image file. Thus, a training set is represented as $\{(\mathbf{G}_i, \mathbf{a}_i[\vec{x}]), i = 1, \ldots, P\}$, where P is the number of training face images and G_i is the fiducial grid of the i-th image.

The average fiducial grid is computed:

$$\bar{\mathbf{G}} = \frac{1}{P} \sum_{i=1}^{P} \mathbf{G}_i \tag{9.11}$$

and each image is warped to this grid using the **Warp** operator.

The creation of the eigenfaces is given in Equation (9.10). Example eigenfaces are shown in Figure 9.4. The features of interest are either the brightest or darkest regions in each image. Thus, image (e) shows a sensitivity to facial hair, (c) shows a sensitivity to lighting, and (b) shows a sensitivity to hair over the forehead, and so on.

Once the eigenfaces are selected, the warped face images are projected into the new space. In this example, there were four images per person, and the mapping of these images into the first three dimensions is shown in Figure 9.5. Each person is represented by a different colored marker. As seen the markers for each person group nicely. This indicates that this eigenspace has the ability to separate the face images. Recognition occurs when a new image is warped and then projected into

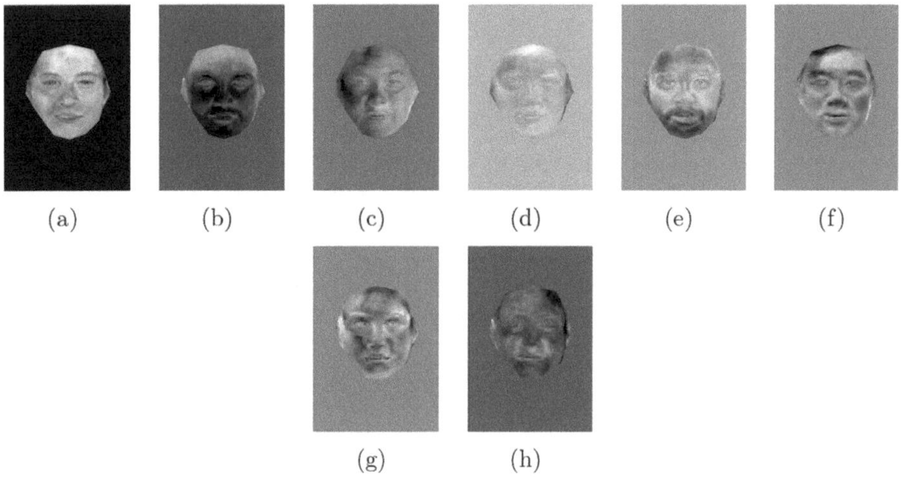

Figure 9.4 Several of the eigenimages.

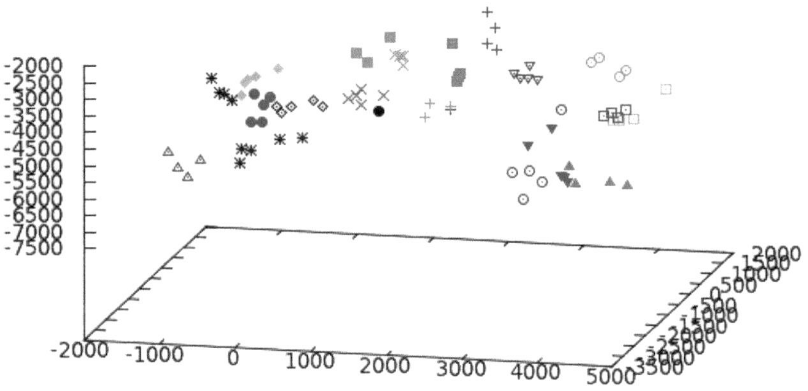

Figure 9.5 Face recognition using eigenfaces.

this space. The classification of the input is the class of the cluster that is nearest to the input in this space.

The eigenimage process is sensitive to first-order information as the projection in the eigenspace is based on dot products. Shape, on the other hand, is higher order and requires knowledge of the relative position of perimeter pixels and so forth. Eigenimages are sensitive only to pixel intensities, and, as such, the location of the data in the eigenspace is a function of pixel intensities. The images on the right side of the chart are brighter on the right portions of their images. The images in the lower middle have intensities that fill the frame. This projection has organized the data based on the location of pixel intensities, and not the feature of pose.

9.1.4 NATURAL EIGENIMAGES

Hancock et al. [10] presented the idea of natural eigenimages, which are created from a large number of random images. Interestingly, the eigenimages of random images are not themselves random, but do show a structure. Figure 9.6 shows the result of a test that started with 20 large images of nature scenes. These images were cut into 256 × 256 blocks resulting in more than 3,000 grayscale images. From this large set of images, the 3,000 eigenimages were created of which the first 100 are shown

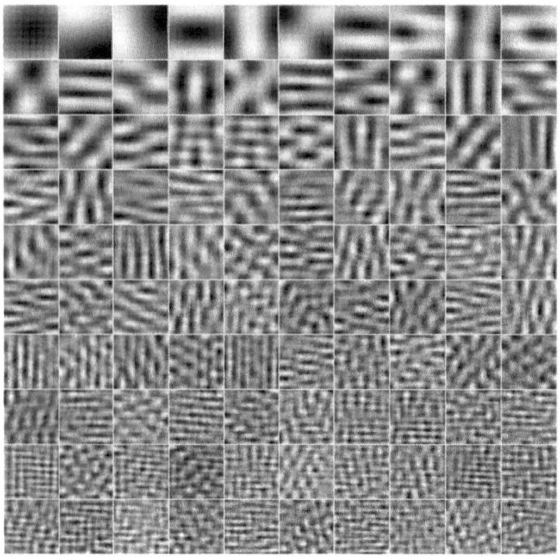

Figure 9.6 Eigenimages of natural scenes.

in Figure 9.6. As seen, there is a definite structure in the results as the frequency of oscillations increases with the eigenimage index. These images are also orthonormal. Furthermore, these results are independent of the input images as long as a few guidelines are followed. Most of the images should have significant structure, and this test cannot be performed on a bunch of images of the sky or something else that is smooth. The second guideline is that it does take a rather large number of images to get these results and this does require some computational resources.

The interesting part of this test is that a completely different set of input images will create basically the same set of natural eigenimages. The implication is that there is a natural structure to images. The usefulness of this internal structure is explored further in Chapter 19.

PROBLEMS

1. Create the eigenimages for the images in Figure 9.7 (Stored as *data/pca.png*).
2. Show that each of the eigenimages from Problem 1 are normal (e.g., the sum of the pixels is 1 or N, where N is the number of pixels).
3. Show that the eigenimages from Problem 1 are all orthogonal to each other.
4. Use the image in Figure 5.7(a) (or another photograph if this is not available). Pick 1,000 random pixels and use these to create 1,000 vectors of length 3 from the RGB values. Compute the covariance matrix.
5. Repeat Problem 4, except use 2,000 random pixels. Write your opinion as to whether or not these two covariance matrices are similar.
6. Choose a person that has many different face images on the web, such as a movie star or a politician. Get ten color images from the web, and place the grayscale versions of the images into frames of the same size with faces about the same size as well. Determine the fiducial grid for each image and save each as a text file. Compute the average fiducial grid.

Figure 9.7 PCA image.

7. Using the information from Problem 6, warp all faces to this grid.
8. Using the results from Problem 7, compute the average face from the data collected in Problem 6.
9. Using data from Problem 1, choose two images of different people. Morph one image to the other in a process that creates 20 intermediate images. Compute the fiducial grids for each of these images as a weighted linear combination of the original grids. Compute the eigenimages of these 20 images.

Part III

Frequency Space Manipulations

10 Image Frequencies

One dimensional signals, such as speech, are often analyzed through their frequencies. Some speaker identification algorithms rely on the identification of a fundamental frequency and overtones that are distinctive to a person's voice. Images are two-dimensional signals, and as such, they also have frequencies that can be analyzed and manipulated. There are many extremely powerful algorithms that perform their operations on the image frequencies rather than the original image. In most cases, these operations would be difficult to accomplish without the analysis of the image frequencies. The most common method of accessing image frequencies is through a Fourier transform. This chapter will review Fourier theory and provide some examples. Subsequent chapters will use Fourier transforms to achieve more advanced processing.

10.1 COMPLEX NUMBERS

The Fourier transform inherently uses complex numbers in the form as $x + \imath y$. Another representation of a complex number is $re^{\imath\theta}$, and these conversions were reviewed in Equations (6.26) and (6.27). The Complex Transformation operator converts all values in an image $\mathbf{d}[\vec{x}]$ from the linear representation to the polar representation as in

$$\mathbf{b}[\vec{y}] = P\mathbf{a}[\vec{x}]. \tag{10.1}$$

Each value in $\mathbf{a}[\vec{x}]$ is in $x + \imath y$ notation, and each element in $\mathbf{b}[\vec{x}]$ is in $r^{\imath\theta}$ notation. The inverse operation is

$$\mathbf{a}[\vec{x}] = P^{-1}\mathbf{b}[\vec{y}]. \tag{10.2}$$

The function **Rect2Polar** shown in Code 10.1 performs Equation (10.1), and the function **Polar2Rect** performs Equation (10.2).

Code 10.1 The **Rect2Polar** and **Polar2Rect** functions

```
# mgcreate.py
def Rect2Polar( data ):
    V,H = data.shape
    answer= np.zeros((V,H,2))
    answer[:,:,0] = np.hypot( data.real, data.imag) #r
    answer[:,:,1] = np.arctan2( data.imag, data.real ) #theta
    return answer

def Polar2Rect( data ):
    V,H,N = data.shape
    answer = np.zeros((V,H),complex)
    answer.real = data[:,:,0] * np.cos( data[:,:,1] )
    answer.imag = data[:,:,0] * np.sin( data[:,:,1] )
    return answer
```

10.2 THEORY

The original theory was developed long before the onset of digital data and so it considers an analog signal $f(x)$. The Fourier transform is described as

$$F(\omega) = \int_{-\infty}^{\infty} f(x)\, e^{-\imath\omega x}\, \mathrm{d}x, \tag{10.3}$$

where ω represents the frequency coordinates and $\imath = \sqrt{-1}$. Recall that

$$e^{\imath\theta} = \cos\theta + \imath\sin\theta, \tag{10.4}$$

and so the integrand is the multiplication of wave functions with $f(x)$. This function is basically the inner product (or projection) of $f(x)$ onto wave functions, and this will indicate the amount of each wave present in the original function. The magnitude of $F(\omega)$ is the magnitude of the ω frequency that is in $f(x)$.

The inverse Fourier transform is

$$f'(x) = \frac{1}{2\pi}\int_{-\infty}^{\infty} F(\omega)\, e^{\imath\omega x}\, \mathrm{d}\omega. \tag{10.5}$$

There is no loss of information and so $f'(x) = f(x)$.

10.3 DIGITAL FOURIER TRANSFORM

In modern times, data is digital, and therefore, it is necessary to review the digital Fourier transform. To distinguish an analog signal from a digital signal, the input has a slightly different form with the use of square brackets. The digital Fourier transform (DFT) of a signal $f[n]$ is

$$F[\omega] = \sum_{n=1}^{N-0} f[n]\, e^{-2\pi\imath n\omega/N}, \tag{10.6}$$

where N is the number of elements in the signal.

The inverse DFT is

$$f'[n] = \frac{1}{N}\sum_{\omega=0}^{N-1} F[\omega]\, e^{2\pi\imath n\omega/N}. \tag{10.7}$$

The philosophy is the same in that the Fourier transform is the projection of the original signal onto a set of frequencies, and the inverse transform is the weighted linear combination of those frequencies.

For two-dimensional signals, the transforms are

$$F[u,v] = \sum_{n=0}^{N-1}\sum_{m=0}^{M-1} f[x,y] e^{-\imath 2\pi(ux/M+vy/N)}, \tag{10.8}$$

and

$$f(x,y) = \frac{1}{MN}\sum_{n=0}^{N-1}\sum_{m=0}^{M-1} F[u,v] e^{\imath 2\pi(ux/M+vy/N)}, \tag{10.9}$$

The operator for a Fourier transform is $\mathfrak{F}$, and the inverse is $\mathfrak{F}^{-1}$.

10.3.1 FFT IN PYTHON

The digital Fourier transform contains a large amount of repetitive computations. The Cooley–Tukey algorithm [7] provides an efficient method of performing the Fourier transforms by eliminating these massive repetitive computations. This fast Fourier transform (FFT) is the algorithm of choice that is used in computer programs.

The *scipy* module contains a package named *fftpack* with several Fourier transform functions. The function **fft** computes the forward Fourier transform, and the function **ifft** computes the inverse transform. These are demonstrated in Code 10.2, which generates a random input vector `a` and computes the transform `A`. Two-dimensional transforms use the functions **fft2** and **ifft2**. Higher dimensional arrays are transformed by the functions **fftn** and **ifftn**.

The efficiency of the FFT is based on eliminating repetitive computations, which occur when the length of the vector is a power of 2. When the length of the data is another value, then the FFT loses its efficiency and there can be undesired spurious effects. So, it is often a good idea to place the data into a frame that has a length that is a power of 2. The same effects apply to images, although it is not necessary for the dimensions to be the same value. For example, the FFT will be efficient for an image with dimensions $(512, 1024)$ because both dimensions are a power of 2. It is almost ten times faster to compute the FFT of an image that has a frame size of (1024,1024) than it is to compute the FFT of an image that has a frame size of (513,513).

The Plop operator will place an image inside a larger frame and is written as

$$\mathbf{b}[\vec{x}] = U_{\vec{w}}\mathbf{a}[\vec{x}], \tag{10.10}$$

where $\vec{w}$ is the frame size of $\mathbf{b}[\vec{x}]$. Many applications would prefer that the frame size of the Fourier transform be the same as the frame size of the input. One manner of performing this is to plop the image into a larger frame, perform the Fourier transform, and then cut out the middle portion of the transform, as in

$$\mathbf{b}[\vec{\omega}] = \square_{\vec{v}_1,\vec{v}_2}\mathfrak{F}U_{\vec{w}}\mathbf{a}[\vec{x}], \tag{10.11}$$

where $\vec{v}_1$ and $\vec{v}_2$ are the upper left and lower right locations of the portion to be cut out. They are designed to extract the center portion of the Fourier transform, and to make $\mathbf{b}[\vec{x}]$ as the same frame size as $\mathbf{a}[\vec{x}]$.

10.3.2 SIGNAL RECONSTRUCTION

Consider the sawtooth signal shown in Figure 10.1(a) which has a regular frequency and a zero sum over each wavelength. The digital Fourier transform is shown in Figure 10.1(b). The horizontal axis is the frequency with low frequencies at each end and high frequencies in the middle. Excepting the $F[0]$ term, this transform is symmetric for real-valued functions $f[n]$. Figure 10.1(c) is a zoomed-in version of Figure 10.1(b) showing the first few elements.

The $F[0]$ is 0 because the original signal was zero-sum. The input function is an odd function, and so only the odd frequencies have nonzero peaks. If the original function were to extend to the left of $\omega = 0$, it would show the same sawtooth behavior but would be a reflection that is out of phase. In other words, $F[\omega] = -F[-\omega]$. Therefore, the original signal can be composed of only

Code 10.2 Forward and inverse FFT

```
>>> import numpy as np
>>> import scipy.fftpack as ft
>>> a = np.random.rand( 32 )
>>> A = ft.fft( a )
```

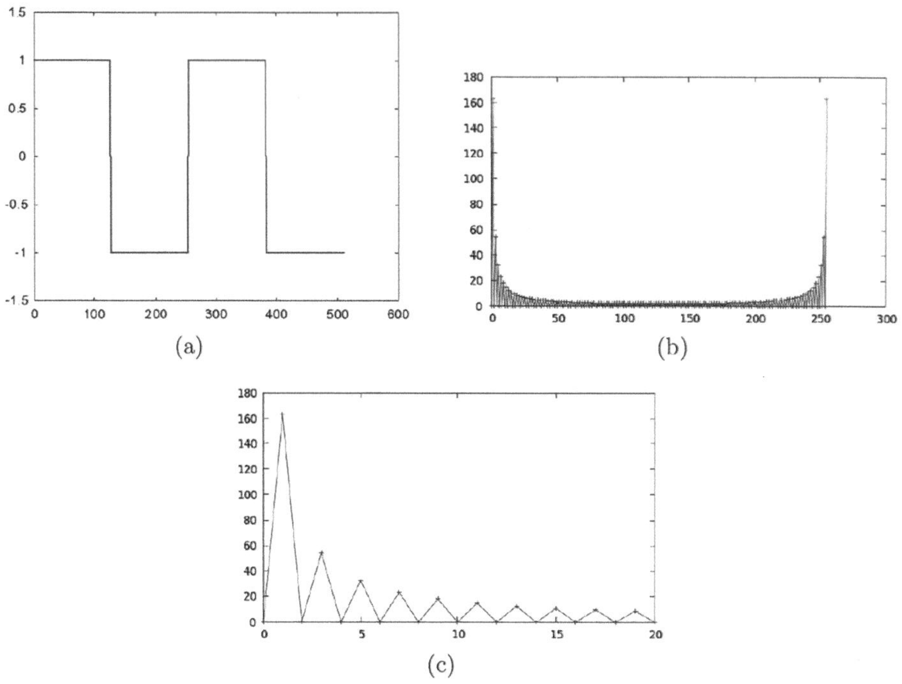

Figure 10.1 (a) The original signal, (b) the digital Fourier transform of the saw tooth, and (c) a zoomed-in version.

odd functions such as the sine, and it cannot contain any even function such as a cosine. Thus, the Fourier transform contains nonzero values for $F[\omega]$ for only odd values of ω.

The inverse transform is shown in Equation (10.7), and it is the multiplication of each frequency magnitude by the wave of that frequency. Basically, it is a weighted linear combination of the frequencies. There should be no loss of information, so $f'[n] = f[n]$. The first three frequencies with the proper magnitudes are shown in Figure 10.2(a). These are sine waves and the magnitudes are defined by the first nonzero values in Figure 10.1(b). Figure 10.2(b) shows the first two frequencies as light lines and their sum as a dark line. Just the addition of these two frequencies begin to show the original sawtooth nature of $f[n]$. However, the sharp edges are not seen because that information is contained in the higher frequencies, which have yet to be used in this reconstruction.

Figure 10.2(c) shows several functions. These are reconstructions using differing number of frequencies. As more frequencies are considered, the reconstructed function $f'[n]$ becomes more similar to $f[n]$. When all of the frequencies are considered, then $f'[n]$ is an exact replicate of $f[n]$.

10.4 PROPERTIES OF A FOURIER TRANSFORM

There are several properties associated with a Fourier transform, and these are useful in image analysis.

10.4.1 DC TERM

The DC term is $F[0]$, which is also the sum of the input signal, and it is called the bias term. This is the average height of the signal. In the case of Figure 10.1(a), the average height is 0, and thus, the first term in the Fourier transform (leftmost term in Figure 10.1(c)) is zero. If the entire input signal was raised by a value of h, then $F[0] = Nh$ (where N is the number of elements) but the other terms are not changed.

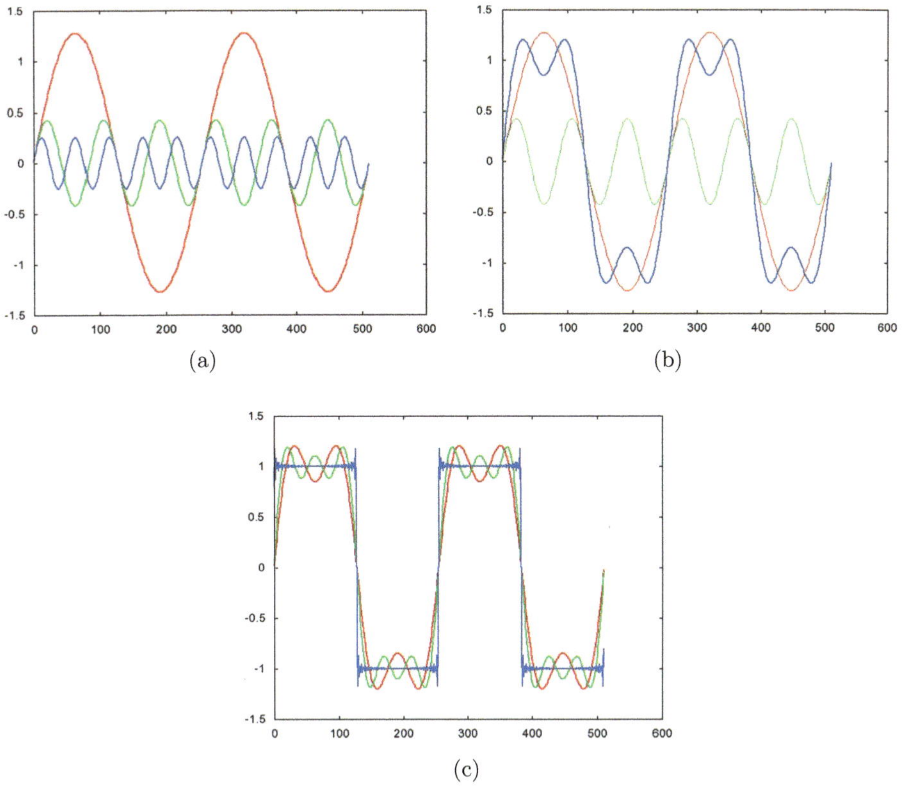

Figure 10.2 (a) The first three nontrivial frequencies. (b) The first two signals (red and green) and the addition of these signals (blue). (c) The addition of three nontrivial frequencies (red), the addition of four nontrivial frequencies (green), and the addition of several frequencies (blue).

Thus, for an image, the first term of the FFT is

$$\mathbf{b}[(0,0)] = \sum_{\vec{x}} \mathbf{a}[\vec{x}]. \tag{10.12}$$

This equation is realized in Code 10.3, which creates a random matrix of 32×32 elements and computes the FFT in line 5. The sum of the input signal is computed in line 6, and it is the same as the DC term of the transform in line 9. Most of the elements in $\mathbf{b}[\vec{\omega}]$ will have complex values. The DC term for an image consisting of real numbers will have only a real component.

10.4.2 CONSERVATION OF ENERGY

Energy is conserved in a Fourier transform for it is merely a transformation. Energy is not added or subtracted in the process. This must be true or the inverse transform will not retrieve the original signal. Parseval's theorem relates the energy of the input space and the frequency space as

$$\sum_{\vec{x}} |\mathbf{a}[\vec{x}]|^2 = \frac{1}{N\mathbf{a}[\vec{x}]} \sum_{\vec{x}} |\mathfrak{F}\mathbf{a}[\vec{x}]|^2. \tag{10.13}$$

Code 10.4 shows the computation of both sides of Equation (10.13) and that they are equivalent.

Code 10.3 The DC term

```
>>> import numpy as np
>>> import scipy.fftpack as ft
>>> np.random.seed( 21434 )
>>> mga = np.random.ranf( (32,32) )
>>> mgb = ft.fft2( mga )
>>> mga.sum()
520.96931772034043
>>> mgb[0,0]
(520.96931772034054+0j)
```

Code 10.4 Conservation of energy

```
>>> (mga*mga.conjugate()).sum()
350.18321610076197
>>> (mgb*mgb.conjugate()).sum()/(32*32)
(350.18321610076197+0j)
```

10.4.3 REPLICATION

The inverse transform returns the original signal without loss of information. If $\mathbf{b}[\vec{\omega}] = \mathfrak{F}\mathbf{a}[\vec{x}]$ and $\mathbf{c}[\vec{x}] = \mathfrak{F}^{-1}\mathbf{b}[\vec{\omega}]$, then $\mathbf{a}[\vec{x}] = \mathbf{c}[\vec{x}]$ is also true. This is shown in Code 10.5, where the inverse transform is computed in line 1. Comparing two arrays in Python can be accomplished with the **allclose** function. This will compare the arrays, element by element, and consider them to be the same if the values are within tight precision, which is set by default to 1.0×10^{-5}. The return of `True` indicates that the two signals are the same.

10.4.4 ADDITION

The Fourier transform of the sum of two functions is the same as the sum of the Fourier transform of the functions. Consider a case in which the function $f(x)$ can be written as the summation of two functions,

$$f(x) = g(x) + h(x).$$

The Fourier transform of $f(x)$ is then

$$F[\omega] = \int (g(x) + h(x))e^{-\imath\omega x}\,\mathrm{d}x = \int g(x)e^{-\imath\omega x}\,\mathrm{d}x + \int h(x)e^{-\imath\omega x}\,\mathrm{d}x. \tag{10.14}$$

These two integrations are just individual Fourier transforms, and so,

$$F[\omega] = G[\omega] + H[\omega]. \tag{10.15}$$

In operator notation, if

$$\mathbf{a}[\vec{x}] = \mathbf{b}[\vec{x}] + \mathbf{c}[\vec{x}], \tag{10.16}$$

Code 10.5 Computing the original image

```
1 >>> mgc = ft.ifft2( mgb )
2 >>> np.allclose( mga, mgc )
3 True
```

then

$$\mathfrak{F}\mathbf{a}[\vec{x}] = \mathfrak{F}\mathbf{b}[\vec{x}] + \mathfrak{F}\mathbf{c}[\vec{x}]. \tag{10.17}$$

10.4.5 SHIFT

A lateral shift of a signal creates a phase shift in Fourier space. If $\mathbf{g}[\vec{\omega}] = \mathfrak{F}\mathbf{f}[\vec{x}]$, then

$$e^{\imath 2\pi\vec{\omega}\cdot\vec{v}}\mathbf{g}[\vec{\omega}] = \mathfrak{F}D_{\vec{v}}\mathbf{f}[\vec{x}]. \tag{10.18}$$

Code 10.6 confirms Equation (10.18) using the bird image for the input $\mathbf{a}[\vec{x}]$. The left side of the equation is computed in lines 7 through 12. Line computes $\mathbf{g}[\vec{\omega}]$. Lines 8 through 11 compute the exponent, and line 12 multiplies the two together. Lines 13 through 16 compute the right side of the equation by applying a shift before the Fourier transform. The amount of shift is defined in line 6. Line 17 determines if the two arrays are the same within a small tolerance. The result of `True` confirms, at least for one image, Equation (10.18).

10.4.6 SCALE

The scale property relates a change of input space to the change of scale in the output space. Basically, if the width of a signal of the input decreases, then the width of the Fourier pattern expands.

The change in scale of a signal by a factor a is represented by $f(ax)$. The Fourier transform is

$$F(\omega) = \int f(ax)\, e^{-\imath\omega x}\, \mathrm{d}x. \tag{10.19}$$

Using $u \equiv ax$, then $\mathrm{d}x = \frac{1}{a}\mathrm{d}u$, and the transform becomes

$$F(\omega) = \frac{1}{a}\int f(u)\, e^{-\imath\omega u/a}\, \mathrm{d}u = \frac{F(\omega/a)}{a}. \tag{10.20}$$

If $a > 1$, then the scale of the input space increases, while the scale in the Fourier space decreases.

Code 10.6 The shifting property

```
1  >>> import numpy as np
2  >>> import imageio
3  >>> import scipy.fftpack as ft
4  >>> mga = imageio.imread('data/bird.jpg',as_gray=True)[:256,:256]
5  >>> V,H = mga.shape
6  >>> shift = 2
7  >>> mgg = ft.fft2( mga )
8  >>> w = np.indices((V,H)).transpose((1,2,0))
9  >>> v = np.array( (0,shift) )/np.array((V,H))
10 >>> wv = w.dot( v )
11 >>> phase = np.exp( -1j*2*np.pi * wv )
12 >>> fmgg = phase * mgg
13 >>> mgf = np.zeros( mga.shape )
14 >>> mgf[:,:shift] = mga[:,-shift:]+0
15 >>> mgf[:,shift:] = mga[:,:-shift]+0
16 >>> fmgf = ft.fft2( mgf )
17 >>> np.allclose(fmgg,fmgf,atol=0.05)
18 True
```

In operator notation, if $\mathbf{b}[\vec{\omega}] = \mathfrak{F}\mathbf{a}[\vec{x}]$, then the scaling property is

$$\frac{\mathbf{b}[\vec{\omega}/a]}{a} = S_a \mathbf{a}[\vec{x}]. \tag{10.21}$$

10.4.7 POWER SPECTRUM

The Fourier transform returns complex values, which include a phase and amplitude. The power spectrum is the power within each term as in,

$$|f(x)|^2 = f(x) f^{\dagger}(x). \tag{10.22}$$

10.5 DISPLAYING THE TRANSFORM

There are three issues that need to be addressed before it is possible to create a viewable Fourier transform image. The first is that the Fourier transform contains complex values. The second is that the DC term is generally several orders of magnitude greater than the other values. The third is that the low frequencies are at the corners and the preference is that they be in the middle of the frame.

A common method of displaying a Fourier transform is

$$\mathbf{c}[\vec{\omega}] = XL(|\mathbf{b}[\vec{\omega}]| + 1). \tag{10.23}$$

The X operator is the quadrant swap. The image is divided into four quadrants. The upper left and lower right quadrants exchange positions, and the other two quadrants exchange positions. The Python function for this is `scipy.fftpack.fftshift`. Since a pixel on the computer screen has only 256 levels of gray, any value that is less than 1/256 of the largest value will appear as black. The solution is to display the log values, and that operator is L. The absolute value converts the complex values to real values, and the addition of 1 will ensure that all log values are positive.

As an example, $\mathbf{a}[\vec{x}] = \mathbf{r}_{\vec{w};\vec{v}_1,\vec{v}_2}[\vec{x}]$, where $\vec{w} = (256, 256)$, $\vec{v}_1 = (96, 112)$, and $\vec{v}_2 = (160, 144)$. The Fourier transform is $\mathbf{b}[\vec{\omega}] = \mathfrak{F}\mathbf{a}[\vec{x}]$, and the image displayed in Figure 10.3 is $\mathbf{c}[\vec{\omega}]$ as computed in Equation (10.23).

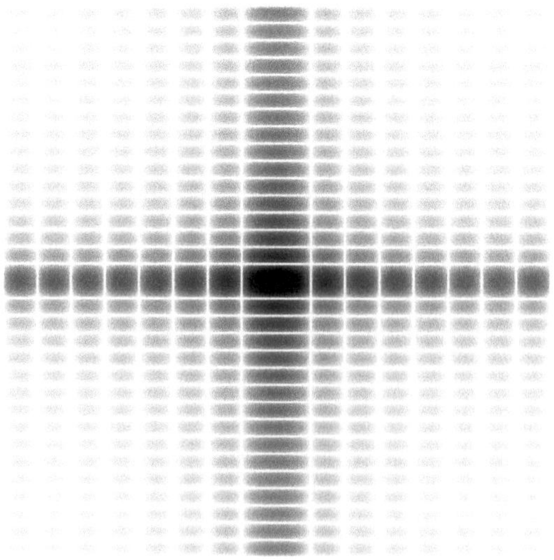

Figure 10.3 The Fourier transform of a rectangle. In this image, darker pixels have a higher intensity.

10.6 SIMPLE SHAPES

The Fourier transforms of images are difficult to visually interpret. So, this section will consider the transform of a few basic shapes to facilitate the understanding of the behavior of the transform.

10.6.1 RECTANGLE

Consider the one-dimensional signal defined by

$$f(x) = \begin{cases} 0 & x < -\frac{1}{2} \\ 1 & -\frac{1}{2} \geq x \geq \frac{1}{2} \\ 0 & x > \frac{1}{2} \end{cases}, \tag{10.24}$$

which is a square wave of amplitude 1 that is centered about $x = 0$. The Fourier transform is

$$F(\omega) = \frac{\sin(\omega/2)}{\omega/2}, \tag{10.25}$$

and is shown in Figure 10.4.

The right-hand side of this equation is a frequent term in the frequency analysis and so it is given the name of a *sinc* function, defined as

$$\texttt{sinc}\ \theta \equiv \frac{\sin(\theta)}{\theta}. \tag{10.26}$$

Figure 10.5(a) shows a simple binary-valued image, which contains a solid rectangle. The two-dimensional Fourier transform is represented in Figure 10.5(b). The horizontal row of values through the middle of the image is similar to the function in Equation (10.24). The horizontal row through the middle of the Fourier transform is representative of the sinc function shown in Figure 10.4 except that the view in Figure 10.5(b) is the log of the absolute values as in Equation (10.23). Likewise, a vertical sample through the middle of the original image is also a function similar to Equation (10.24) except that the scale is different. The vertical sample through the middle of the Fourier transform also shows the sinc behavior but on a different scale. The spacing between the peaks in the sinc function is inversely related to the width of the rectangle. This is a common trait, in that if an object becomes larger in the input space, then the pattern of the Fourier transform becomes smaller.

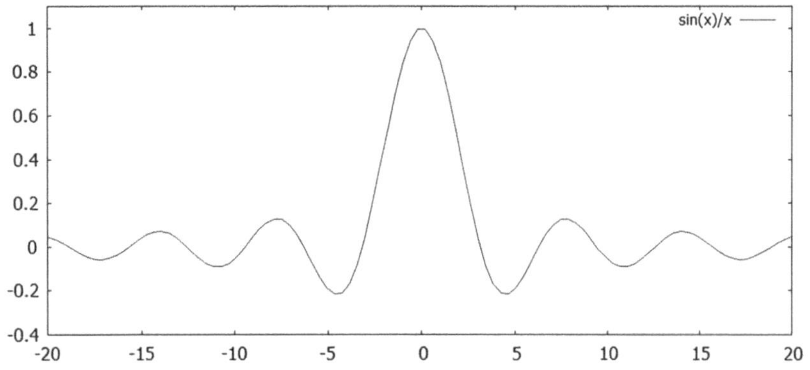

Figure 10.4 The Fourier transform of a rectangle function is a sinc function.

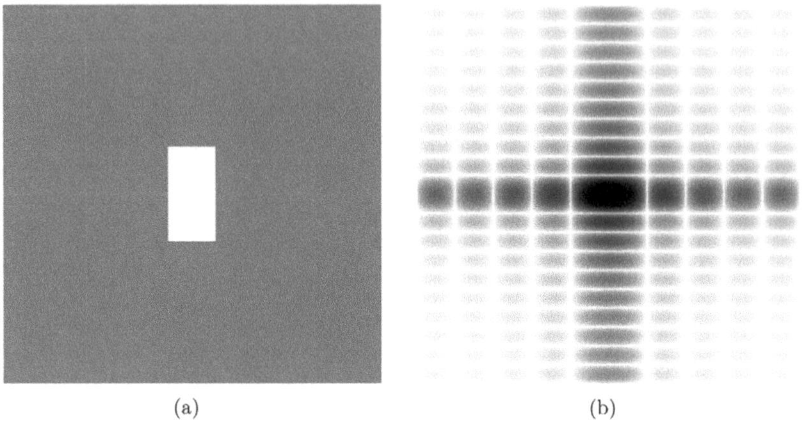

Figure 10.5 (a) The original binary-valued image and (b) the Fourier transform of this image.

10.6.2 CIRCLE

The operator for the creation of a circle is $o_{\vec{w};\vec{v},r}$, where $\vec{w}$ is the frame size, $\vec{v}$ is the location of the center of the circle, and r is the radius. In this example, $\vec{w} = (256, 256)$, $\vec{v} = (128, 128)$, and $r = 16$. The Fourier transform (as displayed by Equation (10.23)) is shown in Figure 10.6. If the radius of the input circle were to be increased, then the difference between the rings in the transform image would be decreased. Near the center of the image, the rings are quite distinct, but this structure decays toward the perimeter of the frame. Recall that this image is the log of the actual values and so these values at the perimeter are all very small. This is also a digital image, and any activity between pixels is not captured in the digital array. So, as the rings get closer together the sampling begins to fail. If the input were an analog signal, then the Fourier transform is a set of rings that become thinner and closer together farther away from the DC term. In the digital transform, the spacing of the rings becomes too close together to be properly sampled.

Figure 10.6 The Fourier transform of a solid circle. In this image, darker pixels have a higher intensity.

10.7 FREQUENCY BANDS

Fingerprints are unique in that most of the information is stored in frequencies instead of intensities. Furthermore, the spacing between the ridges is about the same throughout the image. Thus, these images are ideal for exploring frequency band filtering. A whorl print pattern is shown in Figure 11.5(a). The image displayed in Figure 10.7(b) is the Fourier transform display image as computed in Equation (10.23). In the center is a bright spot that is associated with the DC term. About that center is a strong primary ring and a weaker secondary ring.

The radius of the primary ring is directly related to the spacing between ridges in the fingerprint. The ring has a width since the spacing between the ridges is not exactly the same throughout the fingerprint. Most of the information about the fingerprint is contained within this ring.

Figure 10.8(a) shows a second print image, which is a scaled version of the first created by

$$\mathbf{b}[\vec{x}] = U_{\vec{w}} S_{0.5} \mathbf{a}[\vec{x}], \tag{10.27}$$

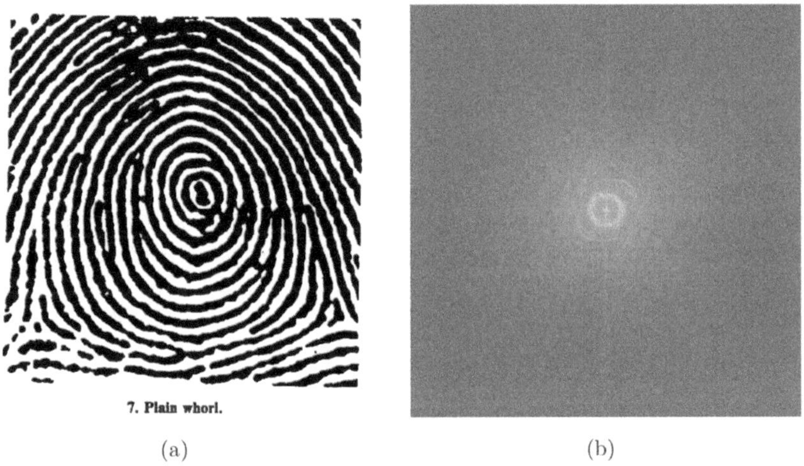

Figure 10.7 (a) A fingerprint and (b) the Fourier transform of the fingerprint.

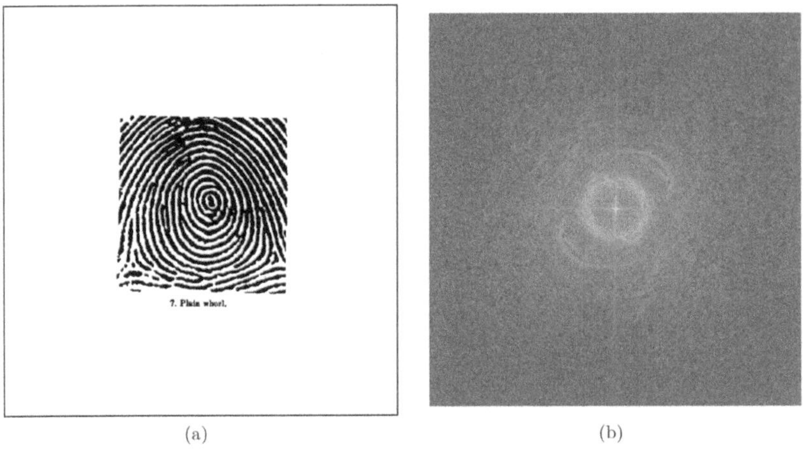

Figure 10.8 (a) The fingerprint image is reduced in scale without reducing the frame and (b) the Fourier transform of the fingerprint.

where $S_{0.5}$ is the Scaling operator that reduces the frame size to 0.5 of its original size in all dimensions. The $\vec{w}$ is the same size as the original fingerprint image. Figure 10.8(b) shows the Fourier transform which is similar in structure but the ring is twice as large since the ridges in the input are now closer together by a factor of 2.

Most of the fingerprint information is contained in the bright ring within in the Fourier image. To support this statement, the rest of the frequencies are removed, and then image is transformed back to the original image space. The process is

$$\mathbf{b}[\vec{x}] = \square_{\vec{v}_1,\vec{v}_2} |\mathfrak{F}^{-1} X\left(\mathbf{m}[\vec{\omega}] \times X \mathfrak{F} U_{\vec{w}} \mathbf{a}[\vec{x}]\right)|, \tag{10.28}$$

where $\mathbf{m}[\vec{\omega}] = \mathbf{o}_{\vec{w};r_2} - \mathbf{o}_{\vec{w};r_1}$. The $\vec{w}$ is the size of the frame, r_2 is a radius that is slightly bigger than the outer part of the white ring, and r_1 is a radius that is slightly smaller than the inner part of the white ring. The $\mathbf{m}[\vec{\omega}]$ is an annular ring mask that encompasses the white ring in the Fourier image. To perform the Fourier transform, the image is placed inside a frame using $U_{\vec{w}}$, where $\vec{w} = (1024, 1024)$. The image is processed but the final operator reduces the frame size back to that of the original image using $\square_{\vec{v}_1,\vec{v}_2}$, where $\vec{v}_1 = (140, 160)$ and $\vec{v}_2 = (885, 863)$.

The multiplication of the mask by the swapped Fourier transform of the image will block out all of the frequencies that are not inside this annular ring. It is removing the frequencies that are not in, or very near, the white band. The quadrants are swapped again before applying the inverse transform. The absolute value of the complex values is obtained before cutting the frame size.

Code 10.7 shows the entire script to execute Equation (10.28). Line 4 loads the data. In this case, the data has a white background and so the intensities are reversed. Lines 7 and 8 create the annular mask. Line 9 computes $X\mathfrak{F}\mathbf{a}[\vec{x}]$, and line 10 finishes the equation. The result is shown in Figure 10.9. Locations where the ridges bifurcate or create a bridge are not within the frequency bands defined by $\mathbf{m}[\vec{\omega}]$. Thus, these features are absent in $\mathbf{b}[\vec{x}]$.

The complement mask is $1 - \mathbf{m}[\vec{\omega}]$, and these are the frequencies other than those in that white band. The filtered image is shown in Figure 10.10(a) and a zoomed in portion is shown in Figure 10.10(b). Some ridge information still remains because the edges of the ridges are sharp. Crisp edges contain high frequency information, which did survive the filtering process. Thus, not all of the ridge information was contained in the white band in Figure 10.7(b). The input image was a binary-valued image, which is a modified version of a real image. In a natural fingerprint image, the edges of the ridges are not sharp and so this type of effect is lessened. In Figure 10.10(b), the intensity of the pixels in the middle of the ridges is similar to the intensity of the valleys. By just looking at this single image, it is not evident as to which part is a ridge and which part is a valley. That information was removed by the masking process.

Code 10.7 The script for Equation (10.28)

```
1  >>> import imageio
2  >>> import scipy.fftpack as ft
3  >>> import mgcreate as mgc
4  >>> adata = 256-imageio.imread( 'data/fing.png', as_gray=True)
5  >>> V,H = 1024,1024
6  >>> adata = mgc.Plop( adata, (V,H) )
7  >>> mask = mgc.Circle((V,H),(V/2,H/2),60)
8  >>> mask -= mgc.Circle((V,H),(V/2,H/2),20)
9  >>> temp1 = ft.fftshift( ft.fft2( adata ))
10 >>> bdata = abs(ft.ifft2(ft.fftshift(mask*temp1)))
11 >>> imageio.imsave('figure.png', bdata[140:885,160:863])
```

Figure 10.9 The output generated by Code 10.7.

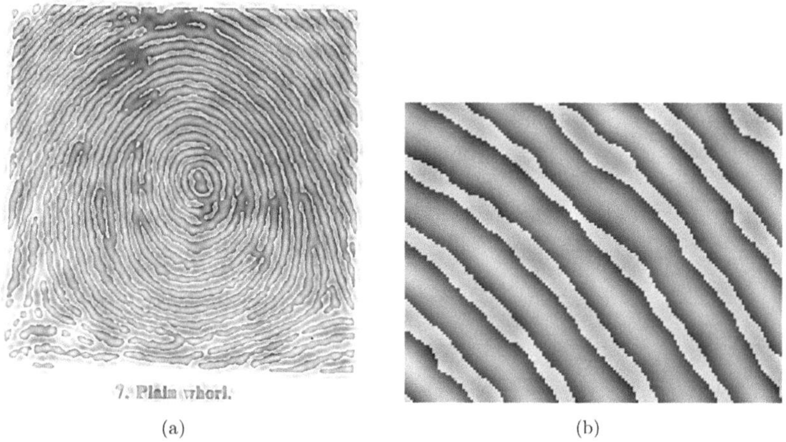

(a) (b)

Figure 10.10 (a) The reconstructed image without the frequencies in the annular ring and (b) a magnified portion of that image.

10.8 WINDOWING

Figure 10.11(a) displays an image of a cell, and the Fourier transform of it is shown in Figure 10.11(b). A strong cross hair pattern is evident in the image of the Fourier transform. As seen in Section 10.6.1, the cross hair pattern is associated with vertical and horizontal lines (edges of the rectangles). However, this appears to be a contradiction because there are no such lines in Figure 10.11(a).

The Fourier theory assumes that the input signal is repetitive. Thus, the input signal can be considered as concatenations of the image in both the vertical and horizontal directions, as seen in Figure 10.12. The presence of vertical and horizontal lines arise from the discontinuity between the top and bottom rows as well as the left-most and right-most columns. The Fourier transform of an image that fills the frame can produce the cross hairs evident in Figure 10.11(b), because of the discontinuities along the edges of the frame.

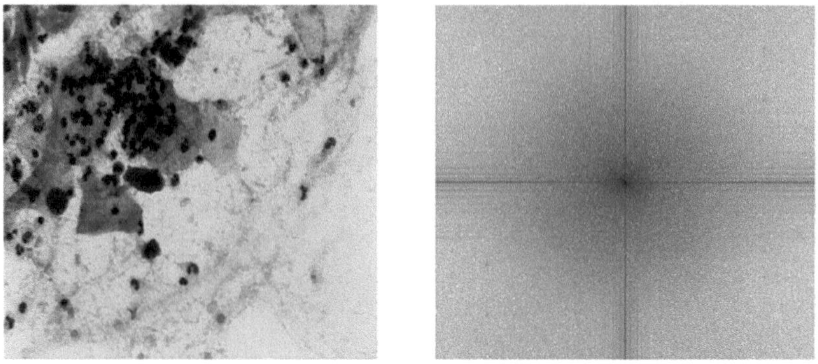

(a) (b)

Figure 10.11 (a) The original image from a pap smear and (b) the Fourier transform of this image shown as a negative so that the darker pixels have a higher magnitude.

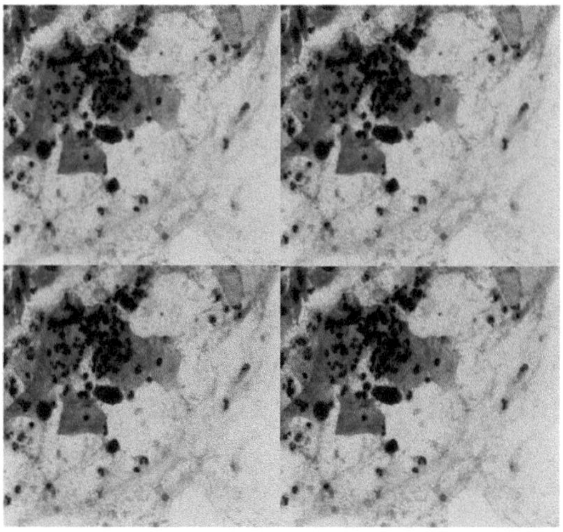

Figure 10.12 Repetitions of the image in both directions.

One solution to this situation is to apply a *window* to the original image. A window is a filter that blocks some of the pixels and allows others to maintain their original value. In this case, the window will block several rows of pixels along each edge of the frame. Given the frame size $\vec{w} = (V, H)$, the filter is $\mathbf{r}_{\vec{w};\vec{v}_1,\vec{v}_2}$, where $\vec{v}_1 = (z, z)$ and $\vec{v}_2 = (V - z, H - z)$. This is a window with a sharp edge will still cause the same issue with the cross hairs, so a Smoothing operator is applied as in $\mathcal{S}_m \mathbf{r}_{\vec{w};\vec{v}_1,\vec{v}_2}$.

Thus, windowing is performed by a simple mask that is rectangular in shape with smoothed edges. Given the grayscale input image, $\mathbf{a}[\vec{x}]$, the entire process is described by

$$\mathbf{b}[\vec{x}] = (X\mathfrak{F}(\mathbf{a}[\vec{x}]) \times \mathcal{S}_m \mathbf{r}_{\vec{w};\vec{v}_1,\vec{v}_2}[\vec{x}], \tag{10.29}$$

where $m = 5$, $\vec{w} = (484, 512)$, $\vec{v}_1 = (20, 20)$, and $\vec{v}_2 = (464, 492)$. Code 10.8 performs all of the steps given the filename, `fname`. Line 5 loads the image. Lines 6 through 8 create the smoothed mask $\mathcal{S}_m \mathbf{r}_{\vec{w};\vec{v}_1,\vec{v}_2}$, and line 9 performs the rest of Equation (10.29). The result is shown in Figure 10.13 using the display Equation (10.23). The vertical and horizontal cross hairs are absent because the harsh wrap-around discontinuities have been removed.

Code 10.8 Creating the mask

```
>>> import numpy as np
>>> import imageio
>>> import scipy.fftpack as ft
>>> import scipy.ndimage as nd
>>> adata = imageio.imread( fname,as_gray=True)
>>> V,H = 1024,1024
>>> mask = np.zeros((V,H))
>>> mask[20:-20,20:-20] = 1
>>> mask = nd.gaussian_filter( mask, 5 )
>>> bdata = ft.fftshift(ft.fft2(adata*mask))
```

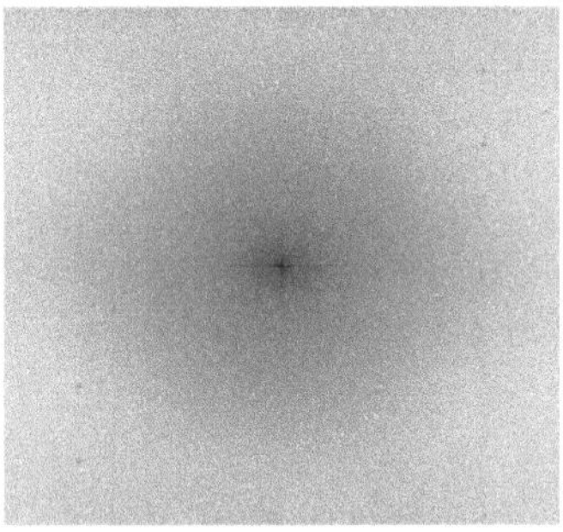

Figure 10.13 The transform after applying a window operation.

There are various types of windows that have been explored which treat the smoothing of the edges with slightly different functions. The Hamming window for one dimension is defined as

$$w(n) = 0.54 - 0.46\cos\left(\frac{2\pi n}{M-1}\right) \qquad 0 \leq n \leq M-1, \tag{10.30}$$

where n is the pixel location. Another alternative is a Hanning window defined by

$$w(n) = 0.5 - 0.5\cos\left(\frac{2\pi n}{M-1}\right) \qquad 0 \leq n \leq M-1. \tag{10.31}$$

The *scipy.signal* module offers several such windows. While analog implementations of these windows have differing results, the digital implementation is not as sensitive. Usually, there are only a few pixels within the region that the window operation affects, and so most windows produce very similar results. For most applications, the use of a window is important but the type of window is not.

These Python functions create windows in one-dimension and it would be up to the user to create a two-dimensional window. Then, there is the task of resizing and locating the window for a particular function. An alternative is the **KaiserMask**, which is a lengthy function and therefore relegated to Code C.5. This creates a two-dimensional matrix that contains a circular region with edges decaying as a Kaiser window. The operator notation is

$$\mathbf{a}[\vec{x}] = \mathbf{k}_{\vec{w};r_1,r_2}[\vec{x}]. \tag{10.32}$$

Code 10.9 shows the call to the function which receives four arguments. The first is the size of the frame, and the second is the center of the masking region. The final two arguments are the radii. In

Code 10.9 Using the **KaiserMask** function

```
>>> mask = mgcreate.KaiserMask( (256,256), (128,128), 40, 50 )
```

this example, the mask is a circle with a center at (128,128). The pixels inside a radius of 40 have a value of 1. The pixels beyond a radius of 50 have a value of 0. The pixels in between the two radii decay from 1 to 0. This mask is useful in both the image and frequency spaces. This mask can be used as a window and replace $\mathcal{S}_m \mathbf{r}_{\vec{w};\vec{v}_1,\vec{v}_2}[\vec{x}]$ in Equation (10.29).

10.9 SUMMARY

The Fourier transform converts an image from its spatial representation to its frequency representation. The transform consists of complex values over a wide range of magnitudes. Thus, the display of the information is often the log of the absolute values. The inverse Fourier transform returns the information back to the image space. These transformations do not lose information and so the consecutive application of the forward and inverse transforms will return the original image. However, the transform does allow access to the frequencies, and thus many applications will alter the frequencies in some manner before returning the information back to image space. Thus, the image is modified and usually providing the user with better access to the pertinent information.

PROBLEMS

1. Write a Python script to prove Equation (10.17).
2. Given an analog signal $f(x)$, prove that

$$e^{-\imath\omega a}F(\omega) = \int f(x-a)e^{-\imath\omega x}\,dx, \tag{10.33}$$

   where a is the amount of shift and $F(\omega)$ is the Fourier transform of $f(x)$.
3. Using $\sin(\theta) = (e^{\imath\theta} - e^{-\imath\theta})/2\imath$, prove that Equation (10.25) is the Fourier transform of Equation (10.24).
4. Start with any image $\mathbf{a}[\vec{x}]$ and compute $\mathbf{b}[\vec{x}] = \mathfrak{F}^{-1}\mathfrak{F}\mathbf{a}[\vec{x}]$. Show the $\sum_{\vec{x}} |\mathbf{a}[\vec{x}] - \mathbf{b}[\vec{x}]|$ is nearly 0.
5. Write a Python script to create Figure 10.5(a) from $\mathbf{a}[\vec{x}] = \mathbf{r}_{\vec{w};\vec{v}_1,\vec{v}_2}$, where $\vec{w} = (256,256)$ is the frame size, and the corners of the rectangle are defined by $\vec{v}_1 = (96,112)$ and $\vec{v}_2 = (160,144)$. Write a Python script to compute $\mathbf{c}[\vec{\omega}] = XL(|\mathfrak{F}\mathbf{a}[\vec{x}]| + 1)$.
6. Use the **mgcreate.Circle** function to create two images $\mathbf{a}[\vec{x}]$ and $\mathbf{b}[\vec{x}]$ with white circles. The radius of the circle in $\mathbf{b}[\vec{x}]$ should be double the radius of the radius of the circle in $\mathbf{a}[\vec{x}]$. Compute the Fourier transforms of each and use Equation (10.23) to create two viewable images. Determine if the structure in $X\mathfrak{F}\mathbf{b}[\vec{x}]$ is twice as big or twice as small as that in $X\mathfrak{F}\mathbf{a}[\vec{x}]$.
7. Create a mask similar to the mask in line 8 of Code 10.8, except that this new mask is created by repeated calls to the **scipy.signal.Hamming** function.
8. Apply the mask created in the previous problem to an image to remove the crosshair pattern.
9. Create $\mathbf{a}[\vec{x}] = \mathbf{r}_{\vec{w};\vec{v}_1,\vec{v}_2}[\vec{x}]$, where $\vec{w} = (256,256)$, $\vec{v}_1 = (96,112)$, and $\vec{v}_2 = (160,144)$. Compute the Fourier transform $\mathbf{b}[\vec{\omega}] = \mathfrak{F}\mathbf{a}[\vec{x}]$ (see Figure 10.3). Create a mask $\mathbf{m}[\vec{x}] = \mathbf{r}_{\vec{w};\vec{v}_1,\vec{v}_2}$, where $\vec{v}_1 = (112,112)$ and $\vec{v}_2 = (144,144)$. Compute $\mathbf{c}[\vec{x}] = \mathfrak{F}^{-1}X(\mathbf{m}[\vec{x}] - X\mathfrak{F}\mathbf{a}[\vec{x}])$.

11 Filtering in Frequency Space

There are many algorithms that rely on the ability to manipulate the frequencies of an image rather than the pixel intensities. In the previous chapter, the Fourier transform was reviewed and a few examples of the effect of manipulating frequencies was presented. This chapter extends that process by exploring frequency-based filters.

11.1 FREQUENCY FILTERING

The Fourier transform organizes the information so that an element in the output array is the intensity of a particular frequency component. This allows for easy access to the individual frequencies, for which there are several powerful tools. Blocking out a selection of frequencies is an effective way to isolate desired information. For example, if the low frequencies are removed from an image, then all that remains is the high frequencies, which are associated with edge information. Thus, such a filtering technique isolates the edges in an image. This is an example of a very simple filter named a high-pass filter since it passes the high frequencies. Low-pass filters, of course, pass the low frequencies and block the high frequencies. This removes the edge information.

A band-pass filter has a minimum and maximum range of frequencies that are kept which is good for isolating objects in an image by size. A wedge filter targets the directions of the edges. All of these filters are shown here with examples and matching Python scripts.

11.1.1 LOW-PASS FILTER

The *low-pass filter* will block the high frequencies and all the low frequencies to pass through the filter. Since the low frequencies are collected near the center of the frame after a swap operation, the low-pass filter is easy to create. A mask is created that has a solid circular region centered in the frame. This mask is multiplied by the frequency array and only the frequencies within this circular region survive the multiplication. The high frequencies are blocked because they are multiplied by zero.

Consider an image $\mathbf{a}[\vec{x}]$ and the operation:

$$\mathbf{b}[\vec{x}] = \mathfrak{F}^{-1}\mathfrak{F}\mathbf{a}[\vec{x}]. \tag{11.1}$$

The Fourier transform does not destroy information, and thus, $\mathbf{b}[\vec{x}] = \mathbf{a}[\vec{x}]$. Now a mask is added to the system, but it is necessary to also swap the quadrants in the Fourier domain. So, the new operation is

$$\mathbf{b}[\vec{x}] = \mathfrak{F}^{-1}X\left(\mathbf{m}[\vec{\omega}] \times X\mathfrak{F}\mathbf{a}[\vec{x}]\right), \tag{11.2}$$

where $\mathbf{m}[\vec{\omega}]$ is this mask. If all elements in $\mathbf{m}[\vec{\omega}]$ were 1, then $\mathbf{b}[\vec{x}] = \mathbf{a}[\vec{x}]$ would still be true. However, in this case, the mask consists of a circular region of pixels with the value of 1 and a background with the values of 0. So, $\mathbf{b}[\vec{x}]$ is now a modified version of $\mathbf{a}[\vec{x}]$. In this case, the mask passes the low frequencies and blocks the high frequencies. Low frequencies are associated with larger solid regions while high frequencies are associated with edges. Thus, the result should be a blurry version of $\mathbf{a}[\vec{x}]$.

As an example, the input image, $\mathbf{a}[\vec{x}]$, is the USAF resolution chart shown in Figure 11.1(a). The original image has black pixels on a white background and this will be reversed before processing. The low-pass filter is $\mathbf{m}[\vec{\omega}] = \mathbf{o}_{\vec{w},\vec{w}/2,r}[\vec{x}]$, where $\vec{w}$ is the size of the frame and r is a radius defined by the user. The process is

$$\mathbf{b}[\vec{x}] = \Re\mathfrak{F}^{-1}X\left(\mathbf{m}[\vec{\omega}] \times X\mathfrak{F}\left(1 - \mathbf{a}[\vec{x}]/256\right)\right), \tag{11.3}$$

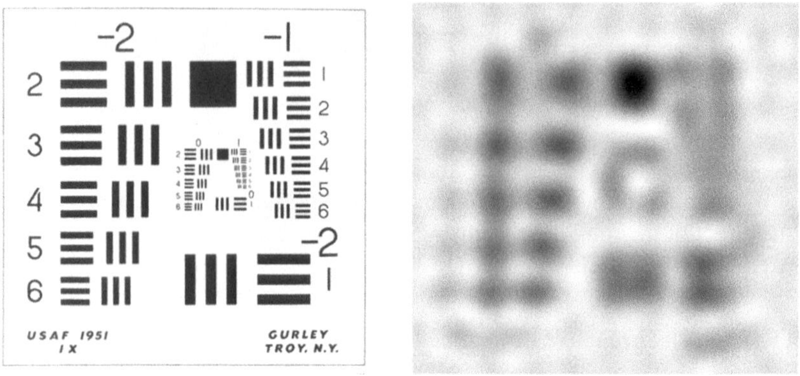

(a) (b)

Figure 11.1 (a) The USAF resolution chart and (b) the chart after a low-pass filtering.

where the $1-\mathbf{a}[\vec{x}]/256$ term scales the input and then reverses the intensity so that the background is black. The $\Re$ operator retrieves the real components of the complex numbers.

Code 11.1 shows the process, where line 4 loads the image and line 5 reverses the bright and dark pixels. The Fourier transform is computed in line 6, and the mask is created in line 8. It has the same frame size as the image and the circle is centered in that frame with a radius of 10. In line 9, this mask is multiplied by the result of line 6 and line 10 displays the image. Since the intensity of the original image was inverted before processing the output intensities are also flipped by using the minus sign in line 10. The result is shown in Figure 11.1(b), and as seen, the regions that were solid in the original have a strong response. The small lines in the original image are merely sharp edges and they appear much weaker after this low-pass filtering process.

11.1.2 HIGH-PASS FILTER

The *high-pass filter* is just the opposite of a low-pass filter, as the masks are the complements of the other. The high-pass filter allows the high frequencies to survive and blocks the low frequencies. So the mask is defined as

$$\mathbf{m}[\vec{\omega}] = 1 - \mathbf{o}_{\vec{w},\vec{w}/2,r}[\vec{x}]. \tag{11.4}$$

In this next example, $r = 150$ is used to pass only the highest frequencies. Code 11.2 shows the creation of the filter and the output image. The absolute value of the output is shown instead of the real component values, because the edges tend to have both positive and negative values. This can be

Code 11.1 An example of a low-pass filter

```
1  >>> import imageio
2  >>> import scipy.fftpack as ft
3  >>> import mgcreate as mgc
4  >>> amg = imageio.imread( 'data/reschartsmall.png', as_gray=True)
5  >>> bmg = 1 - amg/amg.max()
6  >>> cmg = ft.fftshift( ft.fft2(bmg))
7  >>> V,H = cmg.shape
8  >>> circ = mgc.Circle( (V,H), (V/2,H/2), 10 )
9  >>> dmg = ft.ifft2( ft.fftshift( circ*cmg))
10 >>> imageio.imsave('figure.png', -dmg.real)
```

Code 11.2 An example of a high-pass filter

```
>>> circ = 1-mgc.Circle( (V,H), (V/2,H/2), 150 )
>>> dmg = ft.ifft2( ft.fftshift( circ*cmg))
>>> imageio.imsave('figure.png', abs(dmg))
```

useful information in some applications, but for this display, only the presence of the edges is sought. The result is shown in Figure 11.2(a), which shows that this process kept the edge information. The interior information, though, is missing. From just viewing Figure 11.2(a), it is possible to determine if the blocks were solid-filled or originally empty.

11.1.3 BAND-PASS FILTER

A *band-pass filter* is an annular ring instead of a solid circle, and it is created by $\mathbf{m}[\vec{\omega}] = \mathbf{o}_{\vec{w},\vec{w}/2,r_2}[\vec{x}] - \mathbf{o}_{\vec{w},\vec{w}/2,r_1}[\vec{x}]$, where $r_2 > r_1$ are the two radii that define the size of the annular ring. The construction of the mask is shown in the first three lines in Code 11.3. The example uses $r_1 = 48$ and $r_2 = 64$, and the result is shown in Figure 11.2(b). The output displays solid-filled blocks for a limited size range as seen on the right side of the image. Changing the values of r_1 and r_2 will change the range of sizes that survive the filtering process.

The final example considers that the mask itself has sharp edges which can produce undesired ringing effects in the filtered image. The easy solution is to smooth the mask before it is applied to the frequencies as in

$$\mathbf{b}[\vec{x}] = \left|\mathfrak{F}^{-1}X\left((\mathcal{S}_{10}\mathbf{m}[\vec{\omega}]) \times X\mathfrak{F}(1-\mathbf{a}[\vec{x}]/256)\right)\right|. \tag{11.5}$$

This will soften the edges in the Fourier space and reduce the negative effects. Code 11.4 creates a band-pass filter and then in line 5 applies the **gaussian_filter** function from the *scipy.ndimage* package.

(a) (b) (c)

Figure 11.2 (a) The result after a hi-pass filter, (b) the result after a band-pass filter, and (c) the result from band-pass filtering with smooth edges.

Code 11.3 An example of a band-pass filter

```
>>> circ1 = mgc.Circle( (V,H), (V/2,H/2), 48 )
>>> circ2 = mgc.Circle( (V,H), (V/2,H/2), 64 )
>>> circ = circ2 - circ1
>>> dmg = ft.ifft2( ft.fftshift( circ*cmg))
>>> imageio.imsave('figure.png', -abs(dmg))
```

Code 11.4 An example of a band-pass filter with soft edges

```
>>> import scipy.ndimage as nd
>>> circ1 = mgc.Circle( (V,H), (V/2,H/2), 36 )
>>> circ2 = mgc.Circle( (V,H), (V/2,H/2), 72 )
>>> circ = circ2 - circ1
>>> circ = nd.gaussian_filter(circ+0.0,10)
>>> dmg = ft.ifft2( ft.fftshift( circ*cmg))
>>> imageio.imsave('dud7.png', abs(dmg))
```

Figure 11.2(c) shows that the band-pass filter was set so that it was sensitive to objects that were the size of the numerals. Compared to Figure 11.2(a), this image shows a strong response for objects of a specific size. There is less response in the background region because of the smoothing performed on the filter.

11.2 DIRECTIONAL FILTERING

A *wedge filter* is so named because the filter is a set of opposing wedges centered in the frame. Inside the wedges are pixels with a value of 1 and when multiplied by a Fourier transform, this filter passes frequencies associated with orientation. Figure 11.3 shows the negative of a wedge filter (where the black pixels have a value of 1 and the white pixels have a value of 0). The 0° line points eastwards much like a graph and positive angles increase in the counter-clockwise direction. The creation of this image is

$$\mathbf{m}[\vec{\omega}] = \mathbf{w}_{\vec{w};\theta_1,\theta_2}, \tag{11.6}$$

where $\vec{w}$ is the frame size and θ_1 and θ_2 are the two angles. This example uses $\theta_1 = 30°$ and $\theta_2 = 40°$.

The **Wedge** function creates the wedge image and it is displayed in Code 11.5. The inputs are the frame size and the two angles in degrees. This function starts with the **indices** (see Section 6.6) function, which creates two matrices of which one increments its values by rows and the others increments its values by columns. These are centered in lines 5 and 6 to create a coordinate system. The next several lines fills in the two triangles. Line 15 ensures that the center point is 0. The (see Section 10.4.1 term can be set to either 1 or 0 from the previous line of code. A serious problem occurs when using these in Fourier space as some filters will pass the DC term and others will not.

Figure 11.3 The negative of the wedge filter.

Code 11.5 The **Wedge** function

```
# mgcreate.py
def Wedge( vh, t1, t2 ):
    ans = np.zeros( vh )
    ndx = np.indices( vh ).astype(float)
    ndx[0] = ndx[0] - vh[0]/2
    ndx[1] = ndx[1] - vh[1]/2
    mask = ndx[0] == 0
    ndx[0] = (1-mask)*ndx[0] + mask*1e-10
    ans = np.arctan( ndx[1] / ndx[0] )
    ans = ans + np.pi/2
    mask = ans >= t1/180.* np.pi
    mask2 = ans < t2/180.* np.pi
    mask = np.logical_and( mask, mask2).astype(int)
    V,H = vh
    mask[V/2,H/2] = 0
    return mask
```

Since wedge filtering is used just for orientation, the preference is that the DC term not be passed and so this line ensures that this is the case.

Consider the image $\mathbf{a}[\vec{x}]$ shown in Figure 11.4(a). The goal is to isolate one set of lines. The process is

$$\mathbf{b}[\vec{x}] = \mathfrak{F}^{-1}X\left(\mathbf{m}[\vec{\omega}] \times X\mathfrak{F}(\mathbf{a}[\vec{x}] < 100)\right). \tag{11.7}$$

The $\mathbf{m}[\vec{\omega}]$ is created by Equation (11.6) with $\theta_1 = 30°$ and $\theta_2 = 40°$. In the original image, the target (the lines) is darker than the background. So, $\mathbf{a}[\vec{x}] < 100$ is used to set the target pixels to 1 and the background to 0. Figure 11.4(b) shows the negative of $\mathbf{b}[\vec{x}]$, which has the lines that are perpendicular to the direction of the wedge filter.

The ends of the lines though are not crisp. Consider the line as a very long rectangle. The long sides of the rectangles are oriented in the same direction as the lines in Figure 11.4(b). The ends, however, are lines in the perpendicular direction. The filter does not pass these edges, and therefore, the ends of the lines in the result are fuzzy.

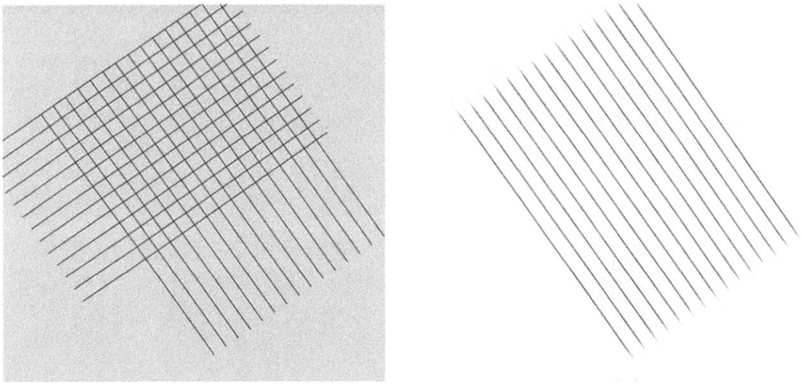

(a) (b)

Figure 11.4 (a) The input image and (b) the negative of the image after passing through a wedge filter.

Code 11.6 An example of line filtering

```
1 >>> adata = imageio.imread( 'data/lines.png', as_gray=True )<100
2 >>> ddata = ft.fftshift( ft.fft2( adata ))
3 >>> V,H = adata.shape
4 >>> mdata = mgc.Wedge( (V,H), 30,40 )
5 >>> bdata = ft.ifft2( ft.fftshift (ddata * mdata ))
6 >>> imageio.imsave('figure.png', -abs(bdata))
```

Code 11.6 executes the steps in Equation (11.7). Line 1 reads the data and applies the threshold. Line 2 applies the Fourier transform and swap functions. Line 4 creates the wedge filter which is applied in line 5. Line 6 shows the negative of the image which is the result shown in Figure 11.4(b).

11.3 FINGERPRINT EXAMPLE

Fingerprint images are interesting in that they consist of lines and flow rather than large contiguous objects. Algorithms that sort the prints into groups rely heavily on the analysis of these flow patterns. This section will use Fourier transforms to detect the flow patterns as an educational device to explore the nature of Fourier transforms. There is no claim here that this is the best method of analyzing fingerprints.

The proposed system uses wedge filters to detect ridge direction. For a single filter, process was described as

$$\mathbf{b}[\vec{x}] = \mathfrak{F}^{-1} X \left(\mathbf{m}[\vec{\omega}] \times X \mathfrak{F} \mathbf{a}[\vec{x}] \right). \tag{11.8}$$

Figure 11.5(a) shows a typical input, and Figure 11.6 shows the result after applying the wedge filer from Figure 11.3. Clearly, the flow perpendicular to the wedges are enhanced in this image.

The goal is to detect flow in all directions, and so this application will required the use of several wedge filters of differing orientations. This set of filters is denoted by $\{\mathbf{m}[\vec{\omega}]\}$, and the i-th filter is $\mathbf{m}_i[\vec{x}]$.

The protocol is then

$$\mathbf{b}_i[\vec{x}] = \mathfrak{F}^{-1} X \left(\mathbf{m}_i[\vec{\omega}] \times X \mathfrak{F}(\mathbf{a}[\vec{x}]) \right), \quad \forall i. \tag{11.9}$$

(a) (b)

Figure 11.5 (a) A fingerprint and (b) the negative of the result after the application of the wedge filter shown in Figure 11.3.

Figure 11.6 A color-coded image for the flow of a fingerprint.

Since there are multiple filters, it is prudent to create a Python function to perform the repetitive operations. Code 11.7 shows the **MaskinF** function, which receives an image and the mask. It then applies the mask and returns the results. The last three lines receive the image, build the filter, and applies the filter.

The process can be repeated several times with different wedge filters to extract ridges flowing in different directions. Each filter has a narrow range of angles and produces a filtered image. Code 11.8 shows the **MultiWedges** function that receives the input image, `indata`, and the number of wedge filters, `nwdjs`. Line 3 computes the `step` which is the number of degrees change between each filter. This is also the degree span of each wedge. Lines 7 and 8 compute the angles for the wedge filter which is created in line 9. The filter is applied in line 10 by calling the **MaskInF** function. The output is a list of filter responses, $\{\mathbf{b}[\vec{x}]\}$, such as the one shown in Figure 11.5(b).

The next task is to display the data, which is cumbersome as there are several images in $\{\mathbf{b}[\vec{x}]\}$. One option is to create a color-coded image, which can be created in a variety of ways. The one chosen here considers the output images as a sequence with the index i. As i increases, the contribution to the red channel increases and the contribution to the blue channel decreases. This is performed in **ColorCode1**, which is shown in Code 11.9. The process is defined as

Code 11.7 The **MaskinF** function

```
# ffilter.py
def MaskinF( indata, fmask ):
    fdata = ft.fftshift( ft.fft2( indata ) )
    fdata *= fmask
    answ = ft.ifft2( ft.fftshift( fdata ))
    return answ

>>> adata = imageio.imread( 'data/fing.png', as_gray=True )
>>> bdata = mgc.Wedge( adata.shape, 30, 45 )
>>> cdata = flt.MaskinF( adata, bdata )
```

Code 11.8 The **MultiWedges** function

```
# ffilter.py
def MultiWedges( indata, nwdjs ):
    step = 180/nwdjs # angle step
    answ = []
    V,H = indata.shape
    for i in range( nwdjs ):
        angle1 = step * i
        angle2 = angle1 + step
        wedj = mgc.Wedge( (V,H), angle1, angle2 )
        answ.append( abs(MaskinF( indata, wedj )))
    return answ
```

Code 11.9 The **ColorCode1** function

```
# color.py
def ColorCode1( mglist ):
    N = len( mglist ) # number of images
    V,H = mglist[0].shape
    answ = np.zeros( (V,H,3) )
    for i in range( N ):
        scale = 256./N * i
        blue = 256. - scale
        answ[:,:,0] += mglist[i] * scale
        answ[:,:,2] += mglist[i] * blue
    mx = answ.max()
    answ /= mx
    return answ
```

$$\mathbf{c}[\vec{x}] = \sum_i \left\{ \begin{matrix} 256/N \\ 0 \\ 256 - 256/N \end{matrix} \right\} \mathbf{b}_i[\vec{x}], \tag{11.10}$$

where N is the number of images in $\{\mathbf{b}[\vec{x}]\}$.

The result is shown in Figure 11.6. The blue and red colors are used to designate flow directions. There are just a few classes of fingerprints (whorl, left loop, right loop, and arch) and two of these have subcategories. The flow direction is unique to each class and thus can be used to determine which class the print is.

11.4 ARTIFACT REMOVAL

The final filtering topic is to demonstrate how filtering can be used to remove undesirable artifacts. Consider the image shown in Figure 11.7, which shows a baseball game with a wire screen between the camera and the action. The goal is to remove the wire screen.

The screen is a regular pattern and with nearly vertical and horizontal lines. Such a pattern will produce large signals along the vertical and horizontal axes of the Fourier transform. The original color image is $\mathbf{a}[\vec{x}]$, and it is converted to gray scale and normalized before the conversion to Fourier space. The process is

$$\mathbf{b}[\vec{\omega}] = X \mathfrak{F} \frac{\mathcal{L}_L U_{2\vec{w}} \mathbf{a}[\vec{x}]}{\bigvee \mathbf{a}[\vec{x}]}, \tag{11.11}$$

Figure 11.7 An original image.

where the $\mathcal{L}$ function performs the conversion to gray scale and the $U_{2\vec{w}}$ function places the original image into the middle of a frame that is twice the size as the original image. This latter step is padding which eliminates some of the effects from the frame edges. The image $\log(|\mathbf{b}[\vec{x}]|+1)$ is shown in Figure 11.8, but this is only the center portion of the image. There are periodic signals along the two axes that are from the frame edge and from the regular pattern of the screen.

The mask $\mathbf{m}[\vec{\omega}]$ should be designed to eliminate the frequencies of the screen without damaging the image. The proper method to do this is to create an image which is nothing but the screen (with the screen pixels set to 1) and then study the pattern in Fourier space. Then, the precise frequencies can be identified and eliminated.

The example shown here is simpler which means that the results could be improved. The mask that is constructed for this task is an image with all of the pixels set to 1 except for four rectangles that remove the periodic signals. This will block the frequencies associated with the screen without too much damage to the image. Once the mask is constructed, the rest of the process follows the protocol that has been explored in previous examples. The final result is

$$\mathbf{c}[\vec{x}] = \Box\Re\mathfrak{F}^{-1}X\left((\mathcal{S}_{10}\mathbf{m}[\vec{\omega}]) \times \mathbf{b}[\vec{\omega}]\right). \tag{11.12}$$

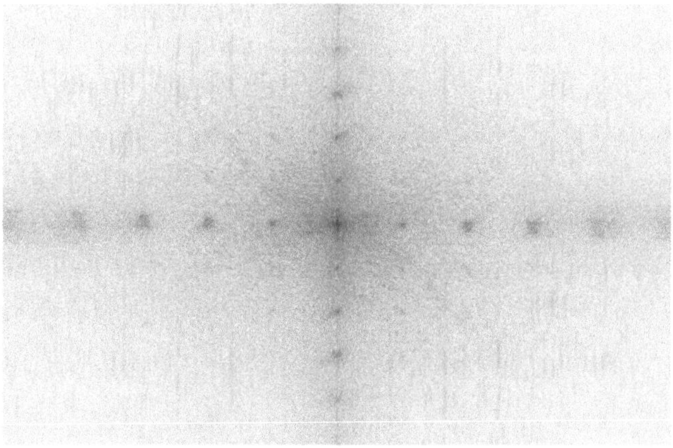

Figure 11.8 The log negative of the baseball image.

Code 11.10 shows the entire process. Line 1 loads the image and converts it to gray scale. Line 3 performs $U_{2\vec{w}}\mathbf{a}[\vec{x}]$, and line 4 completes the computation of Equation (11.11). The creation of the mask runs from lines 5 through 10 with the final step applying $\mathcal{S}_{10}\mathbf{m}[\vec{\omega}]$ to soften the edges of the mask. This will reduce ringing in the final result. Line 11 applies $\mathfrak{F}^{-1}X$, and the windowed real part of the image is extracted in line 12. Only the central portion of the image is extracted. This selection is the same size as the original image and is basically removing the rows and columns that were added by the $U_{2\vec{w}}$ operator.

The result is shown in Figure 11.9, and as seen, the screen is no longer present. There are two noticeable effects that bear mentioning. The first is that the edges of the people and playing field are not as sharp. The mask that was used was a bit sloppy and destroyed more frequencies than necessary. Better results are possible if more care is exercised on which frequencies are removed. The second is that the pixelated camera and regularity of screen spacing created soft interference lines. These are visible in the original image and tend to run from about second base to the bottom of the frame. While the filtering removed the screen, it did not remove interference lines. Their spacing and orientations are different than the screen and so their main frequencies were left unscathed.

Code 11.10 Removal of the screen from the baseball image

```
1  >>> adata = imageio.imread( 'data/baseball.jpg', as_gray=True)
2  >>> V,H = adata.shape
3  >>> temp = mgc.Plop( adata, (2*V,2*H) )
4  >>> bdata = ft.fftshift( ft.fft2 ( temp/adata.max() ))
5  >>> mdata = np.ones( (2*V,2*H))
6  >>> mdata[V-50:V+50,:666] = 0
7  >>> mdata[V-50:V+50,-666:] = 0
8  >>> mdata[:450,H-50:H+50] = 0
9  >>> mdata[-450:,H-50:H+50] = 0
10 >>> mdata = nd.gaussian_filter( mdata, 10 )
11 >>> temp = ft.ifft2( ft.fftshift(mdata * bdata ))
12 >>> cdata = temp.real[V-V/2:V+V/2,H-H/2:H+H/2]
```

Figure 11.9 The result from Equation 11.12.

11.5 SUMMARY

Filtering frequencies is a very powerful technique of which there are many applications. The common approach is to apply a Fourier transform to the image, modify the frequencies, and to use the inverse Fourier transform to convert the image back to its original space. Then, it is possible to see the changes from filtering. This chapter explored a few common frequency filtering techniques. The next chapter will extend this idea into the creation of tailored filters.

11.6 PROBLEMS

1. Use $\mathbf{a}[\vec{x}] = Y(\texttt{data/bird.jpg})$. Compute

$$\mathbf{b}[\vec{x}] = \mathfrak{F}^{-1}X\left(X\mathfrak{F}U_{\vec{w}}\mathbf{a}[\vec{x}] \times \mathbf{o}_{\vec{w};r}[\vec{x}]\right),$$

where $\vec{w} = (1024, 1024)$ and $r = 30$.

2. Use $\mathbf{a}[\vec{x}] = Y(\texttt{data/bird.jpg})$. Compute

$$\mathbf{b}[\vec{x}] = \mathfrak{F}^{-1}X\left(X\mathfrak{F}U_{\vec{w}}\mathbf{a}[\vec{x}] \times (1 - \mathbf{o}_{\vec{w};r}[\vec{x}])\right),$$

where $\vec{w} = (1024, 1024)$ and $r = 50$.

3. Use $\mathbf{a}[\vec{x}] = Y(\texttt{data/bird.jpg})$. Create a band pass filter, $\mathbf{m}[\vec{\omega}]$, where the inner diameter is 20 and the outer diameter is 30. Compute

$$\mathbf{b}[\vec{x}] = \mathfrak{F}^{-1}X\left(X\mathfrak{F}U_{\vec{w}}\mathbf{a}[\vec{x}] \times \mathbf{m}[\vec{\omega}]\right),$$

where $\vec{w} = (1024, 1024)$.

4. Load the image *geos.png* into $\mathbf{a}[\vec{x}]$. Create a mask for the Fourier space to be multiplied by $\mathfrak{F}\mathbf{a}[\vec{x}]$. The result should have a large intensity region at the location of the smaller circle but a smaller response in the area of the larger circle. A typical result is shown in Figure 11.10.
5. Use the $\mathbf{a}[\vec{x}] = Y(\texttt{lines.png})$ from Figure 11.4(a). Create a filter that isolates the lines that were excluded in Figure 11.4(b) and excludes the lines that were kept in Figure 11.4(b).
6. Use $\mathbf{a}[\vec{x}] = \mathcal{L}_L Y(\texttt{data/bird.jpg})$. Compute

$$\mathbf{b}[\vec{x}] = \mathfrak{F}^{-1}X\left(X\mathfrak{F}U_{\vec{w}}\mathbf{a}[\vec{x}] \times (\mathbf{m}_1[\vec{\omega}] + \mathbf{m}_2[\vec{\omega}])\right),$$

where $\vec{w} = (1024, 1024)$, $\mathbf{m}_1[\vec{\omega}]$ is a low-pass filter with radii 10, and $\mathbf{m}_2[\vec{\omega}]$ is a hi-pass filter with a radius of 50.

7. Write the image operator notation to create the filter $\mathbf{m}_1[\vec{\omega}] + \mathbf{m}_2[\vec{\omega}]$ in the previous problem.

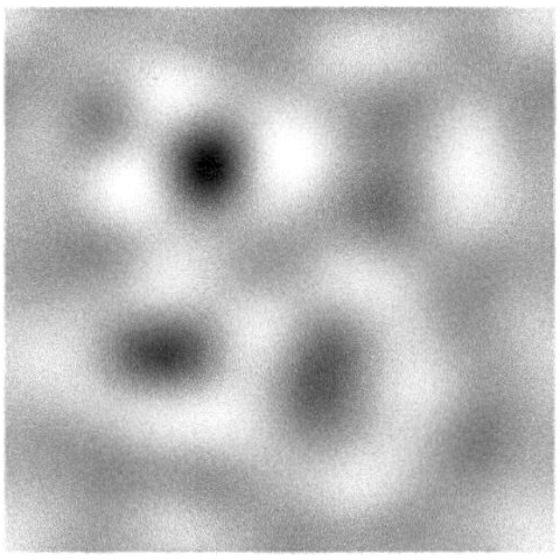

Figure 11.10 Typical result for problem 4.

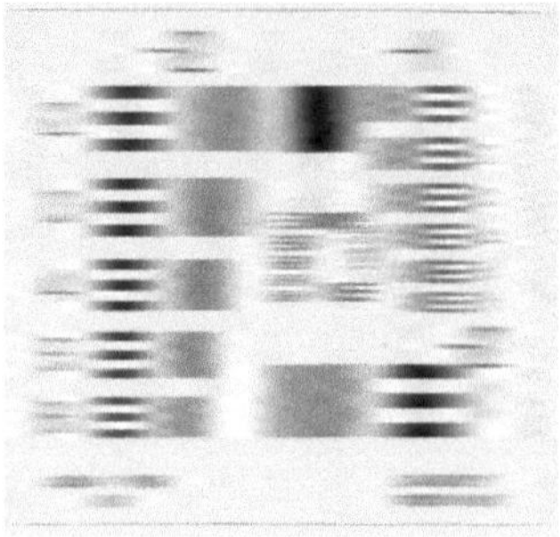

Figure 11.11 Typical result for problem 11.

8. Use $\mathbf{a}[\vec{x}] = Y(\texttt{data/baseball.jpg})$ as the input. Repeat the process of removing the net for this color image. The result should also be a color image.
9. Use $\mathbf{a}[\vec{x}] = \mathcal{L}_L Y(\texttt{data/baseball.jpg})$ as the input. Build a filter to remove the vertical lines in the netting but not the horizontal lines.
10. Write the image operator notation for the previous problem, including operator notation for the construction of the filter.
11. Use the Navy resolution chart as the input. Create an image which is the negative of this chart so that the background is black. Create a filter such that the response is the horizontal lines survive but the vertical lines are smeared. A typical result is shown in Figure 11.11.
12. Use $\mathbf{a}[\vec{x}] = \square_{\vec{v}_1,\vec{v}_2} \mathcal{L}_L Y(\texttt{data/bird.jpg})$, where $\vec{v}_1 = (0,0)$ and $\vec{v}_2 = (512,512)$. Create an image that is the first 512 rows and 512 columns. Create a filter that blocks the DC term. Show that the sum of the result after applying this filter is nearly 0. Show this image.
13. Define $\mathbf{a}[\vec{x}] = \mathbf{o}_{\vec{w};\vec{v},r}[\vec{x}]$, where $\vec{w} = (512,512)$, $\vec{v} = (256,256)$, and $r = 160$. Create and apply a filter that will leave the horizontal edges of the circle as sharp edges but the vertical edges will be fuzzy.
14. Write the operator notation for the creation of the filter in the previous problem. Write the operator notation for the application of the filter in the previous problem.
15. Use $\mathbf{a}[\vec{x}] = Y(\texttt{data/ib3logo.png})$t. Apply a filter that produce sharp edges for all angles except those near 45°.
16. Write the operator notation for the creation of the filter in the previous problem. Write the operator notation for the application of the filter in the previous problem.

12 Correlations

In the previous chapter, image frequencies were manipulated to achieve specific goals. The next step is to modify the frequencies of image in accordance with the frequencies in another image. This chapter pursues this idea by beginning with a review of correlation theory, which may seem out of place. However, the theory will be followed by relating this theory to Fourier filtering. This chapter will present some examples, venture into composite filtering, and then end with some negative issues related to correlations as applied to image analysis.

12.1 JUSTIFICATION AND THEORY

Consider the simple example shown in Figure 12.1(a) that depicts an attempt to find a specific signal embedded in a larger signal. The smaller signal (red) exists somewhere in the larger signal (green). A correlation will consider all possible shifts of the target signal (red) with the data signal (green). One of the features of the target signal is that it is zero-sum. When it is placed as shown, the values of the target signal are multiplied with the corresponding values of the data signal. If there is no match, then the collection of multiplied values will contain both positive and negative results. Thus, the summation would be somewhat close to zero. However, when the two signals align, then the negative values will be multiplied by negative values; thus, the summation will contain only positive values. For this particular relative shift between the two signals, the correlation signal should have a peak. There may be other relative shifts in which the two signals have a significant amount of similarity, and so subsidiary peaks are also possible, particularly if the original signal has a repetitive structure. The correlation of these two signals is shown in Figure 12.1(b).

Locations of the large peaks in the correlation signal indicate relative shifts of the signals for potential alignment. Consider the largest peak in Figure 12.1(b). The distance from the location of the peak to the center of the graph is the distance that needs to be shifted to align. The two signals, now aligned by that shift distance are shown in Figure 12.1(d). As seen, the target is now positioned to be part of the signal that is similar to the target.

Figure 12.1(b) shows the correlation result. Each value is the sum of multiplied values of a specific shift. A large spike in this signal indicates that there is a shifted position of the target signal that matches the content in the data signal. The location of this spike is directly related to amount of shift of the target signal. In these examples, there is a spike near $x = 175$. This indicates the shift required of the target signal to find a matching location in the data signal.

12.2 THEORY

The correlation of the two functions, $f(x)$ and $g(x)$, is

$$c(u) = \int_{-\infty}^{\infty} f(x+u)g^{\dagger}(x)\,\mathrm{d}x, \tag{12.1}$$

where $g^{\dagger}(x)$ represents the complex conjugate fo $g(x)$. The function $f(x)$ is shifted and multiplied by the function $g(x)$. The integration sums the multiplied values. The two functions shown in Figure 12.1(a) would be $f(x)$ and $g(x)$ and the result shown in Figure 12.1(b) would be $c(u)$.

A correlation in two-dimensions is quite similar in nature:

$$c(u,v) = \int_{-\infty}^{\infty}\int_{-\infty}^{\infty} f(x+u,y+v)g^{\dagger}(x,y)\,\mathrm{d}x\,\mathrm{d}y. \tag{12.2}$$

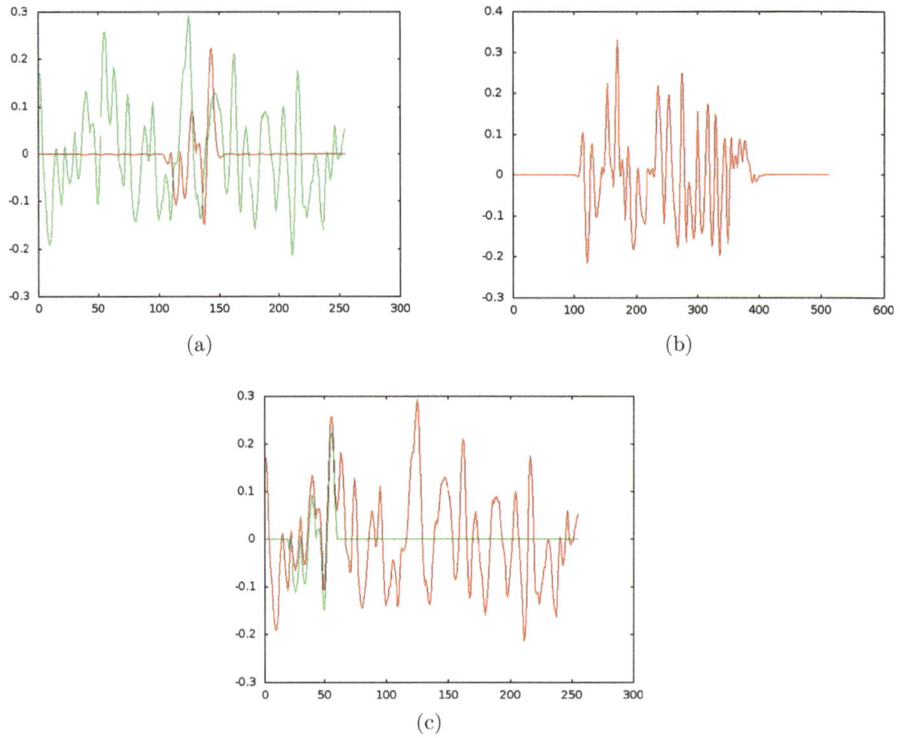

Figure 12.1 (a) The original signal in green and the target signal in red, (b) the correlation, and (c) the alignment of the two signals.

The operator notation for a correlation is $\otimes$, and thus, Equation (12.1) or (12.2) is written as

$$\mathbf{c}[\vec{u}] = \mathbf{a}[\vec{x}] \otimes \mathbf{b}[\vec{x}]. \tag{12.3}$$

The output, $\mathbf{c}[\vec{u}]$, is denoted in a new space by $\vec{u} \in \mathbf{U}$.

The correlation of a function with itself is called a *auto-correlation*. This is a useful procedure for finding presence of regular patterns in a signal. The correlation of two different functions is also called a *cross-correlation*.

12.2.1 COMPUTATIONS IN FOURIER SPACE

For large two-dimensional arrays, the computations of a correlation using Equation (12.2) can be slow. If, for example, an image was $N \times N$ pixels, then the number of mult-adds (multiplications and additions) goes as N^4.

A more efficient method of computing the correlation is to employ Fourier transforms. Consider the Fourier transform of a correlation:

$$C(\alpha) = \int_{-\infty}^{\infty} c(u) e^{-\imath u \alpha} \, \mathrm{d}u. \tag{12.4}$$

By employing Equation (12.1), this becomes

$$C(\alpha) = \int_{-\infty}^{\infty} \int_{-\infty}^{\infty} f(x+u) g^{\dagger}(x) \, \exp\left[-\imath u \alpha\right] \, \mathrm{d}x \, \mathrm{d}u. \tag{12.5}$$

Using the substitution $z = x + u$, the equation becomes

$$C(\alpha) = \int_{-\infty}^{\infty} \int_{-\infty}^{\infty} f(z) g^{\dagger}(x) \, e^{-\imath \alpha (z-x)} \, \mathrm{d}x \, \mathrm{d}z. \tag{12.6}$$

This can be rewritten as

$$C(\alpha) = \int_{-\infty}^{\infty} f(z) e^{-\imath \alpha z}\, \mathrm{d}z \int_{-\infty}^{\infty} g^{\dagger}(x)\, e^{\imath \alpha x}\, \mathrm{d}x = F(\alpha) G^{\dagger}(\alpha), \tag{12.7}$$

where $F(\alpha)$ is the Fourier transform of $f(x)$ and $G(\alpha)$ is the Fourier transform of $g(x)$. This equation computes the Fourier transform of the correlation and so an inverse transform needs to be applied to obtain the correlation.

In operator notation, this is

$$\mathbf{c}[\vec{\alpha}] = \mathfrak{F}^{-1}\left((\mathfrak{F}\mathbf{a}[\vec{x}]) \times (\mathfrak{F}\mathbf{b}[\vec{x}])^{\dagger}\right) = \mathbf{a}[\vec{x}] \otimes \mathbf{b}[\vec{x}] \tag{12.8}$$

The digital correlation for a signal of length N requires N^2 mathematical operations. The digital Fourier transform for a signal of length N requires $N\log_2(N)$ operations. The correlation using the Fourier methods requires three Fourier transforms and the elemental multiplication of two signals. Thus, the number of operations is $3N\log_2(N) + N$. Consider a case in which an image is 1024×1024. Then, N would be larger than 10^6 and the original correlation method would require 10^{12} operations. The Fourier method, though, would only require 6.4×10^7 operations. For large arrays, the computational savings is several orders of magnitude, and very easy to employ.

12.3 IMPLEMENTATION IN PYTHON

There are two possible methods to compute a correlation in Python. The first uses a brute force method, or the digital equivalent of Equation 12.2. This works well if one of the input functions is small in size. The second method uses the Fourier transforms, which is better suited for the cases in which both inputs are images.

12.3.1 BRUTE FORCE

The *scipy.signal* package contains a few functions that can perform the brute force correlations. For a one-dimensional signal, the function is **correlate**, and for a two-dimensional signal, the correct function is **correlate2d**. This function will receive two arrays and the second array should be small in size. Code 12.1 shows a simple example of this function. Line 4 reads in an image and converts it to gray scale. Line 5 creates a 5×5 array in which all elements are 1.0. Line 6 performs the correlation, and line 7 will be discussed shortly.

Figure 12.2(a) shows the original image, and Figure 12.2(b) shows the result after the correlation. In this case, each pixel in the output space is the average overall pixels in a 5×5 region. So, this process smoothens the image.

It should be noted that the size of `cdata` is slightly bigger than `adata` or $N\mathbf{c}[\vec{u}] > N\mathbf{a}[\vec{x}]$. The reason is that all possible shifts of the kernel do allow it to extend beyond the frame of the original image. This, though, can be controlled with the optional argument `mode`, which has three possible

Code 12.1 Smoothing through a correlation with a small solid block

```
1 >>> import numpy as np
2 >>> import imageio
3 >>> import scipy.signal as ss
4 >>> adata = imageio.imread( 'image.png', as_gray=True )
5 >>> bdata = np.ones( (5,5) )/25.0
6 >>> cdata = ss.correlate2d( adata, bdata )
7 >>> cdata = ss.correlate2d( adata, bdata, mode='same' )
```

Figure 12.2 (a) The original image and (b) the image after correlation with a small solid block (Code 12.1).

states. The default state is "full." The option that will produce an output which is the same size of the input is "same." The third option is "valid" which is used to control the size of the output if there is padding in the input.

The small matrix is called a *kernel*, and there are several kernels with known and useful behaviors. Consider the kernel

$$\begin{pmatrix} -1 & -1 & -1 & -1 & -1 \\ -1 & 0 & 0 & 0 & -1 \\ -1 & 0 & 16 & 0 & -1 \\ -1 & 0 & 0 & 0 & -1 \\ -1 & -1 & -1 & -1 & -1 \end{pmatrix}, \tag{12.9}$$

which is a zero-sum matrix and is also called an *on-center off-surround* matrix. Using the same process as in Code 12.1 excepting the kernel, the negative of the absolute value of the result is shown Figure 12.3. In this case, the result is an edge enhanced image. The value of a single pixel in the output space is corresponding pixel in the input space multiplied by 16 subtracted by the values of next neighbors. This is the process used in Section 13.3.

12.3.2 METHOD BASED ON FOURIER TRANSFORMS

Code 12.2 shows the **Correlate1D** function, which performs a one-dimensional correlation using Fourier transforms. This receives two vectors of the same length and computes the correlation. The

Figure 12.3 The result of applying the kernel in Equation (12.9).

Code 12.2 The **Correlate1D** function

```
# correlate.py
import scipy.fft as ft
def Correlate1D( A, B ):
    a = ft.fft(A)
    b = ft.fft(B)
    c = a * b.conjugate( )
    C = ft.ifft( c );
    C = ft.fftshift(C);
    return C
```

Code 12.3 The **Correlate2DF** function

```
# correlate.py
def Correlate2DF( A, B ):
    c = A * b.conjugate( )
    C = ft.ifft2( c )
    C = ft.fftshift(C)
    return C
```

equivalent function for two dimensions is **Correlate2D**. The only difference in the two functions besides the name is that the latter uses **ft.fft2d** and **ft.ifft2d**.

Finally, the function **Correlate2DF** shown in Code 12.3 is used in cases where the input and filter are already in Fourier space. In some cases, an image is to be correlated with several filters and so it is prudent to compute the Fourier transform of the input just once and feed it to this function for each filter. Section 12.5.1 reviews the Fractional Power Filter, which is constructed in Fourier space from the Fourier transforms of images. Thus, both the data and the filter are already in Fourier space, and **Correlate2DF** can be used to finish the correlation computation between the inputs and filter.

12.3.3 EXAMPLE – GEOMETRIC SHAPES

The first example is the correlation of a simple shapes as performed in Code 12.4. The input image is created in lines 3 through 8. Lines 4 through 5 create three rectangles, and lines 7 and 8 create

Code 12.4 Correlating shapes

```
1  >>> import numpy as np
2  >>> import correlate as crr
3  >>> adata = np.zeros( (512,512) )
4  >>> adata[100:130,80:120] = 1
5  >>> adata[300:350,100:210] = 1
6  >>> adata[80:210,300:400] = 1
7  >>> adata += mgc.Circle( (512,512), (150,180), 40 )
8  >>> adata += mgc.Circle( (512,512), (350,300), 60 )
9  >>> bdata = mgc.Circle( (512,512), (256,256), 60 )
10 >>> bdata -= mgc.Circle( (512,512), (256,256), 55 )
11 >>> cdata = crr.Correlate2D( adata, bdata )
```

two circles. These are white shapes on a black background, and the negative of this image is shown in Figure 12.4(a).

The second image that is created in lines 9 and 10 is an annular ring that has the same radius as one of the circles in the original image. This is not a solid circle but just the perimeter or annular ring. The correlation of two solid objects with binary values tends to create a flat-top response. Consider a case in which the two images are solid squares and one square is smaller than the other. At one particular shift, the smaller square is completely inside the other square. The correlation value for this shift is the inner product the value of which is the number of pixels in the smaller square. Now, the smaller square is shifted one pixel and it is still completely inside the larger square, thus the correlation value is the same. The correlation of binary valued, solid shapes tends to produce a correlation signal with a flat-top response. If the two squares are the same size, there is only one correlation value that is the maximum, but the neighboring correlation values are only slightly less.

In this example, the second image is an annular ring of the same size as one of the objects. When they are perfectly aligned, there is a large signal, but when they are slightly misaligned the correlation signal decays rapidly thus producing a sharper peak. The correlation of these two images is performed in line 11 and the negative of the result is shown in Figure 12.4(b). In the lower right, there is a round shape with a sharp peak (dark spot). This is the strongest signal in the correlation and thus indicates a strong match between the object at that location in the input and the filter. Basically, this filter found a circular shape of a specific radius.

12.3.4 EXAMPLE – BOAT ISOLATION

In this example, the individual boats in Figure 12.5 will be isolated and labeled. The main problem in this task is that the boats and the dock touch and that the range of pixel intensities of the boats is beyond the range of the dock. So, it is not possible to separate the dock from the boats using a threshold. The construction of the dock is known, and in this task, the image is considered on only one scale and orientation. A correlation is used to locate the dock, and then it can be subtracted from the original image. Isolation of the boats is easy once the dock is removed.

There are four steps in this process. The first is to load the original image and create the dock mask. The second step is to locate the mask in the original image, and the third step is to subtract the dock. The final step is to clean the remaining components of the image to create isolated boats.

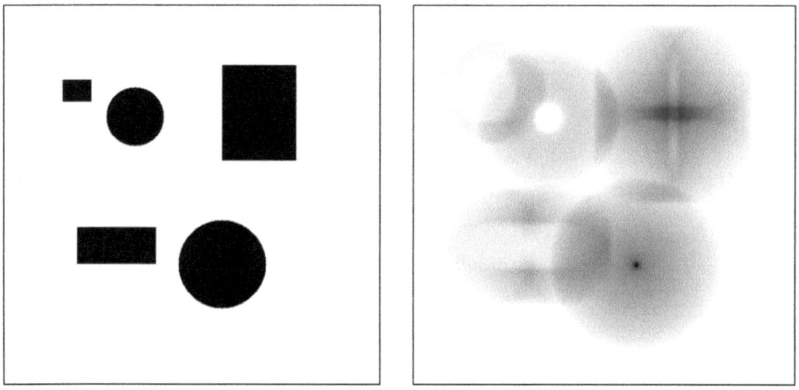

(a) (b)

Figure 12.4 (a) The original signal and (b) the correlation of the original image.

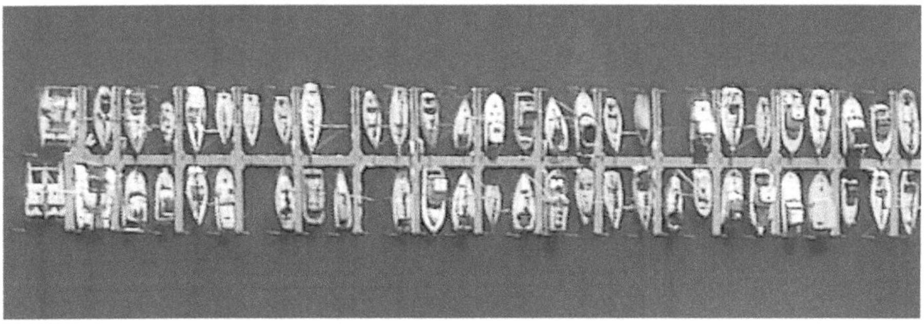

Figure 12.5 An original image extracted from a much larger image. Image copyright DigitalGlobe. All rights reserved. Image courtesy of Satellite Imaging Corporation.

The original image is a color image extracted from a satellite image which is designated as $\mathbf{a}[\vec{x}]$. This image is converted to grayscale as

$$\mathbf{b}[\vec{x}] = \mathcal{L}_L \mathbf{a}[\vec{x}]. \tag{12.10}$$

It is loaded as `orig` in line 15 of Code 12.5.

The dock dimensions are known (perhaps from previous satellite images or other intelligence). The mask of the dock is created from simple rectangles. The function **BuildDock** shown in Code 12.5 creates this mask. Line 3 creates the frame, and line 4 creates the horizontal rectangle. There are two vertical rectangles on the left that are not aligned and so they are created separately. The rest are created from lines 7 through 9. The list `crn` contains the horizontal left edge of each rectangle. The loop creates rectangles that are 84 pixels high and 4 pixels wide. This dock needs to be centered in the frame before being used in a correlation and so line 11 finds the center of mass and line 12 centers the dock in the frame. The negative of the image is shown in Figure 12.6.

The grayscale boat image is represented by $\mathbf{b}[\vec{x}]$, and the dock mask is represented by $\mathbf{m}[\vec{x}]$. The intent is to locate the position of the dock in the original image and then to subtract the dock, which will separate the boats from other pixels in the same intensity range. The process of locating a boat

Code 12.5 Loading and creating the necessary images

```
1  # boats.py
2  def BuildDock():
3      dock = np.zeros((189,545))
4      dock[39:46,0:508] = 1
5      dock[0:39,9:14] = 1
6      dock[46:84,0:5] = 1
7      crn = [30,66,100,136,172,206,242,281,315, 349,384,420,456,492]
8      for c in crn:
9          dock[0:84,c:c+4] = 1
10     V,H = dock.shape
11     v,h = nd.center_of_mass( dock )
12     dock = nd.shift( dock + 0., (V/2-v, H/2-h) )
13     return dock
14
15 >>> orig = imageio.imread( 'data/boats1.png', as_gray=True)
16 >>> dock = boats.BuildDock2()
```

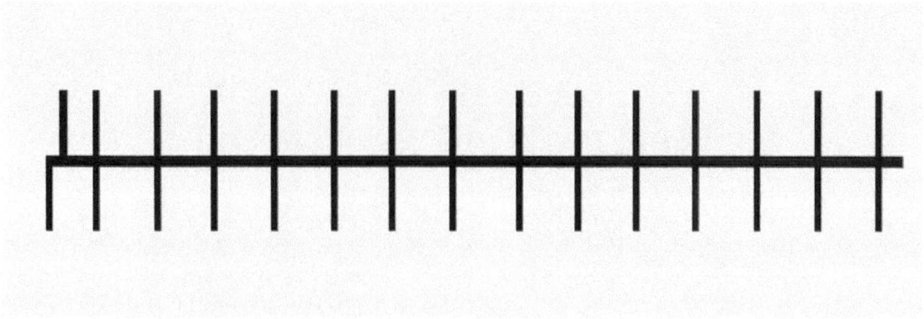

Figure 12.6 The boat dock mask.

is performed by correlating the dock mask with the image. The peak of the correlation is directly related to the location of the center of the dock in the original image. As long as the dock is centered in the mask, then the amount of shift required to align the mask to the dock in the image is the distance between the center of the frame and the location of the correlation peak. The location of the peak is extracted by the $A_\vee$ operator, which is applied to the correlation of the image with the dock mask as in

$$\vec{v} = A_\vee \left(\mathbf{b}[\vec{x}] \otimes \mathbf{m}[\vec{x}] \right). \tag{12.11}$$

The process is shown in Code 12.6 as the **LocateDock** function. Line 3 performs the correlation, and lines 4 and 5 determine the location of the peak. In this example, the peak is located at (91,281).

Figure 12.7 shows the dock positioned over the image. The method by which this image is created begins with the definition of the vector $\vec{c}$ as the center of the image frame. The process is defined as

$$\mathbf{d}[\vec{x}] = \mathbf{b}[\vec{x}] + 300\Gamma_{>0.1} D_{\vec{v}-\vec{c}} \mathbf{m}[\vec{x}]. \tag{12.12}$$

The mask is shifted by $\vec{v} - \vec{c}$ which positions the dock. A small threshold is applied as the **shift** function does install small artifacts if the input images has integer values. The result is multiplied by a large number and then added to the original. This image is merely to determine if the dock is correctly placed and will not be used in further computations.

The **Overlay** function shown in Code 12.7 performs Equation (12.12). Lines 5 shifts the dock, line 6 applies the threshold, and line 7 finishes the computation and returns the result of which the negative is shown in Figure 12.7.

The third step in the process is to subtract the dock from the original image, and the fourth step is to isolate the boats. Both steps can be performed by

$$\mathbf{f}[\vec{x}] = \Gamma_{>140} \left(\mathbf{b}[\vec{x}] - 200\, D_{\vec{v}-\vec{c}} \lhd \Gamma_{>0.1} \mathbf{m}[\vec{x}] \right). \tag{12.13}$$

Code 12.6 The **LocateDock** function

```
1  # boats.py
2  def LocateDock( orig, dock ):
3      corr = crr.Correlate2D( orig, dock)
4      V,H = corr.shape
5      v,h = divmod( abs(corr).argmax(), H )
6      return v,h
7
8  >>> vh = boats.LocateDock( orig, dock )
9  >>> vh
10 (93, 281)
```

Figure 12.7 The dock positioned in the original image.

Code 12.7 The **Overlay** function

```
# boats.py
def Overlay( orig, dock, vh ):
    v,h = vh
    V,H = dock.shape
    ndock = nd.shift( dock, (v-V/2, h-H/2) )
    ndock = ndock > 0.1
    return orig + 300*ndock

>>> olay = boats.Overlay( orig, dock, vh )
```

Once again, a small threshold is applied to the dock before the dilation process is applied as in $\lhd\Gamma_{>0.1}\mathbf{m}[\vec{x}]$. This is to ensure that all of the background pixels are exactly 0. If there is even a small bit error, then the dilation will not work properly. This result is shifted by $D_{\vec{v}-\vec{c}}$, so that dock is centered in the frame and then multiplied by a large value. This is subtracted from the original, thus making the dock pixels much darker than anything else in the image. The threshold $\Gamma_{>140}$ passes only those pixels that are brighter than 140, which excludes all of the ocean pixels and dock pixels. The surviving pixels belong only to the boats as seen in Figure 12.8.

The process is contained within the **SubtractDock** function in Code 12.8. Lines 5 through progressively apply the operators in Equation (12.13).

Not all of the boats in the image are contiguous segments. Some of the boats have separate segments due to vastly different pixel values for the boat deck and cabin canopy. The final step is to

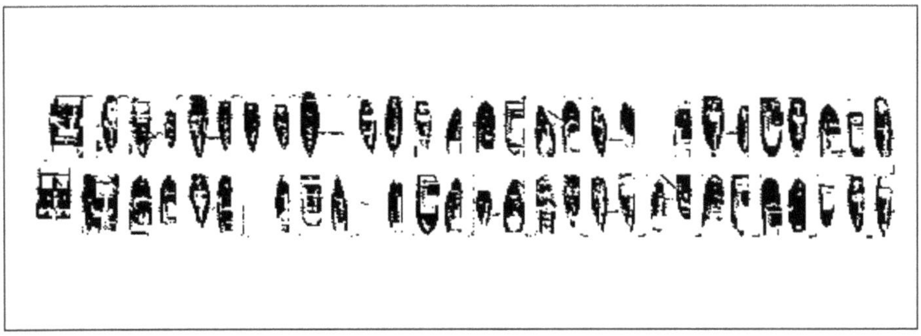

Figure 12.8 The isolated objects.

Code 12.8 The **SubtractDock** function

```
# boats.py
def SubtractDock( orig, dock, vh ):
    v,h = vh
    V,H = dock.shape
    dock2 = nd.binary_dilation( dock>0.1, iterations=1)
    ndock = nd.shift( dock2+0., (v-V/2, h-H/2) )
    ndock = (orig -200*ndock)>140
    return ndock

>>> isoboats = boats.SubtractDock( orig, dock, vh )
```

Code 12.9 The **IDboats** function

```
1  # boats.py
2  def IDboats( sdock ):
3      mask = np.ones( (7,1) )
4      answ = nd.binary_dilation( sdock, structure = mask )
5      lbl, cnt = nd.label( answ )
6      return lbl
7
8  >>> answ = boats.IDboats( isoboats )
```

remove spurious pixels and to unite boats that might be shown in more than one contiguous segment. This step relies on the idea that all of the boats are facing vertically in the image. The dilation process is modified so that there is a dilation in one direction that is different than the dilation in the other direction. In this case, the desire is to dilate in the vertical direction, which would connect separated boat segments. This dilation must be large enough to cross the gaps between segments but small enough to not bridge over the dock.

The process starts with the definition of **P**, which is a matrix that is 7×1 and all of the values are 1. The description is

$$\mathbf{g}[\vec{x}] = \mathcal{I} \lhd_{\mathbf{P}} \mathbf{f}[\vec{x}]. \tag{12.14}$$

The dilation $\lhd_{\mathbf{P}}$ will connect segments that are within seven pixels of each other in the vertical direction, thus uniting the boat segments. The final step is to label the components. These final steps are performed in the **IDboats** shown in Code 12.9. Line 3 creates **P**, line 4 performs $\lhd_{\mathbf{P}}$, and line 5 performs the isolation $\mathcal{I}$. The result in Figure 12.9 displays each isolated segment with a different pixel intensity. Isolation of a single object is performed by $\mathbf{g}[\vec{x}] \stackrel{?}{=} i$, where i is an integer that is the pixel intensity of the desired object.

12.4 COMPOSITE FILTERING

Correlation filters are somewhat useful but in many real applications the target can have several presentations. In the previous example, the search was limited such that the dock could have only a single orientation and magnification. A more realistic case would consider any orientation and a range of scales.

It would be possible to create a filter for every possible orientation/scale combination but that would be a prohibitive computation. A better solution is to use composite filtering which creates a

Figure 12.9 The final result.

filter from several training images. Thus, one single filter could identify the location of a target over a range of orientations and scales. Once the filter is built, the computation time to identify the dock is the same as that of a single correlation.

12.5 SDF AND MACE

The composite filter theory starts the SDF (synthetic discriminant function). The desire is to create a filter, $\vec{h}$, that solves,

$$\vec{c} = \mathbf{X}\vec{h}, \tag{12.15}$$

where $c_i = 1 \ \forall i$ and the columns of $\mathbf{X}$ are the target data vectors. Extrapolation of this system to two-dimensional data is straightforward as it is possible to vectorize the image, placing all of its pixels in a single column vector. If $\mathbf{X}$ were square, then the solution would simply require the computation of the inverse of $\mathbf{X}$. However, when used for images $\mathbf{X}$ is extremely rectangular. Therefore, the pseudo-inverse is employed and the filter is computed by

$$\vec{h} = \mathbf{X}\left(\mathbf{X}^{\dagger}\mathbf{X}\right)^{-1}\vec{c}. \tag{12.16}$$

The center pixel of the correlation between the filter $\vec{h}$ and any of the inputs $\vec{x}_i$ will produce c_i. This value is also $\vec{h} \cdot \vec{x}$. Thus, $\vec{c}$ is called the *constraint vector* as it constrains a single value in the correlation, but the rest of the values are unconstrained.

It is possible that correlation values other than the center have a higher magnitude. In these cases, the peak may not be accurately identified. A proposed solution is to simultaneously reduce the total correlation energy as well as constrain the center value. This led to the minimum average correlation energy (MACE) filter.

The MACE is computed by

$$\hat{\vec{h}} = \mathbf{D}^{-1/2}\mathbf{X}\left(\hat{\mathbf{X}}^{\dagger}\mathbf{D}^{-1}\hat{\mathbf{X}}\right)^{-1}\vec{c}, \tag{12.17}$$

where

$$D_{i,j} = \frac{\delta_{i,j}}{N}\sum_k |\hat{x}_i^{(k)}|^2, \tag{12.18}$$

and $x_i^{(k)}$ represents the i-th element of the Fourier transform of the k-th vector. The MACE must be computed with the data in Fourier space and this is designated by the carets.

The MACE filter builds filters that produce a very sharp correlation peak when correlated with a training input. However, if the correlation is with the filter and another vector that is not used in training, the height of the correlation peak is small. This is the case even if the correlation input vector is mostly similar to a training vector.

12.5.1 FRACTIONAL POWER FILTER (FPF)

The original filter was robust but had large side lobes. The MACE had low side lobes and was too sensitive to changes in the input. The solution was to find a filter that was a mixture of the two.

12.5.1.1 Theory

The matrix **D** is a diagonal matrix, and if the power term in Equation (12.18), where 0 instead of 2, then **D** would be the Identity matrix. In that case, Equation (12.17) would become Equation (12.16). The only difference between the two filters is the power term, and thus, the FPF [5] uses Equation (12.17) with

$$D_{i,j} = \frac{\delta_{i,j}}{N} \sum_k |\hat{x}_i^{(k)}|^\alpha, \tag{12.19}$$

where $0 \leq \alpha \leq 2$.

The operator $\mathcal{Q}_{\vec{c},\alpha}\left\{\mathbf{a}_i[\vec{x}]\right\}$ represents the process of creating an FPF filter. The vector $\vec{c}$ is the constraint vector and is an optional argument. If it is not present, then it is assumed that all of the elements in this vector are set to 1, which coincides with the original use of the filter. The α is the fractional power value between 0 and 2, and the images inside the curly braces represent the training data. Thus, the SDF can be computed by

$$\mathbf{b}[\vec{x}] = \mathcal{Q}_0\left\{\mathbf{a}_i[\vec{x}]\right\}. \tag{12.20}$$

The FPF is constructed on the Fourier transform of the images, and thus, it is defined as

$$\mathbf{b}[\vec{\omega}] = \mathcal{Q}_\alpha\left\{\mathfrak{F}\mathbf{a}_i[\vec{x}]\right\}. \tag{12.21}$$

This filter is in Fourier space and thus has the coordinates defined by $\vec{\omega}$. In some applications, there is an efficiency to be gained by leaving the filter in the Fourier space instead of converting it back to the original image space.

Code 12.10 shows the **FPF** function. It receives the training data as rows in a matrix named `data`. For images, it is necessary to use the **ravel** command to convert the two-dimensional information

Code 12.10 The **FPF** function

```
# fpf.py
def FPF( data, c, fp):
    (N,Dim )= data.shape
    D = ( np.power( abs( data), fp )).sum(0)
    D = D / N
    ndx = (abs(D) < 0.001 ).nonzero()[0]
    D[ndx] = 0.001 * np.sign(D[ndx]+1e9)
    Y = data / np.sqrt(D)
    Y = Y.transpose()
    Yc = Y.conjugate().transpose()
    Q = Yc.dot( Y )
    if N == 1:
        Q = 1./Q
    else:
        Q = np.linalg.inv( Q )
    Rc = Q.dot( c)
    H = Y.dot( Rc ) / np.sqrt(D)
    return H
```

Code 12.11 Testing the **FPF** function

```
1 >>> data = np.random.ranf( (3,10) )
2 >>> c = np.ones(3)
3 >>> h = fpf.FPF( data,c,0)
4 >>> h.dot( data[0] )
5 1.0
6 >>> h.dot( data[1] )
7 1.0
8 >>> h.dot( data[2] )
9 1.0
```

into a long one-dimensional vector that is then placed in a row of `data`. Each row in `data` is a training input. If the fractional power is anything other than 0, then the data must be in Fourier space. The second input is the constraint vector `c` and usually all of the elements are set to 1. The length of this vector is the same as the vertical dimension of `data`. The third input is the fractional power term. Lines 4 and 5 compute D, and lines 6 and 7 are used to handle very small values. Equation (12.18) has three instances of $\mathbf{D}^{1/2}\mathbf{X}$, and so lines 8 through 10 define $Y = \mathbf{D}^{1/2}\mathbf{X}$, which reduces the computation to $\hat{\vec{h}} = \hat{\mathbf{Y}}(\hat{\mathbf{Y}}^{\dagger}\hat{\mathbf{Y}})^{-1}\vec{c}$. Line 11 defines $\mathbf{Q}$, which is the terms inside the parenthesis in Equation (12.18). Lines 12 through 15 invert $\mathbf{Q}$ with line 13 used for the case when $\mathbf{Q}$ is 1×1. Lines 16 and 17 finish the computation.

Code 12.11 shows a very simple case for the SDF. Line 1 creates random data in a 3×10 matrix. This is equivalent to three input vectors of length ten. The constraint vector is created in line 2, and the SDF filter is created in line 3. The rest of the lines show that the dot product (which is also the center of a correlation result) is 1.0 for all three inputs. This corresponds to the three values in the constraint vector.

12.5.1.2 Manipulating α

The value of α ranges from 0 to 2, and when applied to images, the behavior is somewhat predictable. The value $\mathbf{D}_{i,i}$ in Equation (12.19) is proportional to the magnitude of frequency i in the training data. However, Equation (12.17) only uses the inverse of $\mathbf{D}$. Thus, the filter has high magnitudes for low magnitude frequencies in the training data. When applied to most types of images, the increase in α corresponds to a reliance of the edge information in the training images. This occurs since the low frequencies are often very large and correspond to solid shapes in the input. Basically, for image of objects, an increase in α is related to a greater sensitivity to the object's edges. As α goes from 0 to 2, the filter trades off generalization for discrimination.

A simple example of the effect of α begins with an image that has various, solid geometric shapes as seen in Figure 12.10. The creation of the filter is

$$\mathbf{h}[\vec{x}] = \mathfrak{F}^{-1}\mathcal{V}_{\vec{w}_1}\mathcal{Q}_{\vec{c},\alpha}\mathcal{C}_m\left\{\mathcal{V}_{\vec{w}_2}\mathfrak{F}\mathbf{b}[\vec{x}]\right\}, \tag{12.22}$$

and the Python script is shown in Code 12.12. Line 5 loads the original data as $\mathbf{a}[\vec{x}]$ and prepares a scaled version by

$$\mathbf{b}[\vec{x}] = 1 - \frac{1}{255} \times \mathcal{L}_L\mathbf{a}[\vec{x}]. \tag{12.23}$$

The process $\mathcal{V}_{\vec{w}_2}\mathfrak{F}\mathbf{b}[\vec{x}]$ computes the Fourier transform of the image and reshapes it to conform the $\vec{w}_2$. In this case, $\vec{w}_2$ represents a column in the matrix with the length of $V \times H$. The **ravel** function in line 8 completes this action. These operators are inside curly braces, which indicates that the process is repeated for each image (if there are more than one). The Concatenation operator $\mathcal{C}_m$ puts all of these vectors into a matrix. In the example, there is only one image, and it is placed into the

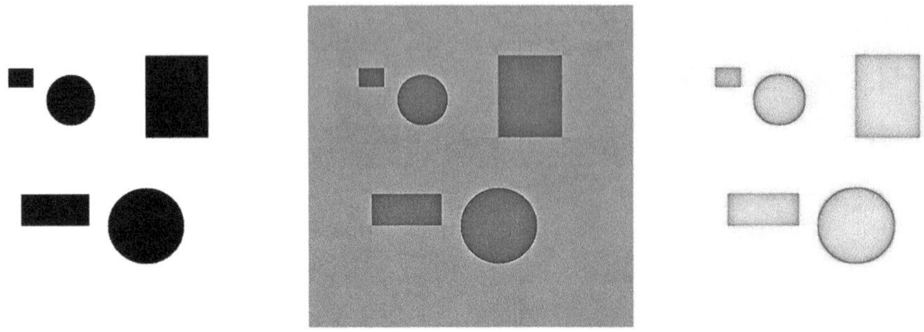

(a) (b) (c)

Figure 12.10 The negative images of the FPFs with different α values. (a) $\alpha = 0$, (b) the real component from $\alpha = 0.3$, and (c) the absolute value of $\alpha = 0.3$.

Code 12.12 Computing an FPF

```
1  >>> import numpy as np
2  >>> import imageio
3  >>> import scipy.fftpack as ft
4  >>> import fpf
5  >>> bdata = 1.0 - imageio.imread( 'data/geos.png', as_gray=True)/255.
6  >>> V,H =  bdata.shape
7  >>> Xt = np.zeros( (1,V*H), complex )
8  >>> Xt[0] = ft.fft2( bdata ).ravel()
9  >>> cst = np.ones( 1 )
10 >>> ffilt = fpf.FPF( Xt, cst, 0 )
11 >>> filt = ft.ifft2( ffilt.reshape( (V,H) ))
```

matrix in line 8. Line 7 creates the matrix and line 8 populates it. If there were more images, then the dimensions in line 7 would change and line 8 would be applied to each image with the index incrementing for `Xt`.

The creation of the filter by $Q_{\vec{c},\alpha}$ is called in line 10 with line 9 creating $\vec{c}$. In this case, $\alpha = 0$. The final two operators, $\mathfrak{F}^{-1}\mathcal{V}_{\vec{w}_1}$ are performed in line 11 where $\vec{w}_1$ represents the original image frame size. Figure 12.10 and (c) show $\Re\mathbf{h}[\vec{x}]$ and $|\mathbf{h}[\vec{x}]|$, respectively, for the case of $\alpha = 0.3$.

When used a correlation, the first filter will generalize well. The input shapes do not have to match those in the training image precisely. However, the correlation peak may be difficult to detect particularly if there is any noise in the input. The second filter will discriminate well. It can better distinguish between to similar shapes because it is more sensitive to shape (the outline or edges of the objects).

12.5.1.3 Example

This example is to identify the location of the numerals in the image of a tachometer shown in Figure 12.11. The goal is to build a single FPF filter that can recognize all of the digits in a single correlation.

The original color image is defined as $\mathbf{a}[\vec{x}]$, and it needs to be processed in the same manner as in Equation (12.23).

To train the FPF, an image of each digit is isolated to create a training set. This set consists of nine images cut out from the original, each 28×28 centered on the one of the digits. This is a sufficient

Figure 12.11 An image of a tachometer.

size to capture all of the pixels of a digit without also capturing pixels from other digits. The set of vectors $\vec{z}_i$ is defined as the location of the center of each digit in the original image. The process of generating these images is

$$\{\mathbf{d}[\vec{x}]\} = U_{\vec{w}_1} \Box_{\vec{z}_i} \mathbf{b}[\vec{x}], \quad \forall i. \tag{12.24}$$

There are nine digits in the image and so the index i runs over the nine choices. For each i, a small segment centered on $\vec{z}_i$ is copied from the image $\mathbf{b}[\vec{x}]$ and then centered in a frame that is the same size, $\vec{w}_1$, as the original image. The result is a set of images $\{\mathbf{d}[\vec{x}]\}$ that are as large as the original image and have the 28×28 image of one of the digits centered in its frame.

Once again, $\vec{c}$ is the constraint vector which now has a length of 9 but all of the elements still have a value of 1.0. Following the previous example, the FPF is computed by Equation (12.22), with $\alpha = 0.8$, and the filter is an edge encouraged, weight linear combination of the digits from the original image.

The final step is to correlate the grayscale image with the filter. If the filter works, then there should be bright peaks at the center location of each digit and these should be the brightest elements in the correlation. The final step is

$$\mathbf{f}[\vec{x}] = \mathbf{b}[\vec{x}] \otimes \mathbf{h}[\vec{x}]. \tag{12.25}$$

Equations (12.23) and (12.24) are performed in the function **LoadTach** shown in Code 12.13. Line 3 loads the image as a gray scale and applies the scale and bias. Lines 5 through 7 are the locations of the center of each digit, and lines 8 and 9 extract these portions of the image. Actually, this is

Code 12.13 The **LoadTach** function

```
1  # tachfpf.py
2  def LoadTach( fname ):
3      bdata = 1 - imageio.imread( fname, as_gray=True)/255.
4      cuts = []
5      vhs = [(303,116), (250, 96), (182,107), (126,150), \
6             (104,222), (116,294), (165,343), (231,362), \
7             (300,341)]
8      for v,h in vhs:
9          cuts.append( bdata[v-14:v+14, h-14:h+14] + 0 )
10     return bdata, cuts
```

only a part of Equation (12.24) as the cutout segments have not been placed into a larger frame yet to save on computer memory.

Code 12.14 shows the **MakeTachFPF** function. Line 5 allocates space for **X**, and the `for` loop started in line 6 puts the cutout images in a larger frame, thus completing Equation (12.24), and then computing their FFTs and placing them in rows of **X**. The rest of the equation is fulfilled in lines 13 through 15. The final filter is multiplied by the number of pixels to scale the correlation output. The process is run as shown in Code 12.15. The last line executes Equation (12.25) completing the correlation of the image with the filter.

The negative of the result is shown in Figure 12.12. Since this is the negative, the darker spots indicate higher correlation values, and as seen, they are located at the locations of the digits. The result is not perfect. As seen in Figure 12.12, there are large correlation signals at the location of all nine targets. However, there are other spurious locations in which there is also a large correlation signal. These would be classified as false positives. The next section discusses methods to reduce the false positives.

12.5.1.4 The Constraints

In the previous section, the constraint vector had a length of N and all of the elements were set to a value of 1, with N being the number of training images. This ensures that $\mathbf{h}[\vec{x}] \cdot \mathbf{a}_i[\vec{x}] = 1$ for all images $\mathbf{d}_i[\vec{x}]$ that were used in training. The center of the correlation, $\mathbf{f}[\vec{x}] = \mathbf{d}_i[\vec{x}] \otimes \mathbf{h}[\vec{x}]$, is likewise constrained to the value of 1. The location of the constrained value is sensitive to the location of the target. Usually, the target is centered in $\mathbf{d}_i[\vec{x}]$, and therefore, it is the center value of the correlation surface that is constrained.

Code 12.14 The **MakeTachFPF** function

```
1  # tachfpf.py
2  def MakeTachFPF( cuts, VH ):
3      V, H = VH
4      NC = len( cuts )
5      X = np.zeros( (NC,V*H), complex )
6      for i in range( NC ):
7          targ = np.zeros( VH )
8          vc, hc = cuts[i].shape
9          targ[V/2-vc/2:V/2+vc/2, H/2-hc/2:H/2+hc/2] = cuts[i] + 0
10         targ = ft.fft2( targ )
11         X[i] = targ.ravel()
12     cst = np.ones( NC )
13     filt = fpf.FPF( X, cst, 0.8 )
14     filt = ft.ifft2( filt.reshape( VH ) )
15     filt *= V*H
16     return filt
```

Code 12.15 Running the functions in the tachometer problem

```
>>> import tachfpf as tach
>>> bdata, cuts = tach.LoadTach( 'data/tach.png')
>>> filt = tach.MakeTachFPF( cuts, bdata.shape )
>>> fdata = crr.Correlate2D( bdata, filt )
```

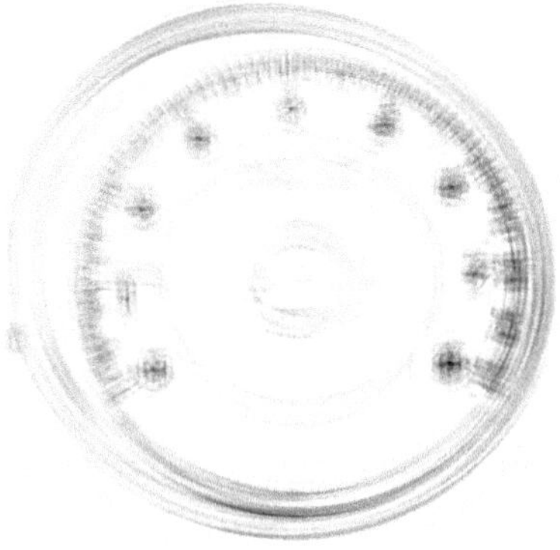

Figure 12.12 The negative of the correlation of the original image with the FPF. Darker pixels indicate a stronger response to the filter. Peaks exist at the center of each numeral.

The correlation of a shifted version of the input is shifted by the same amount as in

$$D_{\vec{v}}\mathbf{f}[\vec{x}] = (D_{\vec{v}}\mathbf{d}_i[\vec{x}]) \otimes \mathbf{h}[\vec{x}]. \tag{12.26}$$

The constrained correlation value is shifted by $\vec{v}$ away from the center of the frame. This effect is seen in the tachometer example. To create the filter, each target (a cutout of the numeral) was placed in the center of an image, and then these images were used to build the filter. The correlation of the tachometer image with the filter produced correlation spikes at the center of each of the numerals. In the training process, the target was in the center of the image, but in the application, the target was shifted by $\vec{v}$ from the center, and so the correlation peak was shifted by the same amount.

In the correlation in Figure 12.12, there are also some large signals that are off-target, particularly in the lower right region. This is quite common when the FPF is applied to a real image. One manner of reducing such false positives is to add new training images and rebuild the FPF. The constraint values of these new inputs are set to values that will then use these images as null-trainers of anti-trainers.

Multiple researchers have explored the effect of values other than 1 in the constraint vector. There are only three values that have been successfully deployed to alter the performance of the filter. These are 0, -1, and $e^{\imath\theta}$. Other values, such as 2, do not significantly alter the performance of the filter, particularly when applied to noisy images. A value of 0 would force the correlation value to be 0 at the center of the target.

In the tachometer example, there were nine training images and the constraint vector was $\vec{c} = (1,1,1,1,1,1,1,1,1,1)$. Now, the goal is reduce one false positive. For the sake of argument, consider the location of the false positive to be $\vec{z}$. A small region about the center of $\vec{z}$ is cut out and centered in a larger frame, $U_{\vec{w}}\square_{\vec{z}}\mathbf{b}[\vec{x}]$. This is added to the training set, but the correlation value is set to 0, thus making $\vec{c} = (1,1,1,1,1,1,1,1,1,1,0)$. The FPF is constructed, and the correlation signal of $\mathbf{b}[\vec{x}] \otimes \mathbf{h}[\vec{x}]$ at $\vec{z}$ become 0, thus eliminating the false positive at that location. For well-behaved systems, the magnitude of the correlation pixels surrounding $\vec{z}$ are also significantly suppressed. However, it is possible that other false positives can be created by this new filter.

Setting a constraint value to 0 is called *null-training*. Another option of suppressing the correlation value is to set the constraint element to -1. This is called *anti-training*. The value of the correlation signal at $\vec{z}$ is now negative, which further suppresses surrounding correlation values. The process of null-training is similar to the filtering ignoring the associated training image, and the anti-training is similar to rejecting the training image. The difference is that the latter still produces a significant magnitude in the output, which is useful in more complicated FPF combinations.

The final option is to set an element in the constraint vector to $e^{\iota\theta}$ which also has a magnitude of 1. The correlation is computed with the use of Fourier transforms, and therefore, the elements are inherently complex. The use of this type of constraint will alter the phase of the correlation values. In the tachometer example, there are correlation spikes at the target, but it is not possible to determine which target is present at each spike. A complex constraint value would alter the phase of the correlation value without altering the magnitude. Thus, by examining the phase, it is possible to determine which target is causing the spike.

These three alternate values for constraint elements are useful for reducing false positives or adding identifier qualities to the correlation output. Performance of the filter is also governed by the quality of the image and the types of frequencies that are in the signal. The construction of a filter may go through several trials before a competent filter is produced.

12.5.1.5 Dual FPFs

Consider a case in which there are several images, $\{\mathbf{a}_i[\vec{x}]; i \in 1,\ldots,N\}$, where i corresponds to a parameter such as rotation, scale, etc. The set of images are generated from a single image through the alteration of a parameter. The FPF is trained on selected values of i. The FPF constrains the correlation peak value for the training images but not for intermediate images. Thus, the value $\mathbf{a}_i[\vec{x}] \cdot \mathbf{h}[\vec{x}] = 1$ and $\mathbf{a}_{i+1}[\vec{x}] \cdot \mathbf{h}[\vec{x}] = 1$. For example, $\mathbf{a}_i[\vec{x}]$ could a 10° rotation of $\mathbf{a}[\vec{x}]$ and $\mathbf{a}_{i+1}[\vec{x}]$ could be a 15° rotation of $\mathbf{a}[\vec{x}]$. The filter would then be trained on these two images but not the images with intermediate rotations. Then, $\mathbf{a}_k[\vec{x}] \cdot \mathbf{h}[\vec{x}] < 1$, where $i < k < i+1$.

Figure 12.13 shows a graph for three different values of α. The horizontal axis corresponds to the changing parameter i, and the vertical axis corresponds to the peak correlation value. In this example, there are four selected training images designated by vertical lines, and the peak values are the constraint values. The intermediate values of i also produce a peak correlation value but the peak value is lower for values of i that are farther away from the training values. This produces a "telephone pole" effect in which the peak correlation value rises nearer the training images and falls farther away from them. There are three examples shown, and there is more sag in the plots for larger values of α. Since larger values of α are more discriminatory, the correlation response degrades faster as the input image differs more from the training images.

Consider a system in which an image has a changing parameter n. A few samples of the image with different values of n are selected for training, and all of the elements of the constraint vector are 1. The peak correlation will then be the same for all of these training images. This is depicted in Figure 12.14, where the horizontal line (constant correlation peak value) crosses the FPF response

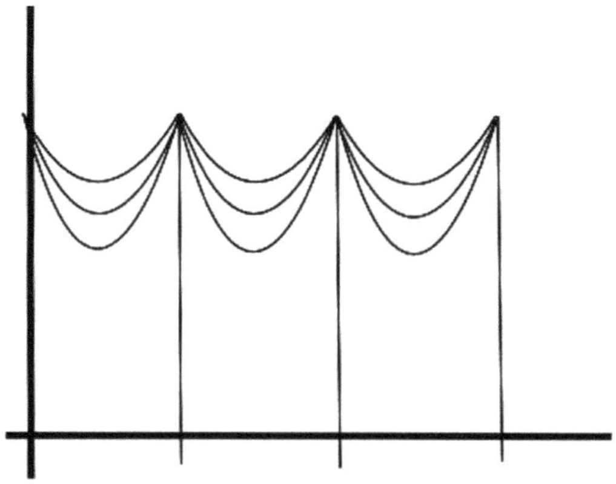

Figure 12.13 Response of the FPF for multiple values of α and for intermediate images.

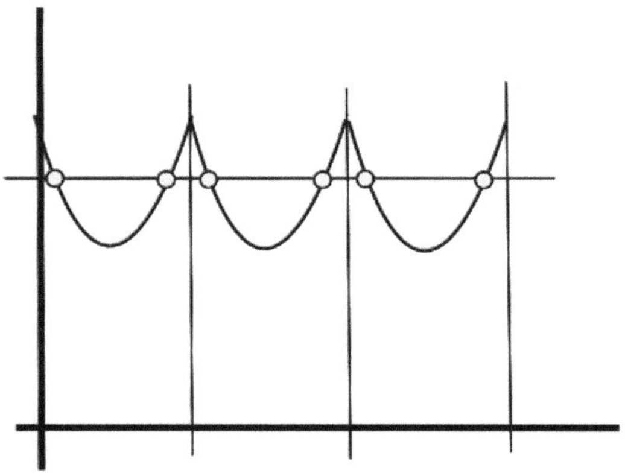

Figure 12.14 The response of three FPFs with a continuum of input images. The horizontal axis corresponds to a gradually changing parameter of the image, and the vertical axis corresponds to the correlation peak height. Vertical lines indicate parameter values used in training the FPF.

curve in multiple places. Thus, if an image with an undetermined value of n is correlated with the FPF, it is not possible to determine the value of n, but it is possible to determine a set of possible values for n. In this example, there are six such possible values.

A properly designed dual-FPF system can determine the value of n with just two filters. These are created with different training image parameters and possibly different values of α. A real case using rotation as the parameter for n is shown in Figure 12.15. Peak correlation values are plotted for two different FPF filters. The first filter is trained from a set of images with a large steps between values of n and the second is trained on a set of data with values of n that are closer together. The selection of the training images is not strict but the goal is that the two filters should react differently for all values of n. Excepting $n = 0$, the filters do not share the same training images or the same difference between two consecutive values of n.

The two FPFs are designated by $\mathbf{b}[\vec{x}]$ and $\mathbf{c}[\vec{x}]$. An input, $\mathbf{a}[\vec{x}]$, with an unknown value of n is then correlated with the two filters and the peak of each is gathered into a vector,

$$\vec{v} = \left\{ \bigvee(\mathbf{a}[\vec{x}] \otimes \mathbf{b}[\vec{x}]), \bigvee(\mathbf{a}[\vec{x}] \otimes \mathbf{c}[\vec{x}]) \right\}.$$

If designed correctly, there should be only one value of n that produces both of those peak values. These two values are compared to the values in the chart, and the one single value of n is determined as the location that best matches the values in $\vec{v}$. This horizontal location is then the estimate of the

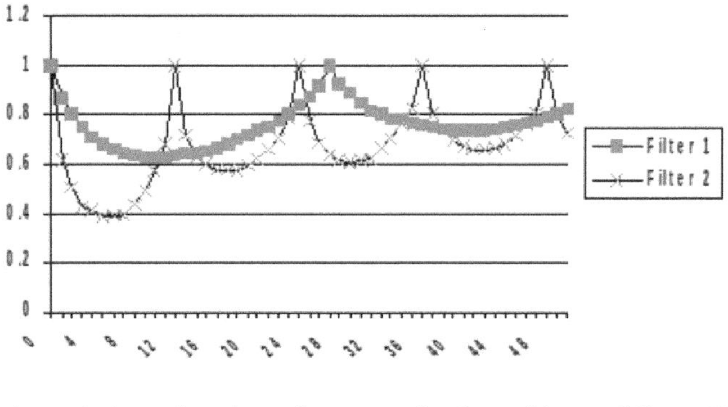

Figure 12.15 Several correlation values for an image varying in n with two different FPFs.

parameter associated with n. In this example, the horizontal value, n, was associated with rotation, and so through these two correlations, the rotation of the input can be estimated.

12.6 RESTRICTIONS OF CORRELATIONS

The FPF is a good tool but it does have limitations that are common among correlation filters. Correlation filters are inherently first order. This means that the filter responds to matches with the input vector. While it is possible to have the filter respond to a particular shape (such as the circle example), it is very difficult to build a filter that will respond to "a circle that does not have a rectangle nearby." In this latter case, the problem is second order and would need a stronger tool.

Correlation filters are translation invariant. In the case of the circle filter, it does not matter where the circle is located in the frame. The correlation spike will be located wherever the target is located. However, there are variances that cause performance decay in the filters. Consider the case of searching for a rectangle of a known ratio. A filter could be constructed to identify this rectangle, but if the query image has a rectangle of a different scale or a different rotation, then the filter may not provide a good response. The correlation filters are not rotation invariant or scale invariant.

There are a few solutions that have been proposed. One option is to construct an FPF with several rotations and scales of the target. Another solution is to transform the data into a different space, such as those created by a polar transformation reviewed in Section 6.7. The advantage of this transformation is that the scale and rotation now become linear shifts in the new space. Since the filters are translation invariant, filters built in the polar space are rotation and scale invariant. The disadvantage of this conversion is that the center of the object must be known, and this information may not be available in some applications.

12.7 SUMMARY

An image correlation is equivalent to the alteration of the image frequencies in accordance with frequencies from a second image. This second image in small kernel correlations has only a few elements. Kernels for enhancing edges or smoothing the input image are easily deployed. The computations for large kernel correlations use Fourier transforms, which reduces the computation time by orders of magnitude. A common implementation is to create a filter from an image that contains a target centered in the frame. The correlation of this with an input that contains the target will produce a correlation surface with a large spike at the location of the target.

A composite filter is created from multiple images and has a control factor that manages the trade-off between generalization and discrimination. This process can also be trained to ignore or reject some targets. Dual composite filters can quantify the changing parameters in a set of images even if that parameter is not well defined.

PROBLEMS

1. Create a vector, $\vec{v}$, which has 48 consecutive 0s, 32 consecutive 1s, and 48 consecutive 0s. Compute the autocorrelation of $\vec{v}$.
2. Create an image that is 256×256 and in the middle is a solid 32×32 square of pixels that have a value of 1. Compute the autocorrelation of this image.
3. Create two images both with a frame size of 256×256. In center of the first image, $\mathbf{a}[\vec{x}]$, create a solid circle of radius 64. In the center of the second image, $\mathbf{b}[\vec{x}]$, create a solid circle of radius 70. (Use **mgcreate.Circle** to generate these images.) Compute the correlation of $\mathbf{a}[\vec{x}]$ and $\mathbf{b}[\vec{x}]$.
4. Repeat the previous problem but replace the solid circles with annular rings of the given radii but with a line width of three pixels. Compute the correlation of $\mathbf{a}[\vec{x}]$ and $\mathbf{b}[\vec{x}]$.

5. Create image $\mathbf{a}[\vec{x}] = \mathbf{r}_{\vec{w};\vec{v}_1,\vec{v}_2} + \mathbf{r}_{\vec{w};\vec{v}_3,\vec{v}_4}$, where $\vec{w} = (256,256)$, $\vec{v}_1 = (112,80)$, $\vec{v}_2 = (144,112)$, $\vec{v}_3 = (112,144)$, and $\vec{v}_4 = (144,176)$. Extract a row of data as $\vec{x} = \mathbf{a}[128]$. Plot $\vec{x}$.
6. Repeat Problem 5, but compute $\mathbf{b}[\vec{x}] = \mathcal{S}_7\mathbf{a}[\vec{x}]$. For this smoothing, use the **ndimage.gaussian_filter** function for smoothing. Extract the row $\vec{x} = \mathbf{b}[128]$ and plot $\vec{x}$. Compare the results to Problem 1.
7. Repeat Problem 5, but compute $\mathbf{b}[\vec{x}] = \mathfrak{F}^{-1}X\left(\mathbf{m}[\vec{\omega}] \times X\mathfrak{F}\mathbf{a}[\vec{x}]\right)$, where $\mathbf{m}[\vec{\omega}]$ is a low-pass filter. Extract the row $\vec{x} = \mathbf{b}[128]$ and plot $\vec{x}$. Compare the results to Problem 2.
8. Repeat Problem 5, but compute $\mathbf{b}[\vec{x}] = \mathfrak{F}^{-1}X\left(\mathbf{m}[\vec{\omega}] \times X\mathfrak{F}\mathbf{a}[\vec{x}]\right)$, where $\mathbf{m}[\vec{\omega}]$ is a high-pass filter. Extract the row $\vec{x} = \mathbf{b}[128]$ and plot $\vec{x}$.
9. Create the kernel
$$\mathbf{b}[\vec{x}] = \frac{1}{25}\begin{pmatrix} 1 & 1 & 1 & 1 & 1 \\ 1 & 1 & 1 & 1 & 1 \\ 1 & 1 & 1 & 1 & 1 \\ 1 & 1 & 1 & 1 & 1 \\ 1 & 1 & 1 & 1 & 1 \end{pmatrix}.$$
Compute $\mathbf{c}[\vec{x}] = \mathbf{a}[\vec{x}] \otimes \mathbf{b}[\vec{x}]$, where $\mathbf{a}[\vec{x}]$ is a grayscale image of your choice. Explain what has occurred to the image.
10. Create the kernel
$$\mathbf{b}[\vec{x}] = \begin{pmatrix} \frac{1}{10} & 0 & \frac{1}{5} & 0 & \frac{1}{10} \\ \frac{1}{10} & 0 & \frac{1}{5} & 0 & \frac{1}{10} \\ \frac{1}{10} & 0 & \frac{1}{5} & 0 & \frac{1}{10} \\ \frac{1}{10} & 0 & \frac{1}{5} & 0 & \frac{1}{10} \\ \frac{1}{10} & 0 & \frac{1}{5} & 0 & \frac{1}{10} \end{pmatrix}.$$
Compute $\mathbf{c}[\vec{x}] = \mathbf{a}[\vec{x}] \otimes \mathbf{b}[\vec{x}]$, where $\mathbf{a}[\vec{x}]$ is a grayscale image of your choice. Explain has occurred to the image.
11. Create $\mathbf{a}[\vec{x}] = \mathbf{o}_{\vec{w},r}$ and $\mathbf{b}[\vec{x}] = \mathbf{o}_{\vec{w},r}$ where $\vec{w} = (256,256)$ and $r = 24$. Compute $\mathbf{c}[\vec{x}] = \mathbf{a}[\vec{x}] \otimes \mathbf{b}[\vec{x}]$. Display $\mathbf{c}[\vec{x}]$.
12. Create $\mathbf{a}[\vec{x}] = \mathbf{o}_{\vec{w},r_1}$ and $\mathbf{b}[\vec{x}] = \mathbf{o}_{\vec{w},r_1} - \mathbf{o}_{\vec{w},r_2}$, where $\vec{w} = (256,256)$, $r_1 = 24$, and $r_2 = 20$. Compute $\mathbf{c}[\vec{x}] = \mathbf{a}[\vec{x}] \otimes \mathbf{b}[\vec{x}]$. Display $\mathbf{c}[\vec{x}]$.
13. Create $\mathbf{a}[\vec{x}] = \mathbf{o}_{\vec{w},r_1} - \mathbf{o}_{\vec{w},r_2}$ and $\mathbf{b}[\vec{x}] = \mathbf{o}_{\vec{w},r_1} - \mathbf{o}_{\vec{w},r_2}$, where $\vec{w} = (256,256)$, $r_1 = 24$, and $r_2 = 20$. Compute $\mathbf{c}[\vec{x}] = \mathbf{a}[\vec{x}] \otimes \mathbf{b}[\vec{x}]$. Display $\mathbf{c}[\vec{x}]$.
14. Correlate the matrix in Equation (12.9) with the grayscale version of the image in "data/man.png" to get a result that is similar to that in Figure 12.2(b).
15. Set $\mathbf{a}[\vec{x}]$ to be the image in "data/man.png". Set $\mathbf{b}[\vec{x}]$ to be the kernel in Equation (12.9). Compute the correlation of the kernel on each of the color channels and then combine the results in a single RGB image.
16. Write the operator notation for the previous problem.
17. Create and FPF filter with $\alpha = 0$ from three images. Each image has a frame size of 256×256. The first image is a solid circle of radius 64, the second image is a solid square with side lengths of 100, and the third is a hollow rectangle of size 80×140 with a line width of five pixels. Show the real components of the FPF.
18. Repeat the previous problem with $\alpha = 0.5$.
19. Write the image operators for the previous problem.
20. Given an image $\mathbf{a}[\vec{x}]$ which has four small hollow squares evenly spaced and in a row horizontally. How many spikes are in $\mathbf{a}[\vec{x}] \otimes \mathbf{a}[\vec{x}]$?
21. Show that the correlation is shift invariant. Create an image $\mathbf{a}[\vec{x}] = \mathbf{r}_{\vec{w};\vec{v}_1,\vec{v}_2}$, where $\vec{w} = (256,256)$ and $\vec{v}_1$, and $\vec{v}_2$ create solid 32×32 square in the middle of the frame. Compute $\mathbf{c}[\vec{x}] = \mathbf{a}[\vec{x}] \otimes D_{\vec{v}}\mathbf{a}[\vec{x}]$, where $\vec{v} = (32,22)$. Find the location of the correlation peak with respect to the center of the frame. Compare that to $\vec{v}$.

22. Create $\mathbf{a}[\vec{x}] = 1 - Y(\texttt{data/geos.png})/255$. Create $\mathbf{b}[\vec{x}] = \mathbf{r}_{\vec{w},\vec{v}_1,\vec{v}_2}$, where $\vec{w} = (512,512)$, $\vec{v}_1 = (230,200)$, and $\vec{v}_2 = (280,310)$. Compute a phase only filter by, $\mathbf{c}[\vec{x}] = \mathfrak{F}^{-1}P^{-1}\left\{\begin{matrix}\varnothing \\ 1\end{matrix}\right\}P\mathfrak{F}\mathbf{b}[\vec{x}]$. Correlate this filter with the original to get $\mathbf{d}[\vec{x}] = \mathbf{a}[\vec{x}] \otimes \mathbf{c}[\vec{x}]$. Determine the location of the peak in $\mathfrak{R}\mathbf{d}[\vec{x}]$.

Part IV

Texture and Shape

13 Edge Detection

Shapes within an image are usually defined by their boundaries, and these boundaries can be detected or enhanced with edge detection or enhancement algorithms. Some texture measures (Section 16.3) are also reliant on the ability to detect edges. This chapter will review some (but certainly not all) of the common edge detection methods.

13.1 EDGES

Edges for grayscale images are generally defined as large or abrupt changes in intensity along a line or curve. These changes would be evident as large values in the first derivative of a signal. Since an image has at least two dimensions, the derivative has a slightly more complicated definition than in the one-dimensional case. A *hard edge* exists when the change is very abrupt perhaps only 2 or 3 pixels wide. A *soft edge* exists when the transition from bright to dark takes several pixels.

As there are several models for enhancing edges, the operator notation is initially generic. Given an image $\mathbf{a}[\vec{x}]$, the edge enhanced is generated via

$$\mathbf{b}[\vec{x}] = E_m \mathbf{a}[\vec{x}], \tag{13.1}$$

where E_m is the edge operator for model m. The following sections display a few of the possible models.

Perhaps, one of the easiest methods of detecting edges is to compare shifted versions of the same image. The Shift operator, $D_{\vec{x}}$, performs this shift, and it is associated with the **shift** function from the *scipy.ndimage* module. However, the function is designed to accommodate shifts that are fractions of a pixel width. In order to do this, the algorithm employs a spline fit that is third-order by default. This can cause some issues in some cases.

Consider the simple example shown in Code 13.1. Lines 1 and 2 create a 5×5 matrix with only one pixel set to 1. Line 3 shifts this matrix one pixel and so the result should also have one pixel set to 1. However, lines 6 and 7 show that there are 10 pixels that are greater than 0. The function uses the spline fit to compute the values of the pixels, and now, nine of the pixels have a nonzero value. These values are very small, but if a threshold was applied to the images as it is in line 6, then the error becomes enormous. So, for some applications, the user may choose to bypass the spline fit by setting the `order` to 0 as shown in line 8.

Code 13.1 Shifting a simple array

```
1  >>> adata = np.zeros((5,5))
2  >>> adata[2,2] = 1
3  >>> bdata = nd.shift( adata, (1,0))
4  >>> (adata>0).sum()
5  1
6  >>> (bdata>0).sum()
7  10
8  >>> bdata = nd.shift( adata, (1,0), order=0)
9  >>> (bdata>0).sum()
10 1
```

The vertical edges of an image can be enhanced by

$$\mathbf{b}[\vec{x}] = |\mathbf{a}[\vec{x}] - D_{\vec{v}} \mathbf{a}[\vec{x}]|, \tag{13.2}$$

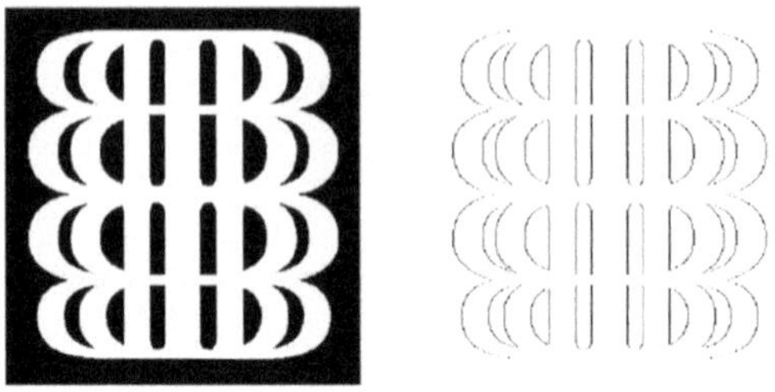

(a) (b)

Figure 13.1 (a) The original image and (b) the vertical edges.

Code 13.2 Extracting the vertical edges

```
1 >>> import imageio
2 >>> import scipy.ndimage as nd
3 >>> amg = imageio.imread('data/ib3logo.png', as_gray=True)
4 >>> bmg = abs( amg - nd.shift(amg,(0,1),order=0))
5 imageio.imsave('image.png', -bmg)
```

where $\vec{v}$ is (0,1) for a horizontal shift. The image in Figure 13.1(a) contains sharp edges in all directions and is used as $\mathbf{a}[\vec{x}]$. The negative of output $\mathbf{b}[\vec{x}]$ is shown in Figure 13.1(b), and notably, the horizontal edges are absent. Code 13.2 displays the steps. Equation (13.2) is performed in line 4.

Shifts in one direction extract edges that are perpendicular to the direction of the shift. Any value of $\vec{v}$ will produce an image with strong edges in one direction and missing edges in perpendicular direction. To capture edges in all directions, two opposing shifts are used as in

$$\mathbf{b}[\vec{x}] = |\mathbf{a}[\vec{x}] - D_{\vec{v}_1}\mathbf{a}[\vec{x}]| + |\mathbf{a}[\vec{x}] - D_{\vec{v}_2}\mathbf{a}[\vec{x}]|, \tag{13.3}$$

where $\vec{v}_1 \perp \vec{v}_2$.

This method is perhaps the simplest but not necessarily the most effective. The generic edge enhancement operator is E_m, where the m is the user-defined model. Thus, one definition is

$$\mathbf{b}[\vec{x}] = E_m\mathbf{a}[\vec{x}] = |D_{\vec{v}}\mathbf{a}[\vec{x}] - \mathbf{a}[\vec{x}]|. \tag{13.4}$$

13.2 THE SOBEL FILTERS

A slightly more sophisticated method of edge enhancement is the Sobel filter. This process is similar to the previous except that shift operations are replaced by small kernel correlations. These are two 3×3 matrices, which are convolved with the original image:

$$\mathbf{g_h}[\vec{x}] = \begin{bmatrix} -1 & 0 & 1 \\ -2 & 0 & 2 \\ -1 & 0 & 1 \end{bmatrix} \tag{13.5}$$

and

$$\mathbf{g_v}[\vec{x}] = \begin{bmatrix} -1 & -2 & -1 \\ 0 & 0 & 0 \\ 1 & 2 & 1 \end{bmatrix}. \tag{13.6}$$

The $\mathbf{g_h}[\vec{x}]$ matrix is sensitive to horizontal edges, and the $\mathbf{g_v}[\vec{x}]$ is sensitive to vertical edges. Given a grayscale image $\mathbf{a}[\vec{x}]$, the process correlates the matrices with the input image as

$$\mathbf{b}[\vec{x}] = \mathbf{a}[\vec{x}] \otimes \mathbf{g_h}[\vec{x}] \tag{13.7}$$

and

$$\mathbf{c}[\vec{x}] = \mathbf{a}[\vec{x}] \otimes \mathbf{g}_\mathbf{v}[\vec{x}]. \tag{13.8}$$

From these two results, it is possible to compute a single edge-enhanced image sensitive to edges in any direction:

$$\mathbf{d}[\vec{x}] = \sqrt{(\mathbf{b}[\vec{x}])^2 + (\mathbf{c}[\vec{x}])^2}, \tag{13.9}$$

and to create an image that indicates the direction of the edge,

$$\mathbf{t}[\vec{x}] = \tan^{-1}\left(\frac{\mathbf{c}[\vec{x}]}{\mathbf{b}[\vec{x}]}\right). \tag{13.10}$$

The *ndimage* module provides a Sobel filter allowing for easy implementation of this method. The two matrices $\mathbf{g_h}$ and $\mathbf{g_v}$ are produced by the function **ndimage.Sobel** with a different axis argument. This is shown in Code 13.3, where lines 3 and 4 convolve the Sobel matrix with the input. Line 5 creates the output by adding the absolute values of the first two arrays. The negative of result is shown in Figure 13.2.

The flow of the edges is obtained by Equation (13.10), and the result is shown in Figure 13.3(a). There are two methods for computing the arctangent in most programming languages. The **arctan2** function is sensitive to which quadrant the angle is in and provides a slightly different result as shown in Figure 13.3(b).

13.3 DIFFERENCE OF GAUSSIANS

Each of the Sobel filters are OCOS (on-center off-surround) shaped. This type of shape is typical of edge extraction filters. In the shift method, the edge value of a pixel was determined by the subtraction of just two pixels. In the Sobel method, the value was determined by the small kernel

Code 13.3 Using the **Sobel** function to create an edge enhancement

```
1 >>> fname = 'data/d88small.png'
2 >>> data = imageio.imread( fname )
3 >>> a = nd.Sobel(data+0.,axis=0 )
4 >>> b = nd.Sobel(data+0.,axis=1 )
5 >>> edj = abs(a) + abs(b)
```

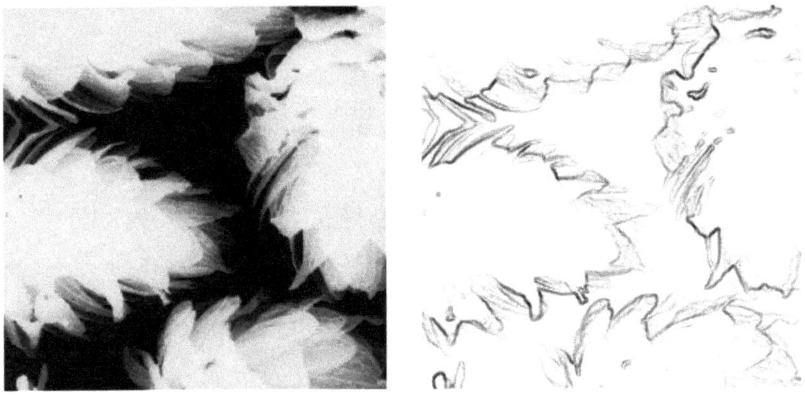

(a) (b)

Figure 13.2 (a) The original Brodatz image and (b) the edge-enhanced image.

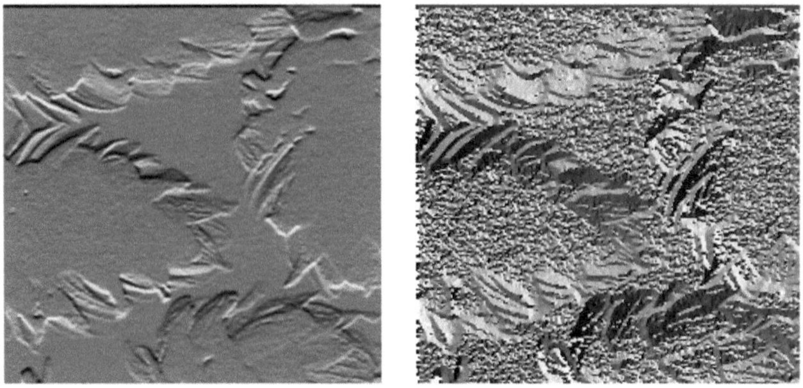

(a) (b)

Figure 13.3 (a) The result of Equation (13.10) using **arctan** and (b) the same equation using **arctan2**.

which included values from nine pixels. The method in this section is similar in nature, but there are more pixels involved in the determination of the edge value of a single pixel.

A Gaussian function with unit area is

$$y = \frac{1}{(2\pi)^{n/2}|\Sigma|^{-1}} \exp\left(-\frac{1}{2}(\vec{x}-\vec{\mu})\Sigma^{-1}(\vec{x}-\vec{\mu})^T\right), \tag{13.11}$$

where Σ is the covariance matrix and $\vec{\mu}$ locates the peak of the function at the center of the frame. The *Difference of Gaussian* (DoG) filter is the subtraction of a wide Gaussian function from a narrower one. The one-dimensional profile of a two-dimensional filter is shown in Figure 13.4. The filter in two-dimensions is circularly symmetric, and the profile from any angle is this profile.

The *scipy.ndimage* module provides the **gaussian_filter** function, which convolves a Gaussian profile with an input image. To coincide with this function, the Gaussian operator is the convolution of a Gaussian profile with the input image. Thus, the creation and application of the filter is simply

$$\mathbf{b}[\vec{x}] = G_{\vec{\sigma}}\mathbf{a}[\vec{x}], \tag{13.12}$$

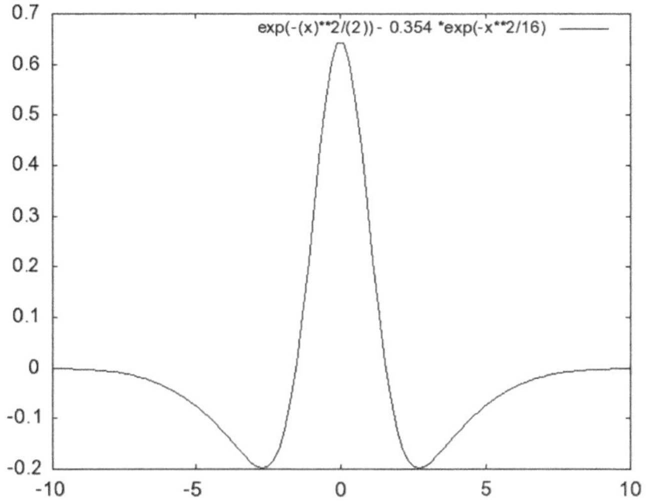

Figure 13.4 Profile of DoG filter.

Code 13.4 Application of the DoG filter

```
1 # edge.py
2 def DoGFilter( amg, sigma1, sigma2 ):
3     b1 = nd.gaussian_filter( amg, sigma1 )
4     b2 = nd.gaussian_filter( amg, sigma2 )
5     answ = b1 - b2
6     return answ
7
8 >>> amg = imageio.imread( 'D88small.png', as_gray=True)
9 >>> bmg = edge.DoGFilter( amg, 2,4)
```

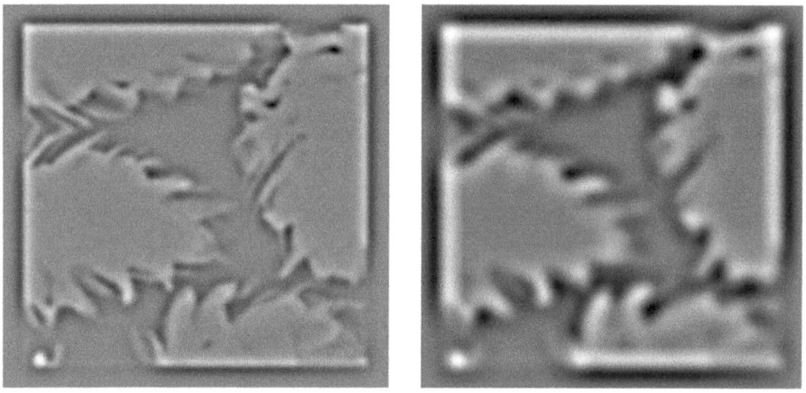

(a) (b)

Figure 13.5 (a) The result of the DoG filter with small sigma values and (b) the result of the DoG filter with larger sigma values.

where $\vec{\sigma}$ is the standard deviations along each axis. If σ is a scalar, then the same standard deviation is applied to all dimensions. This process has already been used in Code 10.8. Equation (13.12) will smooth an image. If the values of $\vec{\sigma}$ are dissimilar, then the smoothing will be greater along one dimension instead of the other.

The DoG filter is the difference between two applications of the Gaussian filter:

$$\mathbf{b}[\vec{x}] = G_{\sigma_1}\mathbf{a}[\vec{x}] - G_{\sigma_2}\mathbf{a}[\vec{x}], \tag{13.13}$$

where $\sigma_2 > \sigma_1$.

The application of the filter is shown in Code 13.4. The function **DoGFilter** receives the input image and two σ values. It then applies two filters to the original image in lines 3 and 4. The difference is computed in line 5. Lines 8 and 9 show the application of the filter. Two tests are shown in Figure 13.5 with filters of different size. Usually, the σ values are quite small.

13.4 CORNERS

As seen in the previous sections, the edges can be enhanced by several methods. Detecting corners is a little more complicated. The approach shown here is the *Harris detector*, which relies on edge detection methods and the eigenvalues of the image structure. The latter requires a more detailed explanation.

The first step in the Harris detector [23] is to create a *structure tensor*. These values are combined algebraically to produce an image, where the bright pixels are at the locations of corners in the image.

The structure tensor for this problem is computed by

$$\mathbf{C} = \begin{pmatrix} \mathcal{S}(E_h\mathbf{a}[\vec{x}])^2 & \mathcal{S}(E_{vh}\mathbf{a}[\vec{x}])^2 \\ \mathcal{S}(E_{vh}\mathbf{a}[\vec{x}])^2 & \mathcal{S}(E_v\mathbf{a}[\vec{x}])^2 \end{pmatrix}, \tag{13.14}$$

where $\mathcal{S}$ is the Smoothing operator and E_m is the Edge operator. The subscripts indicate the direction of the edges that are extracted. Thus, $E_v\mathbf{a}[\vec{x}]$ is the vertical edges, and $E_{vh}\mathbf{a}[\vec{x}]$ are at a 45° angle. The tensor **C** contains four numbers for each pixel, and these values are based upon the edge enhancements in various directions.

For each pixel, there is a 2×2 matrix, and the eigenvalues of this matrix are indicative of the type of flow in the image. If both eigenvalues are small, then there is no flow in either direction. If one of the eigenvalues is large and the other small, then there is a flow in a single direction which is a trait of an edge. If both eigenvalues are large, then there is flow in both directions and that is a characteristic trait of a corner.

The Harris measure is then a combination of the eigenvalues:

$$R = \det(C) - \alpha \texttt{trace}^2(C) = \lambda_1 \lambda_2 - \alpha(\lambda_1 + \lambda_2)^2. \tag{13.15}$$

Computing the determinant and trace for all of the small matrices can be done en masse as shown in the function **Harris** shown in Code 13.5. Lines 3 and 4 extract the edges using the Sobel function. Lines 7 through 9 perform the smoothing on the three different elements in the structure matrix. Line 10 computes the determinant for all matrices, and line 11 computes the trace for all matrices. These are combined in line 12 as prescribed to create the Harris measure.

The Harris detector is represented by $\mathcal{H}$ and so the operation is simply,

$$\mathbf{r}[\vec{x}] = \mathcal{H}\mathbf{a}[\vec{x}].$$

The creation of the Harris operator begins with the Sobel filters:

$$\mathbf{b}[\vec{x}] = (G_v\mathbf{a}[\vec{x}])^2$$

$$\mathbf{c}[\vec{x}] = (G_h\mathbf{a}[\vec{x}])^2$$

Code 13.5 The **Harris** function

```
1  # edge.py
2  def Harris( indata, alpha=0.2 ):
3      Ix = Sobel( indata, 0 )
4      Iy = Sobel( indata, 1 )
5      Ix2 = Ix**2;      Iy2 = Iy**2
6      Ixy = abs(Ix * Iy)
7      Ix2 = gaussian_filter( Ix2, 3 )
8      Iy2 = gaussian_filter( Iy2, 3 )
9      Ixy = gaussian_filter( Ixy, 3 )
10     detC = Ix2 * Iy2 - 2 * Ixy
11     trC = Ix2 + Iy2
12     R = detC - alpha * trC**2
13     return R
```

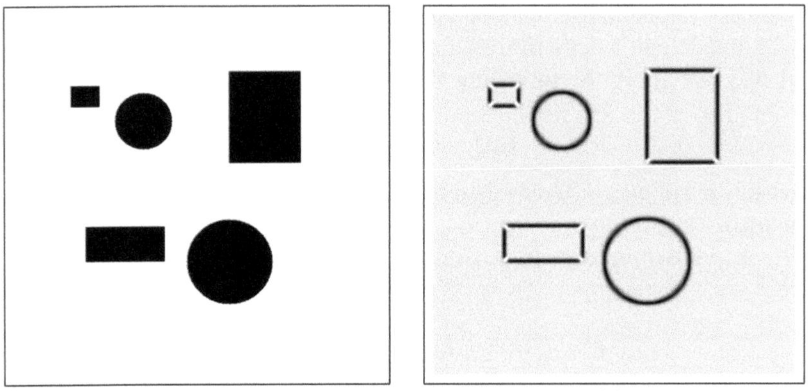
(a) (b)

Figure 13.6 (a) The original image, and (b) the response of the Harris detector.

Code 13.6 Applying the Harris detector to simple geometric shapes

```
>>> mg = imageio.imread( 'data/geos.png', as_gray=True)
>>> R = edge.Harris( mg )
```

$$\mathbf{d}[\vec{x}] = |\mathbf{b}[\vec{x}] \times \mathbf{c}[\vec{x}]|.$$

Each of these is smoothed:

$$\mathbf{f}_1[\vec{x}] = \mathcal{S}\mathbf{b}[\vec{x}]$$
$$\mathbf{f}_2[\vec{x}] = \mathcal{S}\mathbf{c}[\vec{x}]$$
$$\mathbf{f}_3[\vec{x}] = \mathcal{S}\mathbf{d}[\vec{x}].$$

The determinant is computed by:

$$\mathbf{g}[\vec{x}] = \mathbf{f}_1[\vec{x}] \times \mathbf{f}_2[\vec{x}] + 2\mathbf{f}_3[\vec{x}],$$

and the trace is computed by,

$$\mathbf{h}[\vec{x}] = \mathbf{f}_1[\vec{x}] + \mathbf{f}_2[\vec{x}].$$

The final result is then

$$\mathbf{r}[\vec{x}] = \mathcal{H}\mathbf{a}[\vec{x}] = \mathbf{g}[\vec{x}] - \alpha\mathbf{h}^2[\vec{x}].$$

A simple example uses an image with simple geometric patterns shown in Figure 13.6(a). Code 13.6 shows the steps to load the image and apply the filter. The result is shown in Figure 13.6(b). The rectangles have sharp corners, and reaction of these to the filter creates bright pixels in the output. Sharp edges that are not corners have been suppressed as seen by the dark pixels. The circles, of course, do not produce any positive response by the filter.

PROBLEMS

1. Create a binary valued image that contains a square annulus (use the function **SquareAnnulus** with the default values from the module *mgcreate* in the provided software). This image is $\mathbf{a}[\vec{x}]$ and apply the following subtraction to get a version of an edge-enhanced image,

$$\mathbf{b}[\vec{x}] = |\mathbf{a}[\vec{x}] - D_{(1,1)}\mathbf{a}[\vec{x}]|.$$

2. Create a binary valued image that contains a square annulus (use the function **SquareAnnulus** with the default values from the module *mgcreate* in the provided software). This image is $\mathbf{a}[\vec{x}]$ and apply the following subtraction to get a version of an edge-enhanced image,

$$\mathbf{b}[\vec{x}] = \mathbf{a}[\vec{x}] - D_{(1,1)}\mathbf{a}[\vec{x}].$$

Note that this is slightly different than the previous problem in that the absolute value is not computed.

3. Use $\mathbf{a}[\vec{x}] = Y(\texttt{data/bananas.jpg})$. Compute the following:

$$\mathbf{b}[\vec{x}] = E_m \begin{Bmatrix} \varnothing \\ \varnothing \\ 1 \end{Bmatrix} \mathcal{L}_{HSV}\mathbf{a}[\vec{x}]$$

$$\mathbf{c}[\vec{x}] = \begin{Bmatrix} 1 \\ \varnothing \\ \varnothing \end{Bmatrix} \mathcal{L}_{HSV}\mathbf{a}[\vec{x}]$$

$$\mathbf{d}[\vec{x}] = \begin{Bmatrix} \varnothing \\ 1 \\ \varnothing \end{Bmatrix} \mathcal{L}_{HSV}\mathbf{a}[\vec{x}]$$

$$\mathbf{f}[\vec{x}] = \mathcal{L}_{RGB} \begin{Bmatrix} \mathbf{c}[\vec{x}] \\ \mathbf{d}[\vec{x}] \\ 1 \end{Bmatrix}$$

$$\mathbf{g}[\vec{x}] = \alpha\mathbf{f}[\vec{x}] + (1-\alpha)\mathbf{b}[\vec{x}] \; / \; \bigwedge \mathbf{b}[\vec{x}]$$

4. Create a binary valued image that contains a square annulus (use the function **SquareAnnulus** with the default values from the module *mgcreate* in the provided software). This image is $\mathbf{a}[\vec{x}]$ and apply the following subtraction to get a version of an edge-enhanced image,

$$\mathbf{b}[\vec{x}] = |\mathbf{a}[\vec{x}] - D_{(2,-2)}\mathbf{a}[\vec{x}]|.$$

5. Write the operator notation that receives a gray image $\mathbf{a}[\vec{x}]$, and then smooth the edge-enhanced version of this image.

6. Use $\mathbf{a}[\vec{x}] = Y(\texttt{data/bananas.jpg})$, and apply the Sobel edge enhancement algorithm to compute the following:

$$\mathbf{b}[\vec{x}] = E_m \begin{Bmatrix} 1 \\ \varnothing \\ \varnothing \end{Bmatrix} \mathcal{L}_{\text{RGB}}\mathbf{a}[\vec{x}]$$

$$\mathbf{c}[\vec{x}] = E_m \begin{Bmatrix} \varnothing \\ 1 \\ \varnothing \end{Bmatrix} \mathcal{L}_{\text{RGB}}\mathbf{a}[\vec{x}]$$

$$\mathbf{d}[\vec{x}] = E_m \begin{Bmatrix} \varnothing \\ \varnothing \\ 1 \end{Bmatrix} \mathcal{L}_{\text{RGB}}\mathbf{a}[\vec{x}]$$

$$\mathbf{f}[\vec{x}] = \mathbf{b}[\vec{x}] + \mathbf{c}[\vec{x}] + \mathbf{d}[\vec{x}]$$

7. Consider $\mathbf{b}[\vec{x}] = E_m\mathbf{a}[\vec{x}]$, where m uses the **DerivEdge** function. The second argument to this function is `(k,k)`, where `k` is the amount of shift equal in both dimensions. Which value of `k` produces the maximum pixel value in $\mathbf{b}[\vec{x}]$? Why?

8. Use the **Checkerboard** function in the provided *mgcreate* module to create a checkerboard pattern $\mathbf{a}[\vec{x}]$. Using the **DerivEdge** function, compute the edge-enhanced image $\mathbf{b}[\vec{x}] = E_m\mathbf{a}[\vec{x}]$ with a shift of (1,1). Repeat the same experiment with a shift of (32,32). Explain why the outputs are different.
9. Given an input image $\mathbf{a}[\vec{x}]$ from the **Homeplate** function in the *mgcreate* module. Compute $\mathbf{b}[\vec{x}]$ as
$$\mathbf{b}[\vec{x}] = E_m\mathbf{a}[\vec{x}]$$
where m is the Sobel edge enhancement algorithm. Create $\mathbf{c}[\vec{x}]$ as,
$$\mathbf{c}[\vec{x}] = \Gamma_{5.5}\mathbf{b}[\vec{x}].$$
10. Use $\mathbf{a}[\vec{x}] = Y(\texttt{data/bananas.jpg})$. Compute the following:
$$\mathbf{b}[\vec{x}] = E_m \left\{ \begin{matrix} 1 \\ \varnothing \\ \varnothing \end{matrix} \right\} \mathcal{L}_{\text{YUV}}\mathbf{a}[\vec{x}]$$
$$\mathbf{c}[\vec{x}] = E_m \left\{ \begin{matrix} \varnothing \\ 1 \\ \varnothing \end{matrix} \right\} \mathcal{L}_{\text{YUV}}\mathbf{a}[\vec{x}]$$
$$\mathbf{d}[\vec{x}] = E_m \left\{ \begin{matrix} \varnothing \\ \varnothing \\ 1 \end{matrix} \right\} \mathcal{L}_{\text{YUV}}\mathbf{a}[\vec{x}]$$
$$\mathbf{f}[\vec{x}] = \mathbf{b}[\vec{x}] + \mathbf{c}[\vec{x}] + \mathbf{d}[\vec{x}]$$
11. Use $\mathbf{a}[\vec{x}] = Y(\texttt{data/bananas.jpg})$. Compute the following:
$$\mathbf{b}[\vec{x}] = E_m \left\{ \begin{matrix} \varnothing \\ \varnothing \\ 1 \end{matrix} \right\} \mathcal{L}_{\text{HSV}}\mathbf{a}[\vec{x}]$$
$$\mathbf{c}[\vec{x}] = \left\{ \begin{matrix} 1 \\ \varnothing \\ \varnothing \end{matrix} \right\} \mathcal{L}_{\text{HSV}}\mathbf{a}[\vec{x}]$$
$$\mathbf{d}[\vec{x}] = \left\{ \begin{matrix} \varnothing \\ 1 \\ \varnothing \end{matrix} \right\} \mathcal{L}_{\text{HSV}}\mathbf{a}[\vec{x}]$$
$$\mathbf{f}[\vec{x}] = \mathcal{L}_{\text{RGB}} \left\{ \begin{matrix} \mathbf{c}[\vec{x}] \\ \mathbf{d}[\vec{x}] \\ \varnothing \end{matrix} \right\}$$
$$\mathbf{g}[\vec{x}] = \alpha\mathbf{f}[\vec{x}] + (1-\alpha)\mathbf{b}[\vec{x}],$$
where α is a value between 0.0 and 1.0. Choose $\alpha = 0.5$ for this problem.
12. Use $\mathbf{a}[\vec{x}] = Y(\texttt{data/bananas.jpg})$, and the Sobel edge enhancement model to compute the following:
$$\mathbf{b}[\vec{x}] = \mathbf{a}[\vec{x}] + E_m\mathcal{L}_L\mathbf{a}[\vec{x}].$$

14 Hough Transforms

The Hough transform is designed to find particular shapes at any location and any orientation within an image. This chapter will consider the detection of lines and annular rings. Hough transformations for other shapes are possible, but, generally, other approaches are preferred. For example, a Hough transform for a circle is the same as a set of correlations with different sized rings. The correlation method is faster and with the use of composite filtering can become even more efficient. Thus, the Hough transform is generally applied to detect simple line shapes.

14.1 DETECTION OF A LINE

Consider an image that contains only a long straight line on a black background. Each pixel is identified by its (x, y) location. The Hough transform maps these points into a new space according to

$$r = x\cos\theta + y\sin\theta, \tag{14.1}$$

where (x, y) are the locations of any point in the input space and (r, θ) is the new location in the output space. This is not a polar transformation and in fact is not a one-to-one mapping. A single point in the input space actually maps to a series of points in the output space. An example is shown in Figure 14.1. The vertical axis in this image is r, and the horizontal image is θ, and as seen, a single point in the input space creates a curve in the output space. Through this single point, there are infinite numbers of lines that can pass through it. This output curve represents all of these possibilities were r is the distance from the origin to the perpendicular point of the line and θ is the angle of the line.

The operator for the Hough transform is H for a line transform and H_O for a circular transform. Thus, the expression to convert an input image $\mathbf{a}[\vec{x}]$ is

$$\mathbf{b}[\vec{y}] = H\mathbf{a}[\vec{x}]. \tag{14.2}$$

The function **LineHough** shown in Code 14.1 performs this transform. It receives two inputs with `data` being the input data and `gamma` as a threshold. In some cases, the data can have low noise values and these are discarded by the threshold value. The output space is created in line 5, and the height depends on the size of the image size. A larger image produces a largervalue of r. However, the horizontal range is fix at 180 pixels so that each column corresponds to 1 degree of rotation. The process starts by creating a replica of the input in line 6. Each point in `work` above, a threshold is considered in the loop starting at line 10. Line 15 finds the location of the largest pixel value. Lines 17 through 20 perform the transformation. Line 21 removes this point from consideration. The iterations continue until there are no pixels in `work` that are above the threshold. The process of creating Figure 14.1 is shown in Code 14.2. The input image is created in lines 4 and 5 and is simply an image with a single point in the middle. Line 6 calls **LineHough**, and the figure shows the negative of the image so that the curve is shown in black.

Consider a case in which the input space has two points instead of one. Each creates a curve as shown in Figure 14.2. However, two points also define a line and so there is one point where both curves intersect. This point defines a line where r is the distance to the original and ρ is the angle of the line.

The creation of this image is shown in Code 14.3. A second point is added and **LineHough** is called again. Code 14.4 shows the case where the input is a perfect line. Each point creates a curve but since all of the input points are on a line, then all of the output curves intersect at a single point.

The location of the output intersection is sensitive to the input line's orientation and location. Code 14.5 shows the case were a line is shifted in space. The Hough transform of the original

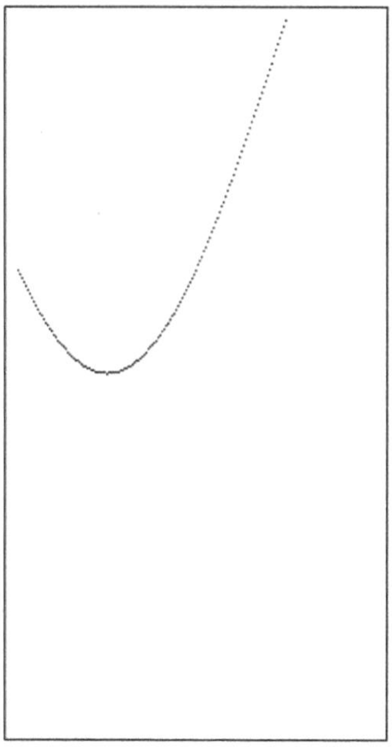

Figure 14.1 The output mapping for a single point in the input space.

Code 14.1 The **LineHough** function

```
# hough.py
def LineHough( data, gamma ):
    V,H = data.shape
    R = int(np.sqrt( V*V + H*H ))
    ho = np.zeros( (R,90), float ) # Hough space
    work = data + 0
    ok = 1
    theta = np.arange( 90 )/180. * np.pi
    tp = np.arange( 90 ).astype(float ) # theta for plotting
    while ok:
        mx = work.max()
        if mx < gamma:
            ok = 0
        else:
            v,h = divmod( work.argmax(), H )
            y = V-v; x = h
            rho = x * np.cos( theta ) + y*np.sin(theta)
            for i in range( len( rho )):
                if 0 <= rho[i] < R and 0<=tp[i]<90:
                    ho[int(rho[i]),int(tp[i])] += mx
            work[v,h] = 0
    return ho
```

Code 14.2 Creating Figure 14.1

```
>>> import numpy as np
>>> import imageio
>>> import hough
>>> adata = np.zeros( (256,256) )
>>> adata[128,128] = 1
>>> bdata = hough.LineHough( adata, 0.5 )
>>> imageio.imsave('figure.png', -bdata)
```

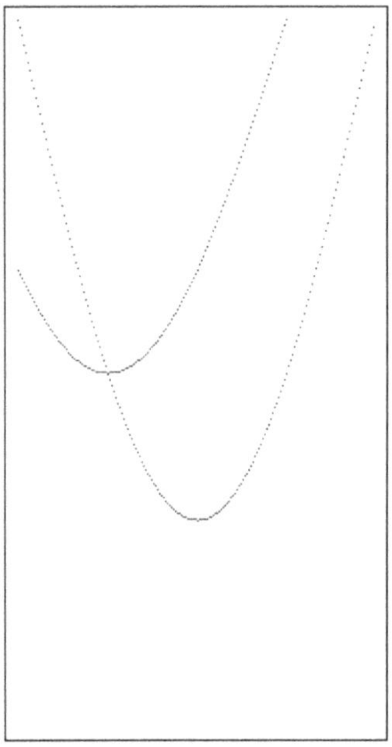

Figure 14.2 The Hough transform of a single point.

Code 14.3 Creating Figure 14.2

```
>>> adata[0,0] = 1
>>> bdata = hough.LineHough( adata, 0.5 )
>>> imageio.imsave('figure.png', -bdata)
```

Code 14.4 Running the Hough transform on an image with a line

```
>>> for i in range( 256 ):
        adata[i,i] = 1
>>> bdata = hough.LineHough( adata, 0.5 )
>>> imageio.imsave('figure.png', -(np.sqrt(bdata)) )
```

line is shown in Figure 14.3(a) and the Hough transform of the line shifted in location is shown in Figure 14.3(b). The location of the peak has shifted downwards. However, there is no horizontal displacement as the angle of the input line is unchanged. The input line is merely shifted and that changes only r, which is displayed as a vertical shift in the output.

Code 14.5 The Hough transform applied to a different image

```
>>> adata = np.zeros( (256,256) )
>>> for i in range( 205 ):
        adata[i,i+50] = 1
>>> bdata = hough.LineHough( adata, 0.5 )
>>> imageio.imsave('figure.png',-(np.sqrt(bdata)) )
```

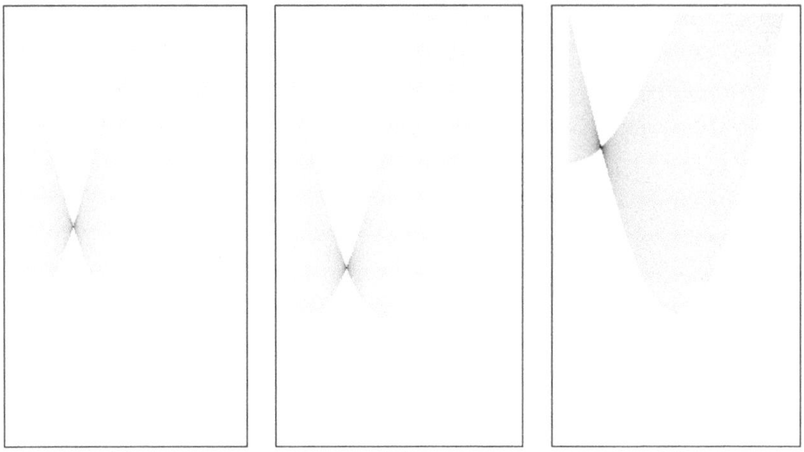

(a) (b) (c)

Figure 14.3 (a) The Hough transform for a straight line, (b) the Hough transformation for a line in a different location, and (c) the Hough transform for a line in a different location and orientation.

Code 14.6 Creating a line that is at a different orientation

```
>>> adata = np.zeros( (256,256) )
>>> for i in range( 256 ):
        adata[i,i*0.5] = 1
>>> bdata = hough.LineHough( adata, 0.5 )
>>> imageio.imsave('figure.png', -(np.sqrt(bdata)) )
```

Code 14.6 creates a line at different location and a different orientation. The output is shown in Figure 14.6(c), and as seen, the intersection point has shifted in both the vertical and horizontal dimensions because the line has a different position and orientation than its predecessor.

14.2 DETECTION OF A CIRCLE

The Hough transform can be applied to any shape, but such an operation may increase the output dimensionality. For example, a circle has a location but also has a radius; thus, there is an extra dimension to accommodate the additional parameter.

Consider the Hough transform for a specific radius R. A single point in the input space creates a circle of radius R in the output space. Thus, three points that lie on a circle of radius R in the input space will create three intersecting rings in the output space as shown in Figure 14.4. The location of the intersection is indicative of the location of the circle in the input space.

If the Hough transform radius does not match the radius of the input circle, then the circles do not intersect a single location. Instead, there is a ring where intersections occur, and this ring

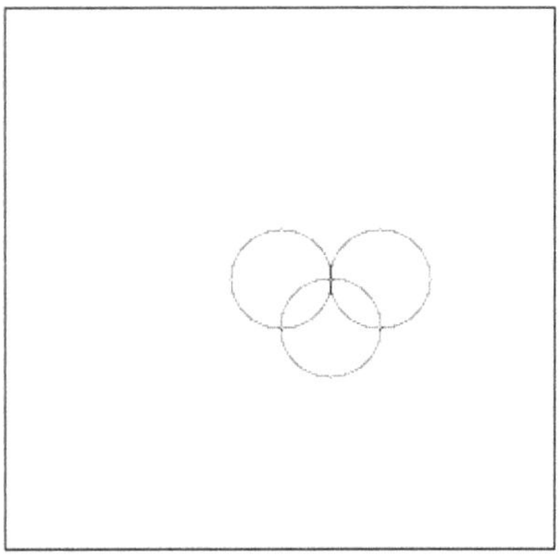

Figure 14.4 The output mapping for a three points that lie of a circle of radius R.

increases in width as the difference between the transform radius and input radius increase. Thus, in a full Hough transform that considers all radii, the output will be two opposing cones that expand in opposite directions in the third dimension of the output space.

An input with a single ring has several points, each of which create a ring in the output space. If the radius of the ring matches the radius of the transform plane, then all of the output rings intersect as shown in Figure 14.5.

The output space for a single radius R is a two-dimensional image which is the correlation of the input image with a ring of radius R. The three dimensional image has several planes which are the correlations with rings of various radii. So, the computation of this output space does not require a new function.

The operator notation is

$$\mathbf{b}[\vec{y}] = H_0\mathbf{a}[\vec{x}]. \tag{14.3}$$

As an example, an image is created using the **RandomRings** function, and an example is shown in Figure 14.6(a). A couple of the rings have a radius near 36, and so a new image is created with the **Ring** function that creates a ring with that radius. The output is shown in Figure 14.6(b), which is just one plane from the full Hough transform. The image shows two peaks, which are indicative of the existence of two rings with the specified radius. The full Hough transform would correlate the

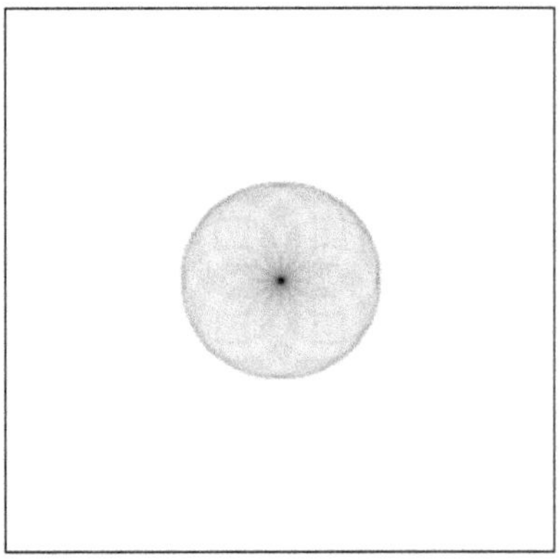

Figure 14.5 The output mapping for an input ring.

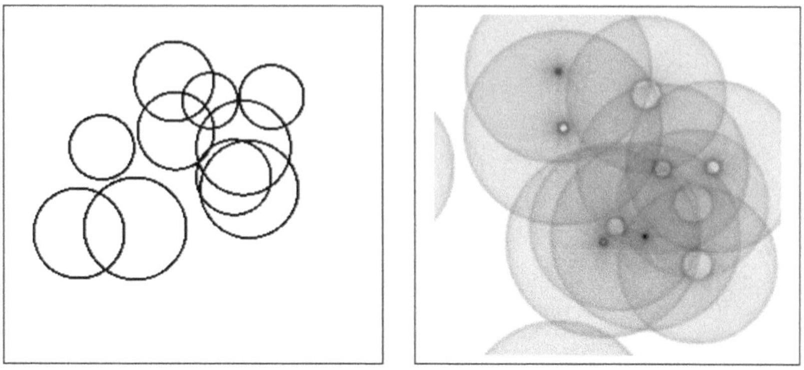

(a) (b)

Figure 14.6 (a) The input image and (b) the Hough transform for a single radius.

Code 14.7 Circle Hough transform applied to multiple rings

```
>>> adata = mgcreate.RandomRings( (256,256), 10, 20, 40 )
>>> cdata = mgcreate.Ring( (256,256), (128,128), 36 )
>>> ddata = crr.Correlate( adata, cdata )
>>> imageio.imsave('figure.png',-(np.sqrt(abs(ddata))) )
# Rotate image
```

image with rings with radii from large to small. Code 14.7 shows the steps to create the two shown images.

14.3 APPLICATION

Consider the image shown in Figure 14.7(a), which shows a man holding a cane. The goal is to detect the presence of a brown cane. This is performed in two steps with the first isolation brown pixels (Figure 14.7(b)) and the second using a Hough transform to determine, if there are brown pixels that form a line.

Since color will be used to isolate pixels, the image is converted to the YIQ color space. A few points on the cane were selected, and the values of the I and Q values are shown in Table 14.1. These help determine the range of values for the color of the cane. In the first step, pixels that have an I value between 12 and 20 and a Q value between 3.5 and 6.5 will be set to 1. This will include most of the pixels on the cane as well as other brown pixels mostly from the man's skin.

The entire algorithm is

$$\mathbf{b}[\vec{y}] = H \prod_{\mathcal{L}} \left\{ \begin{matrix} \varnothing \\ 12.0 \\ 3.5 \end{matrix} \right\} < \mathcal{L}_{\text{YIQ}} \mathbf{a}[\vec{x}] < \left\{ \begin{matrix} \varnothing \\ 20.0 \\ 6.5 \end{matrix} \right\}. \tag{14.4}$$

The input image is $\mathbf{a}[\vec{x}]$, which is converted to YIQ color space. The Y channel is not used, and lower and upper thresholds are applied to the other two channels. These two binary images are multiplied together so that only the pixels that pass both thresholds survive. The linear Hough transform is applied, and the output image is shown in Figure 14.8. As seen, there is a peak in the output space which indicates the presence of the line. The peak is in column 7, which corresponds to the angle of the cane which is 7° from the vertical.

(a)

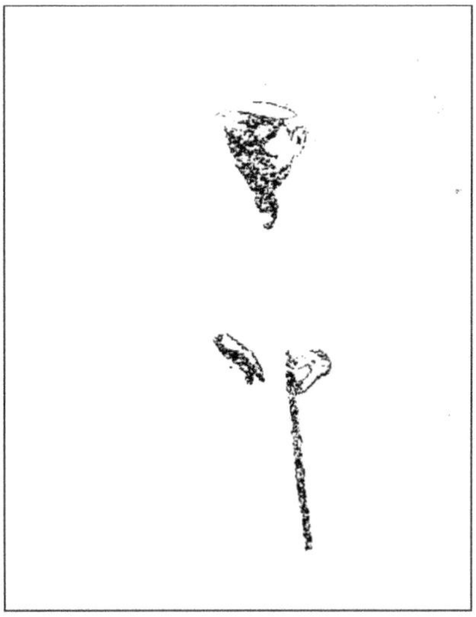

(b)

Figure 14.7 (a) The input image and (b) the isolated brown pixels.

Table 14.1
Data from Selected Pixels

Coordinate	I	Q
(400,310)	13.13	5.52
(452,314)	12.81	5.84
(511,322)	14.33	5.95
(547,327)	19.0	4.30
(573,332)	19.6	4.00
(593,330)	12.30	5.01

The entire process is shown in Code 14.8. The **colorsys.rgb_to_yiq** function converts one RGB pixel to the YIQ format. Using the **numpy.vectorize** function provides a new process that will apply the function to all pixels as in line 4. Lines 5 through 8 show the steps to obtain the I and Q values from one pixel. This step was repeated for all selected pixels in Table 14.1. Line 9 applies the threshold, and line 10 applies the Hough transform. The peak of the resultant image is found to determine the location in Hough space of the indicative signal. Thus, the presence, location, and orientation of the brown cane are determined.

14.4 SUMMARY

The Hough transform is designed to find specific geometric shapes such as a line or a curve. By far, the most common use is to detect lines in an image. The Hough transform can find lines of various thickness, any orientation, and any length above a minimum. The Hough output is (r, θ) space in which a curve is draw for every point from the line in the input space. For a perfectly straight, thin line, all of the curves intersect at one point, which provides information as to the location and angle of the line. Lines with imperfections in linearity or thickness will create overlapping curves in a small region, which still provides information as to the nature of the line in the image. The chapter used a real image example to demonstrate the effectiveness of the Hough transform.

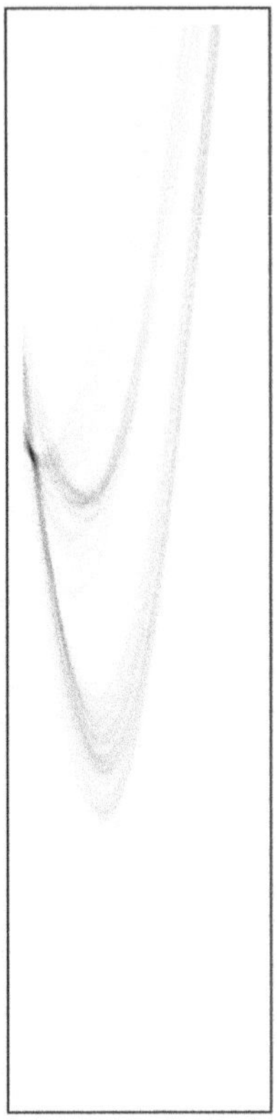

Figure 14.8 The output Hough space for the sitting man.

Code 14.8 The detection of the cane

```
>>> import colorsys
>>> adata = imageio.imread( 'data/man.png')
>>> rgb_to_yiq = np.vectorize( colorsys.rgb_to_yiq )
>>> bdata = rgb_to_yiq( adata[:,:,0], adata[:,:,1], adata[:,:,2] )
>>> bdata[1][400,310]
13.1336
>>> bdata[2][400,310]
5.5232
>>> cdata = (bdata[1]>12)*(bdata[1]<20) * (bdata[2]>3.5)*(bdata[2]<6.5)
>>> ddata = hough.LineHough( cdata, 0.5 )
>>> ddata.shape
(846, 180)
>>> divmod( ddata.argmax(), 180)
(342, 7)
```

PROBLEMS

1. Create $\mathbf{a}[\vec{x}] = \mathbf{r}_{\vec{w};\vec{v}_1,\vec{v}_2}$ where $\vec{w} = (512, 512)$, $\vec{v}_1 = (200, 254)$, and $\vec{v}_2 = (300, 258)$. Compute $\mathbf{b}[\vec{y}] = H\mathbf{a}[\vec{x}]$. Find the location of the peak in $\mathbf{b}[\vec{y}]$.
2. Create $\mathbf{a}[\vec{x}] = \mathbf{r}_{\vec{w};\vec{v}_1,\vec{v}_2}$, where $\vec{w} = (512, 512)$, $\vec{v}_1 = (200, 198)$, and $\vec{v}_2 = (300, 202)$. Compute $\mathbf{b}[\vec{y}] = H\mathbf{a}[\vec{x}]$. Find the location of the peak in $\mathbf{b}[\vec{y}]$.

3. Create $\mathbf{a}[\vec{x}] = \mathbf{r}_{\vec{w};\vec{v}_1,\vec{v}_2}$, where $\vec{w} = (512,512)$, $\vec{v}_1 = (200,254)$, and $\vec{v}_2 = (300,258)$. Compute $\mathbf{b}[\vec{y}] = H\mathcal{R}_\theta \mathbf{a}[\vec{x}]$, where $\theta = 25°$.
4. Create $\mathbf{a}[\vec{x}] = \mathbf{r}_{\vec{w};\vec{v}_1,\vec{v}_2} + \mathbf{r}_{\vec{w};\vec{v}_3,\vec{v}_4}$, where $\vec{w} = (512,512)$, $\vec{v}_1 = (200,100)$, $\vec{v}_2 = (300,104)$, $\vec{v}_3 = (200,300)$, and $\vec{v}_4 = (300,304)$. Compute $\mathbf{b}[\vec{y}] = H\mathbf{a}[\vec{x}]$. Find the location of the two peaks in $\mathbf{b}[\vec{y}]$.
5. Create $\mathbf{a}[\vec{x}] = \mathbf{r}_{\vec{w};\vec{v}_1,\vec{v}_2}$, where $\vec{w} = (512,512)$, $\vec{v}_1 = (200,200)$, and $\vec{v}_2 = (300,300)$. Compute $\mathbf{b}[\vec{y}] = H\mathbf{a}[\vec{x}]$. Explain why there are not sharp peaks in $\mathbf{b}[\vec{y}]$.
6. Create $\mathbf{a}[\vec{x}] = \mathbf{r}_{\vec{w};\vec{v}_1,\vec{v}_2}$, where $\vec{w} = (512,512)$, $\vec{v}_1 = (200,254)$, and $\vec{v}_2 = (300,258)$. From that create, $\mathbf{b}[\vec{x}] = \mathbf{a}[\vec{x}] + \mathcal{R}_{20°}\mathbf{a}[\vec{x}] + \mathcal{R}_{40°}\mathbf{a}[\vec{x}] + \mathcal{R}_{80°}\mathbf{a}[\vec{x}]$. Compute $\mathbf{c}[\vec{y}] = H\mathbf{b}[\vec{x}]$. List the location of all major peaks.
7. Use $\mathbf{a}[\vec{x}] = Y(\texttt{data/pens.png})$ (Figure 15.1). In this image, there are three items that are linear (two pens and a ruler). Using a Hough transform, isolate the pens using color thresholds and apply a Hough transform. Find the two peaks that are associated with the pens.
8. Use $\mathbf{a}[\vec{x}] = Y(\texttt{data/ib3logo.png})$. Compute $\mathbf{b}[\vec{y}] = H\mathbf{a}[\vec{x}]$. How many peaks in $\mathbf{b}[\vec{y}]$ are associated with the vertical lines in $\mathbf{a}[\vec{x}]$?
9. Create an image with frame size (512,512). Place in this image a solid circle with a radius of 50 at location (200,200). Place another solid circle with a radius of 50 at location (350,350). Erase the left half of this second circle. Compute the Hough circle transform for a radius of 50.
10. Write the operator notation for the previous problem
11. Write a program that uses the Hough circle transform to detect the presence of the circle about the clock face in the image "data/clock.png."

15 Noise

Noise is the term applied to artifacts in the image that inhibit the ability to detect the targets. Noise can come in many forms, although the term is most commonly used for variations in intensity from sources other than the objects in the image. This chapter will review some types of noise and the methods by which the noise can be reduced.

15.1 RANDOM NOISE

Random noise is the addition of random values as in

$$\mathbf{b}[\vec{x}] = \mathbf{a}[\vec{x}] + \mathbf{n}[\vec{x}], \tag{15.1}$$

where $\mathbf{n}[\vec{x}]$ is an array that has the same frame size as the original image $\mathbf{a}[\vec{x}]$. The range of the values in $\mathbf{n}[\vec{x}]$ may depend on the image collecting system and/or the values of the associated pixels in $\mathbf{a}[\vec{x}]$. In the early days of couple-charged device (CCD) cameras, a bright intensity on one pixel tended to bleed over into several pixels in the same row or column. Thus, the noise was also dependent on pixel intensities of the surrounding pixels.

Gaussian noise is similar in theory except that the values in $\mathbf{n}[\vec{x}]$ are governed by a Gaussian distribution. The values are then random but within the guidelines of a distribution, and so the effect has a similar grainy appearance to that of random noise.

Consider the image shown in Figure 15.1, which consists of a student's lab work as they were tasked to use their smart phone in an pendulum experiment [18]. Of interest in the image is the laptop and table which are mostly smooth in the original image.

Code 15.1 adds noise to an original image, which is loaded as `data`. Line 5 shows that the max value of any pixel is 255, which is needed into order to properly scale the noise. In this case, 10% noise is added and thus the random array in line 5 is multiplied by 25.6. The noise is added in line 7. Replacing the **ranf** function with the **normal** generates Gaussian noise.

A portion of grayscale version of the original image is shown in Figure 15.2(a), and Figure 15.2(b) shows the same portion with noise added. The second image is grainier, which is particularly noticeable in the regions that were originally smooth. Increasing the percentage of noise increases the grainy nature of the image.

One simple method of reducing noise is to smooth the image. This works because the noise of one pixel is independent of the noise in the neighboring pixels. Thus, the average of the noise over a region should be close to a constant. Since $\mathbf{a}[\vec{x}] + k$ does not change the appearance of the image, the smoothing operation reduces the noise. Smoothing is very similar to averaging and so the graininess is reduced. However, smoothing also deteriorates the edges of the objects in the image.

The process of smoothing is simply,

$$\mathbf{b}[\vec{x}] = \mathcal{S}_{\alpha=2}\mathbf{a}[\vec{x}], \tag{15.2}$$

where α controls the amount of smoothing, and in this example $\alpha = 2$. Code 15.2 shows one method of smoothing in Python using the **cspline2d** function. The second argument to the function is the smoothing parameter. The result of the smoothing operation is shown in Figure 15.2(c). As seen, the graininess is greatly reduced but the edges are slightly more blurred.

15.2 SALT AND PEPPER NOISE

Salt and pepper noise describes the noise in which some pixels are completely white or completely black. This was common in the early days of CCD cameras as some pixels would fail. In this case,

Figure 15.1 An original image.

Code 15.1 Adding random noise

```
>>> import imageio
>>> import numpy as np
>>> data = imageio.imread( 'data/altroom.png', as_gray=True)
>>> data.max()
255.0
>>> noise = (np.random.ranf( data.shape )-0.5) * 25.6
>>> sdata = data + noise
```

(a) (b) (c)

Figure 15.2 (a) A portion of the original image, (b) the image with 10% noise, and (c) the image after smoothing.

most of the pixels have the original values and some have a value of either 0 or 255. Smoothing will not be as effective here as the average noise about any pixel is not zero.

Creating the salt noise is shown in the first few lines in Code 15.3. Line 1 creates an array of random numbers that have values between 0 and 1. However, these are compared to a small factor such as 0.01. Thus, in this case, 1% of the pixels will be changed. Line 2 creates a new array `bdata`, which has the salt noise. The first term, `(1-r)*data`, maintains the original values for the 99% of

Code 15.2 Smoothing in Python

```
>>> import scipy.signal as ss
>>> bdata = ss.cspline2d( sdata,2 )
```

Code 15.3 Salt noise

```
>>> r = np.random.ranf( data.shape )<0.01
>>> bdata = (1-r)*data + r*255
>>> cdata = nd.grey_erosion(bdata,2)
>>> ddata = nd.grey_dilation(cdata,2)
```

the unaffected pixels. The array `r` is binary valued with 1% of the pixels set to 1. Thus, `1-r`, has 99% of the values set to 1.

One method of reducing this type of noise is

$$\mathbf{d}[\vec{x}] = \lhd_2 \rhd_2 \mathbf{b}[\vec{x}], \tag{15.3}$$

where $\rhd$ and $\lhd$ are the erosion and dilation operators (see Section 6.5). The erosion operator replaces the large value with the lowest neighbor value. This removes the salt noise, but it also makes bright regions slightly smaller as the perimeters also decay. Therefore, a dilation is applied to regrow bright regions to their original size. Since the salt noise was completely eliminated by the erosion, it does not return with the dilation.

Figure 15.3(a) shows the image after the salt noise is applied, and Figure 15.3(b) shows the image after the erosion and dilation. As seen, the salt noise is removed, and all of the objects are the original size.

This method works well but does have a small disadvantage. Edges in the output image that are not vertical or horizontal may have slightly exaggerated stair step borders. For the case of pepper noise, the dilation operator is applied first to fill in the black pixels with the brightest neighbor. To shrink bright areas back to their original size, the erosion operator is subsequently applied.

Salt and pepper noise is removed by

$$\mathbf{d}[\vec{x}] = \lhd_2 \rhd_4 \lhd_2 \mathbf{b}[\vec{x}]. \tag{15.4}$$

If the erosion operator is applied first, then the salt pixels are removed but the pepper pixels are larger. Thus, the erosion operator needs to have twice the extent as the original. A final erosion operator is then applied to get all shapes back to their original size.

The process could also apply the dilation first followed by the doubled erosion and final dilation. If the number of salt and pepper pixels is the same, then either method produces a similar result. If either salt or pepper is significantly dominant in the image, then it should be removed first.

(a) (b)

Figure 15.3 (a) After adding white pixels and (b) after erosion and dilation.

15.3 CAMERA NOISE

Early CCD cameras had issues that infused a lot of noise into the images. Decades of improvements though have eliminated many issues. However, some types of noise are still evident.

A lot of processing occurs in the capture of a digital image. For example, in some digital cameras, a signal is sent from the camera to the object when the shutter button is pressed. The returned signal helps with determining the focus and the bias and scale of the detected intensities. The image received is not actually the raw data. Modern cameras will even perform color correction, blur correction, face finding, and more.

15.4 COLORED NOISE

The term *colored noise* is applied for random noise in the Fourier plane. For an RGB image, this random noise can appear as blotchy colors. The reason is that some of the noise is applied to the lower frequencies, which correspond to larger regions in the input image.

Consider the original image shown in Figure 15.4(a). The remaining paint on the house is white. The same image under the influence of colored noise is shown in Figure 15.4(b).

Regions of the image have changes in color due to the random alteration of the original frequencies. This process is described by

$$\mathbf{b}[\vec{x}] = \mathfrak{F}^{-1} \frac{1}{1 + \alpha \mathbf{n}[\vec{x}]} \mathfrak{F} \mathbf{a}[\vec{x}]. \tag{15.5}$$

The process is shown in Code 15.4.

15.5 COMPARISON OF NOISE REMOVAL SYSTEMS

This section will compare some of the common methods applied to random noise. The target image is shown in Figure 15.5(a). It is chosen because it has smooth regions (the road) and textured regions

(a) (b)

Figure 15.4 (a) An original image, and (b) the image after colored noise is applied.

Code 15.4 Applying colored noise

```
>>> bdata = np.zeros((V,H,3),complex)
>>> for i in range( 3 ):
        r = np.random.ranf( (V,H) )
        fdata = ft.fft2( data[:,:,i] )
        fdata = fdata/(1 + 1*r)
        bdata[:,:,i] = ft.ifft2( fdata )
```

(a)

(b)

Figure 15.5 (a) The original image and the (b) after the noisehas been applied.

(the leaves). Noise is added via the **AddNoise** function shown in Code 15.5 and the result is shown in Figure 15.5(b).

15.5.1 SMOOTHING

The application of smoothing to remove noise was presented in Equation (15.2) and Code 15.2. The result is shown in Figure 15.6, and as seen, the objects are a little blurrier and the noise has not been completely removed.

Code 15.5 The **AddNoise** function

```
# deer.py
def AddNoise( fname ):
    adata = imageio.imread(fname,as_gray=True)
    noise = np.random.ranf( adata.shape )-0.5
    bdata = adata + 84*noise # 33 percent
    return bdata
```

Figure 15.6 The deer image after smoothing.

15.5.2 LOW-PASS FILTERING

Since the noise is random, the noise in consecutive pixels is independent. This also means that the noise is in the highest frequencies. Thus, a low-pass filter can be applied to remove the highest of frequencies. This type of filtering was discussed in Section 11.1.1.

The grayscale input is defined as $\mathbf{b}[\vec{x}]$. The mask is an image that is the same frame size as the input with a solid circle centered in the frame. The radius of the circle is r, which the user can define to suit their needs. The radius is dependent on the frame size, and the qualities that are desired.

The process is defined as

$$\mathbf{d}[\vec{x}] = \Re \mathfrak{F}^{-1} X \left(\mathbf{o}_r[\vec{x}] \times X \mathfrak{F} \mathbf{b}[\vec{x}] \right). \tag{15.6}$$

Code 15.6 displays the function **Lopass** , which performs these functions. The mask is created in line 4 and multiplied by the swapped, Fourier transform in line 6. The image is returned back to the image space after another swap, and finally, the real components are returned. The result using $r = 128$ is shown in Figure 15.7.

15.5.3 EROSION AND DILATION

Consecutive erosion and dilation operators are very useful for removing salt and pepper noise, but as this example demonstrates, this process is less effective on removing random noise. The process is described in Equation (15.4).

Code 15.7 shows the **ErosionDilation** function, which applies the erosion and dilation operators. The result is shown in Figure 15.8, and as seen, the noise was not significantly reduced.

Code 15.6 The **Lopass** function

```
1 # deer.py
2 def Lopass( bdata ):
3     V,H = bdata.shape
4     circ = mgc.Circle( (V,H), (V/2,H/2), 128 )
5     fbdata = ft.fftshift( ft.fft2( bdata ) )
6     fddata = fbdata * circ
7     ddata = ft.ifft2( ft.ifftshift( fddata ))
8     return ddata.real
```

Figure 15.7 The deer image after applying a low-pass filter.

Code 15.7 The **ErosionDilation** function

```
# deer.py
def ErosionDilation( bdata ):
    b1 = nd.grey_erosion( bdata, 2 )
    b2 = nd.grey_dilation( b1, 4 )
    b3 = nd.grey_erosion( b2, 2 )
    return b3
```

Figure 15.8 The deer image after the application of erosion and dilation operators.

15.5.4 MEDIAN FILTER

The *median filter* is a method that computes the local average for each pixel. It is very similar to smoothing. The Python *scipy.ndimage* module does contain a median filter function, so implementation is easy. Code 15.8 shows the application of this filter. The second argument controls the extent of the averaging. A single value uses a square extent but it is also possible to define a rectangular or

Code 15.8 Applying a median filter

```
>>> import scipy.ndimage as nd
>>> cdata = nd.median_filter(bdata,4 )
```

Figure 15.9 The deer image after the application of a median filter.

other shape for the region that is used to compute the average. Users should consult the *scipy* manual for a full display of the options that are available. The result of this filter is shown in Figure 15.9.

15.5.5 WIENER FILTER

The *Wiener Filter* attempts to minimize the noise in the Fourier space. The theory starts with the alteration of the original input by both a point spread function, $\mathbf{h}[\vec{x}]$, and additive noise, $\mathbf{n}[\vec{x}]$. Given the original image $\mathbf{a}[\vec{x}]$, the image received by the detector is modeled as

$$\mathbf{b}[\vec{x}] = \mathbf{h}[\vec{x}] \otimes \mathbf{a}[\vec{x}] + \mathbf{n}[\vec{x}], \tag{15.7}$$

where $\mathbf{h}[\vec{x}]$ is a point spread function perhaps a very thin Gaussian function. The $\mathbf{b}[\vec{x}]$ is the additive noise. Consider the filter $\mathbf{g}[\vec{\omega}]$, that satisfies

$$\mathfrak{F}\mathbf{a}[\vec{x}] = \mathbf{g}[\vec{\omega}] \times \mathfrak{F}\mathbf{b}[\vec{x}]. \tag{15.8}$$

The filter that performs this optimization is defined as

$$\mathbf{g}[\vec{\omega}] = \frac{\mathfrak{F}\mathbf{h}[\vec{x}]}{|\mathfrak{F}\mathbf{h}[\vec{x}]|^2 + \mathbf{f}[\vec{\omega}]/S}, \tag{15.9}$$

where $\mathbf{f}[\vec{\omega}]$ is the power spectrum of the input and S is the power spectrum of the estimated noise. However, for random noise, this value is just a scalar defined as $S = VH\sigma^2$, where V and H are the vertical and horizontal frame size and σ^2 is the standard deviation of the pixels in the received image. The power spectrum is

$$\mathbf{f}[\vec{\omega}] = |\mathfrak{F}\mathbf{a}[\vec{x}]|^2. \tag{15.10}$$

The *scipy.signal* module contains the **wiener** function as displayed in Code 15.9. Again, there are multiple options for the user to employ and consulting the *scipy* manual is informative. The result is shown in Figure 15.10 and as seen does a better job than some of the other filters.

Code 15.9 Applying a Wiener filter

```
>>> import scipy.signal as ss
>>> cdata = ss.wiener( bdata )
```

Figure 15.10 The deer image after the application of a Wiener filter.

15.6 OTHER TYPES OF NOISE

The work in this chapter was mostly applied to random noise. However, there are other types of noise: one example is shown in Figure 11.7. As the netting is in the way of the ball field, this is also a type of noise. *Structured noise* is a pattern that interferes with the image. The example in Section 11.4 removes these artifacts through Fourier filtering. This requires that the structure of the noise be known so that the correct frequencies can be removed.

Clutter is usually the presence of other objects in the image that interfere with the presentation of the target. For example, if the target is an image of a vehicle and there is a tree between the viewer and the target, then it interferes with the viewing of the target. The car is then presented to the viewer as two distinct segments separated by significant presence of the tree.

Glint is the bright reflection of the illuminating source. An example is shown in Figure 15.11, which shows two balloons in a room with several light sources. Several bright spots are seen on the balloons due to the reflection of the illuminating source. This reflection is so bright that the camera can no longer record the color of the balloon material in that region.

Shadows also provide a source of noise. A vehicle driving along a tree-lined lane will have several shadow patterns. The problem can be severe as the intensity of the car changes drastically between regions in the shade and in the direct light. If edge enhancing algorithms are applied to this type of image, then there are edges that arise from the boundary of the shadows rather than the objects in the image. One possible method of alleviating this issue is to convert the image to a different color space (HSV, YUV, etc.) in order to separate intensity from the hue. However, if the range of values between shaded and brightly lit areas is severe, then the hue viewed in the image will also be affected by this type of noise.

15.7 SUMMARY

Noise is a general term that accounts for artifacts that disrupt the purity of an image. Random noise adds a grainy texture to the image. An image with random pixels incorrectly set to fully ON or OFF

Figure 15.11 Reflecting balloons.

has salt and pepper noise. Colored noise are random fluctuations in the frequencies of the image. Several other types of noise exist as well. This chapter reviewed some of the types of noise and a few popular methods used to remove the noise.

PROBLEMS

1. Use $\mathbf{a}[\vec{x}] = \mathcal{L}_L Y(\text{'}\texttt{clock.png}\text{'})$. Create $\mathbf{b}[\vec{x}]$ by adding 40% random noise to $\mathbf{a}[\vec{x}]$. Compute the error between the two images by

$$E = \sqrt{\sum_{\vec{x}} \frac{(\mathbf{a}[\vec{x}] - \mathbf{b}[\vec{x}])^2}{(256)^2 VH}}, \tag{15.11}$$

   where V and H are the vertical and horizontal dimensions.
2. Use $\mathbf{a}[\vec{x}] = \mathcal{L}_L Y(\text{'}\texttt{clock.png}\text{'})$. Create $\mathbf{b}[\vec{x}]$ by adding 40% random noise to $\mathbf{a}[\vec{x}]$. Compute the error using Equation (15.11).
3. Use $\mathbf{a}[\vec{x}] = \mathcal{L}_L Y(\text{'}\texttt{clock.png}\text{'})$. Create an image $\mathbf{b}[\vec{x}]$ that is the same as $\mathbf{a}[\vec{x}]$ except that 2% of randomly selected pixels are set to the max value (255). Measure the error by Equation (15.11).
4. Using the image from problem 3 as $\mathbf{b}[\vec{x}]$, compute $\mathbf{c}[\vec{x}] = \lhd_2 \rhd_2 \mathbf{b}[\vec{x}]$. Measure the error between $\mathbf{a}[\vec{x}]$ and $\mathbf{c}[\vec{x}]$ using Equation (15.11) by replacing $\mathbf{b}[\vec{x}]$ with $\mathbf{c}[\vec{x}]$.
5. Use $\mathbf{a}[\vec{x}] = \mathcal{L}_L Y(\text{'}\texttt{clock.png}\text{'})$. Create an image $\mathbf{b}[\vec{x}]$ that is the same as $\mathbf{a}[\vec{x}]$ except that 2% of randomly selected pixels are set to the min value (0). Compute $\mathbf{c}[\vec{x}] = \rhd_2 \lhd_2 \mathbf{b}[\vec{x}]$. Measure the error between $\mathbf{a}[\vec{x}]$ and $\mathbf{c}[\vec{x}]$ using Equation (15.11) by replacing $\mathbf{b}[\vec{x}]$ with $\mathbf{c}[\vec{x}]$.
6. Use $\mathbf{a}[\vec{x}] = \mathcal{L}_L Y(\text{'}\texttt{clock.png}\text{'})$. Create $\mathbf{b}[\vec{x}]$ by adding 1% salt and 1% pepper noise to $\mathbf{a}[\vec{x}]$. Compute $\mathbf{c}[\vec{x}] = \lhd_2 \rhd_4 \lhd_2 \mathbf{b}[\vec{x}]$. Measure the error between $\mathbf{a}[\vec{x}]$ and $\mathbf{c}[\vec{x}]$ using Equation (15.11) by replacing $\mathbf{b}[\vec{x}]$ with $\mathbf{c}[\vec{x}]$.
7. Use $\mathbf{a}[\vec{x}] = \mathcal{L}_L Y(\text{'}\texttt{clock.png}\text{'})$. Create $\mathbf{b}[\vec{x}]$ by adding 1% salt and 3% pepper noise to $\mathbf{a}[\vec{x}]$. Compute $\mathbf{c}[\vec{x}] = \lhd_2 \rhd_4 \lhd_2 \mathbf{b}[\vec{x}]$. Compute $\mathbf{d}[\vec{x}] = \rhd_2 \lhd_4 \rhd_2 \mathbf{b}[\vec{x}]$. Measure the error between $\mathbf{a}[\vec{x}]$ and $\mathbf{c}[\vec{x}]$ using Equation (15.11) by replacing $\mathbf{b}[\vec{x}]$ with $\mathbf{c}[\vec{x}]$. Measure the error between $\mathbf{a}[\vec{x}]$ and $\mathbf{d}[\vec{x}]$ in the same manner. Which method of removing noise performed better?
8. Use $\mathbf{a}[\vec{x}] = \mathcal{L}_L Y(\text{'}\texttt{clock.png}\text{'})$. Create $\mathbf{b}[\vec{x}] = \mathbf{a}[\vec{x}] + 0.1\mathbf{q}[\vec{x}]$. Create $\mathbf{c}[\vec{x}] = \mathcal{S}_2 \mathbf{b}[\vec{x}]$. Measure the error between $\mathbf{a}[\vec{x}]$ and $\mathbf{c}[\vec{x}]$ using Equation (15.11) by replacing $\mathbf{b}[\vec{x}]$ with $\mathbf{c}[\vec{x}]$.

9. Use $\mathbf{a}[\vec{x}] = \mathcal{L}_L Y(\text{'}\texttt{clock.png}\text{'})$. Create $\mathbf{b}[\vec{x}] = \mathbf{a}[\vec{x}] + 25 * \mathbf{q}_{\vec{w}}$. Clean $\mathbf{b}[\vec{x}]$ using a median filter.
10. Use $\mathbf{a}[\vec{x}] = \mathcal{L}_L Y(\text{'}\texttt{clock.png}\text{'})$. Create $\mathbf{b}[\vec{x}]$ by adding 1% salt and 1% pepper noise to $\mathbf{a}[\vec{x}]$. Clean $\mathbf{b}[\vec{x}]$ using a median filter.
11. Use $\mathbf{a}[\vec{x}] = \mathcal{L}_L Y(\text{'}\texttt{clock.png}\text{'})$. Create $\mathbf{b}[\vec{x}]$ by adding 1% salt and 1% pepper noise to $\mathbf{a}[\vec{x}]$. Compute
$$\mathbf{d}[\vec{x}] = \mathfrak{F}^{-1}(\mathfrak{F}\mathbf{b}[\vec{x}] \times (0.1\mathbf{q}[\vec{x}] + 0.9))$$
Clean $\mathbf{b}[\vec{x}]$ using a median filter.
12. Use $\mathbf{a}[\vec{x}] = \mathcal{L}_L Y(\text{'}\texttt{clock.png}\text{'})$. Create $\mathbf{b}[\vec{x}] = \mathbf{a}[\vec{x}] + 25 * \mathbf{q}_{\vec{w}}$. Clean $\mathbf{b}[\vec{x}]$ using a Weiner filter.
13. Use $\mathbf{a}[\vec{x}] = \mathcal{L}_L Y(\text{'}\texttt{clock.png}\text{'})$. Create $\mathbf{b}[\vec{x}]$ by adding 1% salt and 1% pepper noise to $\mathbf{a}[\vec{x}]$. Clean $\mathbf{b}[\vec{x}]$ using a Weiner filter.
14. Use $\mathbf{a}[\vec{x}] = \mathcal{L}_L Y(\text{'}\texttt{clock.png}\text{'})$. Create $\mathbf{b}[\vec{x}]$ by adding 1% salt and 1% pepper noise to $\mathbf{a}[\vec{x}]$. Compute
$$\mathbf{d}[\vec{x}] = \mathfrak{F}^{-1}(\mathfrak{F}\mathbf{b}[\vec{x}] \times (0.1\mathbf{q}[\vec{x}] + 0.9))$$
Clean $\mathbf{d}[\vec{x}]$ using a Weiner filter.

16 Texture Recognition

Texture is the quality of region of an image that can include smoothness, regularity of a pattern, or variations in the pattern. Texture for a single pixel requires information from the surrounding pixels and therefore is a higher-ordered calculation. Many different approaches to texture have been proposed. This chapter will review some of the methods and organize these methods into classes according to their principles.

16.1 DATA

The Brodatz [6] data set is a long standing set of texture-based images. This data set is available through many different websites [26]. The data set consists of 111 large images of which small samplings of three images are shown in Figure 16.1. The digital images can be obtained from Ref. [26] as 640×640 images. These are quite large, and commonly, researchers will cut these large images into smaller subimages. Some subimages will be used for training and the rest will be reserved for testing.

16.2 EDGE DENSITY

Some definitions of texture rely on the density of edges within a local region. Smooth regions have few, if any, edges, and therefore, the density of edges should be small. This technique, though, does not consider and structure within the target region. Of course, there are many different approaches to use edge densities to contrive a texture metric, and a few are reviewed here.

16.2.1 STATISTICAL METHOD

It is easy to apply a threshold to pixel intensities; thus, one approach to texture recognition is to convert the regional texture to intensity levels. Thus, if a region is smooth then in the new image the pixels in that region are dark. Whereas, if a region is rough then the pixels in that region should be bright. This is just a general overview, for it is plausible that an application would prefer regions of regularity to be bright and regions of confused intensities to be dark. There are many possibilities that are defined by the application, but as a whole, the process is to convert the texture into pixel intensities.

Perhaps the simplest is to consider the ratio of local mean to the local standard deviation. The mean of an image is the average of all of the pixel values, whereas the local mean is the average over a small region. In this case the local mean output is also an image with the same dimensions as the input. The operator is $\mathcal{M}_r$ where r is defined by the user. It could be as simple as a small radius, and thus, the value of an output pixel is the average of all pixels in the input image, a distance r away from the pixel. The user may define other selection criteria and modify the definition of r. The local average of an input image $\mathbf{a}[\vec{x}]$ is

$$\mathbf{b}[\vec{x}] = \mathcal{M}_r \mathbf{a}[\vec{x}]. \tag{16.1}$$

An example using $r = 2$ as applied to the bird image (Figure 2.1(a)) is shown in Figure 16.2(a). As seen, this is just a smoothed version of the original image.

The standard deviation for a set of data represented by x_i is

$$\sigma = \sqrt{\frac{1}{N}\sum_{i=1}^{N}(x_i - \mu)^2}, \tag{16.2}$$

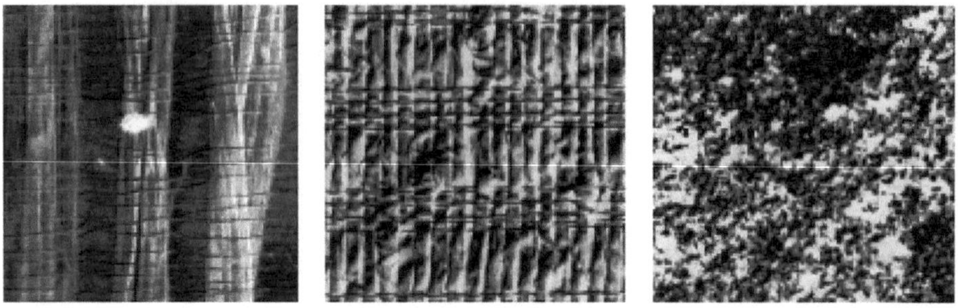

(a) (b) (c)

Figure 16.1 (a) Brodatz 50, (b) Brodatz 80, and (c) Brodatz 100.

where N is the number of data points and μ is the average. The local standard deviation performs over a small region for each pixel. The result is shown in Figure 16.2(b). The brightness of the pixels is associated with the variation in the pixels within a local region. So, smooth regions will have darker output pixels. The operator is $\mathcal{T}_r$, and thus, the notation for computing the local standard deviation is

$$\mathbf{b}[\vec{x}] = \mathcal{T}_r \mathbf{a}[\vec{x}]. \tag{16.3}$$

A simple, and somewhat effective, method of converting texture to intensity is to compute the ratio of the local mean to the local standard deviation

$$\mathbf{b}[\vec{x}] = \frac{\mathcal{M}_r \mathbf{a}[\vec{x}]}{\mathcal{T}_r \mathbf{a}[\vec{x}]}. \tag{16.4}$$

Code 16.1 shows this approach with line 6 using the **scipy.ndimage.gaussian_filter** function to smooth the image. Line 7 computes the local standard deviation, and line 8 computes the ratio.

Results for $r = 2$ and $r = 8$ are shown in Figure 16.3. The intent is to create a system in which each type of texture is associated with a single pixel intensity level in the output. As seen, there are differences depending on the value of r, which was set in line 5. In the case of $r = 8$, there is a bright spot on the bird's body, but it does not fill the body. As r increases, so does the effects at the edges of objects.

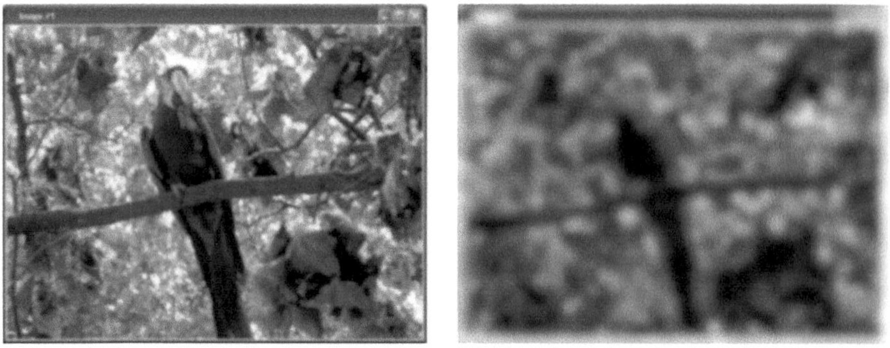

(a) (b)

Figure 16.2 (a) The local mean and (b) the local standard deviation.

Code 16.1 Simple texture measure through the ratio of the mean and standard deviation

```
>>> import numpy as np
>>> import imageio
>>> import scipy.ndimage as nd
>>> amg = imageio.imread('data/bird.jpg', as_gray=True)
>>> r = 2
>>> bmg = nd.gaussian_filter(amg,r)
>>> cmg = np.sqrt(nd.gaussian_filter( (amg-bmg)**2, r) )
>>> dmg = nd.gaussian_filter(amg,2) / cmg
```

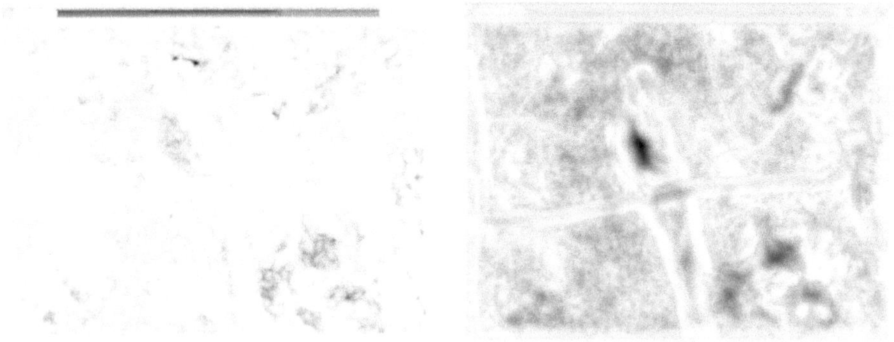

(a) (b)

Figure 16.3 (a) The conversion of texture to intensity for $r = 2$ and (b) for $r = 8$.

16.2.2 THE METHOD OF ROSENFELD AND THURSTON

The *edge density* method was introduced by Rosenfeld and Thurston in the early days of image processing [27]. The smoothed version of an edge-enhanced image is

$$\mathbf{b}[\vec{x}] = \mathcal{S}_2 |\mathbf{a}[\vec{x}] - D_{(1,1)}\mathbf{a}[\vec{x}]|. \tag{16.5}$$

The shift of (1,1) enhances the edges that are at a 45° angle. Code 16.2 shows the process in a single line of Python script. The **shift** function obviously performs the shift, and once again, the **gaussian_filter** function is used to smooth the image. The output `bdata` has a measure of texture for each pixel. The histogram of these values is computed in line 2 using the **histogram** function from the *scipy.ndimage* module. This function produces two vectors. The first is captured as `cts`, which is the y value of each bin, and `bins` is the x value of the bins. In this case, the bins are equally spaced from 0 to 255 incrementing by a value of 1, so only `cts` has important information.

The histograms of Equation (16.5) for images D2 and D90 from the Brodatz collection are shown in Figure 16.4. The original images are shown in Figure 16.5. The first image is more homogeneous than the second, and this leads to the different shapes in the histograms. A perfectly homogeneous image would produce a Gaussian distribution.

Code 16.2 Compute the edge density

```
1 >>> bdata = nd.gaussian_filter( abs( adata - nd.shift(adata,(1,1))), 2)
2 >>> cts,bins = np.histogram( bdata,255,(0,255))
```

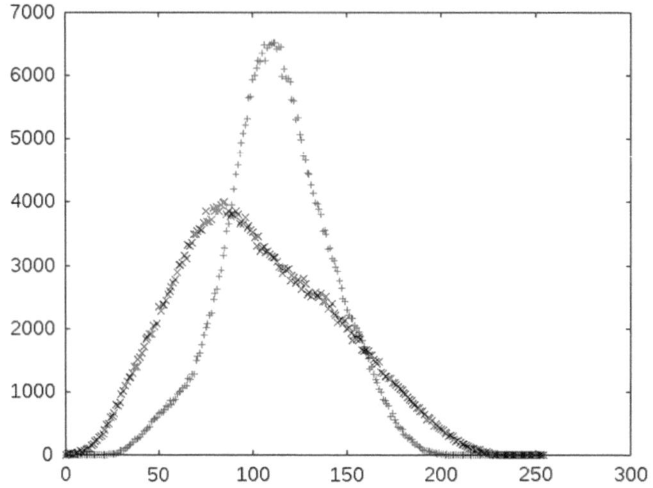

Figure 16.4 The distribution of data points for Brodatz images 2 and 90.

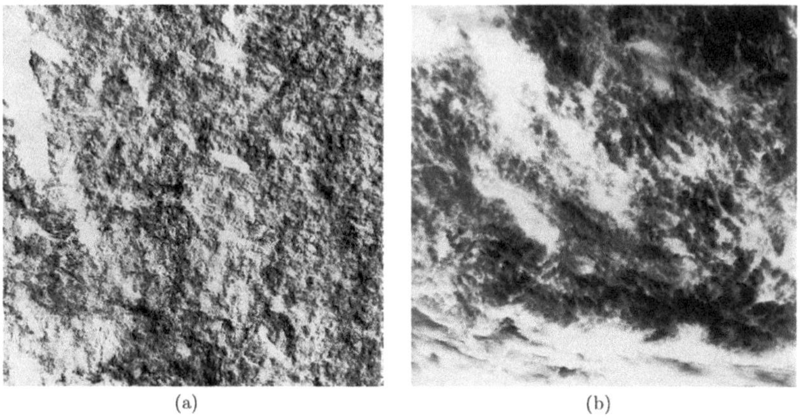

Figure 16.5 (a) Brodatz image 2 and (b) Brodatz image 90.

There are four moments of a Gaussian distribution of which the first two are quite well known. The first is the mean, which is related to sum of the x values of the data, and the second is the standard deviation which is sum of the x^2 values (with the average removed). The third moment is *skew*, which measures the lack of symmetry in the distribution. In the example above, one of the distributions looks like a bell curve, and therefore, the skew is near 0. The other is quite lopsided and so has a large magnitude of skew. The fourth moment is *kurtosis*, which measures the nature of the curve. Once again, a perfect bell curve, would have a value near 0. A distribution that differs from the bell curve shape has a large kurtosis magnitude. For example, a graph that looks like a mesa would have a high kurtosis value.

Code 16.3 considers the entire process. The image is read in line 2, and the texture metric is measured in line 3. The following lines compute the four metrics. For Brodatz image D2, the four moments are (119, 24.7, 0.0441, −0.221). The same process was applied to image D90, and the four values are (103, 41.2, 0.317, −0.521). As seen, the skew value is much larger which is reflected in the distribution.

Code 16.3 Measuring the four moments

```
>>> import scipy.stats as st
>>> adata = imageio.imread( 'brodatz/D2.gif')
>>> bdata = nd.gaussian_filter( abs( adata - nd.shift(adata,(1,1))), 2)
>>> bdata.mean()
119.14822021484375
>>> bdata.std()
24.674943919451326
>>> st.skew(bdata.ravel())
0.044139225640128586
>>> st.kurtosis(bdata.ravel())
-0.2209047798138375
```

The four moments are computed in the function **FourMoments** shown in Code 16.4. The computations are the same as in Code 16.3; but in this function, the four moments are stored in a vector named `vec`. The process is repeated for all files in the given directory `indir`. The names of the files are read in line 3, and an empty list is created in line 4. As the vectors are computed, they are appended to this list, which is converted to a matrix in line 14. This will facilitate further analysis. The main caveat is that the input directory should contain nothing other than the images to be analyzed. There is no safeguard if the program attempts to load a different file as an image. However, this can easily be rectified with a few lines of code.

The function is called in Code 16.5 in line 1. This function returns two items. The first is the matrix, and the second is the list of file names. The first row in the matrix corresponds to the first file in this list and so on. Before a simple analysis can be performed, it is necessary to normalize the data. The numerical range of the first two moments is much larger than that of the last two. So, in simple comparisons, the first two moments would dominate. The solution is to normalize the data so that all moments are on the same scale. This is performed in line 2, and the normalized data is stored as `statsn`.

A very simple method of comparison is to pick one image at random and the compare its state vector to the state vectors of all of the other images. In this case, the comparison is chosen to be the

Code 16.4 The **FourMoments** function

```
1  # texture.py
2  def FourMoments( indir, r=2 ):
3      fnames = os.listdir( indir )
4      stats = []
5      for fn in fnames:
6          adata = imageio.imread( indir+'/'+ fn, as_gray=True )
7          bdata = nd.gaussian_filter( abs( adata - nd.shift(adata,(1,1))), r)
8          vec = np.zeros( 4 )
9          vec[0] = bdata.mean()
10         vec[1] = bdata.std()
11         vec[2] = st.skew( bdata.ravel() )
12         vec[3] = st.kurtosis( bdata.ravel() )
13         stats.append( vec )
14     stats = np.array( stats )
15     return stats, fnames
```

Code 16.5 Beginning the comparison of textures

```
1 >>> stats, fnames = txu.FourMoments( '../data/brodatz')
2 >>> statsn = stats/stats.mean(0)
3 >>> dists = np.sqrt(((statsn-statsn[5])**2).sum(1))
4 >>> ag = dists.argsort()
5 >>> ag[:5]
6 array([ 5, 21, 68, 56, 29])
7 >>> np.take(fnames, ag[:5] )
8 array(['D100.gif', 'D15.gif', 'D87.gif', 'D76.gif', 'D22.gif'],
9       dtype='<U8')
```

Euclidean distance. The target image is selected to be the sixth vector and is therefore at `statsn[5]`. Line 3 subtracts this from the entire matrix, and this subtracts the target vector from all of the rows in the matrix. These values are squared and then summed across the rows using `sum(1)`. The result is now a vector that has N elements, where N is the number of images. The square root is computed, thus completing the calculation of the Euclidean distance for all rows. This one line computes the Euclidean distance of the target to all of the vectors.

Line 4 uses **argsort** to determine the sort order of the distance values. The first five are printed and of course, the first one is five. This is the comparison of the target vector with itself, and so, it should have a distance of 0. The next best vector is #21. The filenames are stored in `fnames`, but there was never a process to sort them, and thus, there is no guarantee that the names are in any meaningful order. The names of the best five images are printed to the screen, and as seen, the target is D100 and the next best match was D15. The first image is shown in Figure 16.1(c), and the second is shown in Figure 16.6.

The images at first do not appear to be alike. One has blob regions, and the other has a bunch of lines. This algorithm only considered the density of edges and not the overall structure. So, the fact that one image has lines was not part of the consideration. Instead, it only looked at the differences in intensities in local regions. Consider a single pixel which has n neighbors that are bright and m neighbors that are dark. If there is a similar pixel in the other image, then their contributions to the

Figure 16.6 Brodatz image 15.

overall calculation are the same. This does not take into account if one of the pixels is part of a line or a blob. So, in viewing the images again, the only similarity that matters is that if there are pixels that have the same types of neighbors. There are many pixelsin one image that are bright with dark neighbors, and there are the same number of similar pixels in the other image. In this respect, the two images do appear to have a lot of similarity. The fact that the lines and blobs were not included in the calculation is the fault of the algorithm, and an indication as to the difficulty of measuring texture.

This is not a unique fault. Many texture algorithms are based on only local differences in the pixels. Computationally, it is expensive to compare each pixel to all of the pixels in each image, and thus, texture algorithms tend to have this feature of only considering local pixel variations. Thus, two versions of the same image with different resolution scales could be considered to be different with some measures of texture.

16.2.3 WAVELET DECOMPOSITION AND TEXTURE

The previous texture systems measure the density of edges in localized regions. The next step in the process is to consider edges across a wider scale; thus, it is necessary to extract edge information across a scale. One approach to extract this information is wavelet decomposition [2].

Decomposition of an image using *wavelets* is performed through iterations of downsampling and edge enhancement. Figure 16.7 shows a simple diagram of a single iteration. The frame is the same size as the original image, and it is divided into four quadrants. A downsampled version of the image will be placed in the upper left quadrant. The other three quadrants will contain a downsampled version of the edge-enhanced image. In the upper right quadrant is the data for the vertical edges, in the lower left is the data for the horziontal edges, and in the lower right is the data for the slanted edges. In practice, it is best if the vertical and horizontal dimensions of the image are powers of 2. Thus, it may be necessary to plop the input image into a larger frame as in

$$\mathbf{a}[\vec{x}] = U_{\vec{w}} Y(\texttt{fileName}). \tag{16.6}$$

The wavelet decomposition is represented by

$$\mathbf{b}[\vec{x}] = \mathcal{W}\mathbf{a}[\vec{x}], \tag{16.7}$$

where $\mathcal{W}$ is the Wavelet Decomposition operator.

The decomposition of the image is an iterative process, in which each iteration downsamples the image, creates three edge-enhanced images, and then places these four images into the larger frame. A single iteration is performed in the function **WvlIteration** shown in Code 16.6. Line 5 creates the array that will contain the output image. Lines 6 and 7 create vectors that only have even numbers.

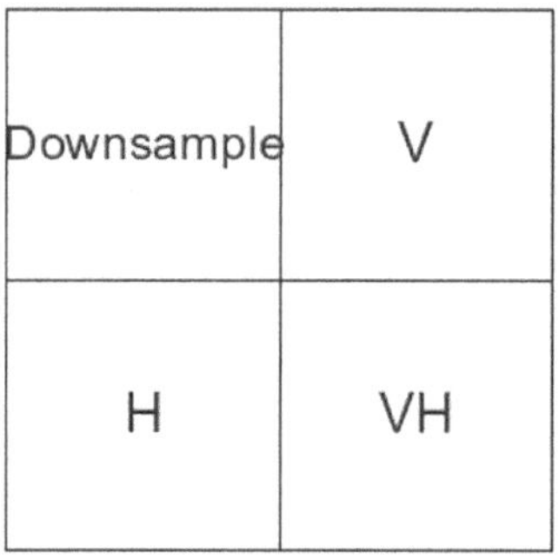

Figure 16.7 Quadrants for wavelet decomposition.

Code 16.6 The **WvlIteration** function

```
1  # wavelet.py
2  def WvlIteration( data ):
3      V,H = data.shape
4      V2, H2 = V//2, H//2
5      answ = np.zeros( (V,H) )
6      vndx = np.arange( 0, V, 2 )
7      hndx = np.arange( 0, H, 2 )
8      a = data[vndx] + 0
9      answ[:V2,:H2] = a[:,hndx] + 0
10     answ[:V2,H2:] = 0.25*abs(nd.sobel(answ[:V2,:H2],0))
11     answ[V2:,:H2] = 0.25*abs(nd.sobel(answ[:V2,:H2],1))
12     temp = nd.sobel(answ[:V2,:H2],1)
13     answ[V2:,H2:] = 0.25*abs(2/3*nd.sobel(answ[:V2,:H2],0) + temp )
14     return abs(answ)
```

These will be used to downsample the image. Lines 8 and 9 create downsampled images, and lines 10 through 12 create three edge-enhanced images. The last one is sensitive to edges at a 45°. The *ndimage* module does not have such a function, and thus, it is created from two Sobel operators and then scaled to match the magnitude of the previous two lines. Each of the edge-enhanced quadrants is also multiplied by 0.25 to keep them on the same scale as the image.

The process is called in Code 16.7, where the data is loaded in line 3. However, lines 4 and 5 show that the size of the image is not a power of two in both axes. Thus, it is placed in proper frame as indicated by the Plop operator in Equation (16.6). The first decomposition iteration is shown in Figure 16.8(a). The upper left quadrant contains the downsampled image, and the other three quadrants contain the edge-enhanced versions for the vertical, horizontal, and 45° directions.

The second iteration operates only on the upper left quadrant of the first iteration. This process repeats with each iteration operating on a smaller image, until one of the dimensions in the downsampled remainder is 2. The iterations are performed by **WaveletDecomp** shown in Code 16.8. The uppercase `V` and `H` are the vertical and horizontal frame sizes, respectively. The lowercase `v` and `h` are the sizes of the current downsampled frame. The `while` loop 'processes the upper left quadrant until one of the lowercase dimensions is reduced to 2. The example output is shown in Figure 16.8(b).

The decomposition image contains edge information across a wide range of scales. As there are many possible metrics that can be applied to the quadrants in the image, the process is broken into two steps. The first is to isolate the quadrants, and the second is to apply the user's selected metrics. The isolation of the quadrants is

$$\mathbf{c}_i[\vec{x}] = \Box_i \mathbf{b}[\vec{x}] \ \ \forall i, \tag{16.8}$$

Code 16.7 Creating an output after a single iteration in wavelet decompostion

```
1  >>> import mgcreate as mgc
2  >>> import wavelet as wvl
3  >>> adata = imageio.imread('data/reschartsmall.png',as_gray=True)
4  >>> adata.shape
5  (500, 512)
6  >>> bdata = mgc.Plop( adata, (512,512))
7  >>> cdata = wvl.WvlIteration(bdata)
8  >>> imageio.imsave('figure.png', -cdata)
```

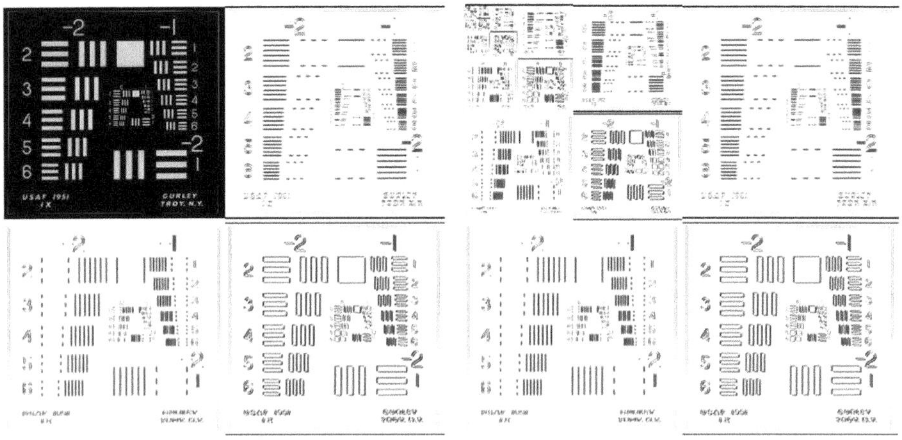

(a) (b)

Figure 16.8 (a) The result after one iteration and (b) the result after all of the iterations.

Code 16.8 The **WaveletDecomp** function

```
# wavelet.py
def WaveletDecomp( data ):
    V,H = data.shape
    ans = data+0.0
    v,h = V,H
    ok = True
    while ok:
        dt = ans[:v,:h] + 0
        b = WvlIteration( dt )
        ans[:v,:h] = b + 0
        v,h = v//2, h//2
        if v<2 or h<2:
            ok = False
    return ans
```

where i is the index over all of the quadrants. Code 16.9 shows the function **GetParts**, which isolates the different quadrants that contain edge-enhanced images. Each scale contains three quadrants that differ in frame size to those in the other quadrants. Thus, the quadrants are stored in a list.

The texture signature is the energy of the edge densities in each subimage within the decomposition image. Each element of the energy density vector, $\vec{v}$, is computed by

$$v_i = \frac{1}{N\mathbf{c}_i[\vec{x}]} \sum_{\vec{x}} (\mathbf{c}_i[\vec{x}])^2, \ \forall i, \tag{16.9}$$

where $N\mathbf{c}_i[\vec{x}]$ is the number of pixels is $\mathbf{b}_i[\vec{x}]$. The function **WaveletEnergies** shown in Code 16.10 receives the output from **GetParts** and measures the energy in each image. These are store sequentially in a vector. The energy is computed in line 6 and contains the scaling factor `(v*h)` to compensate for the difference in quadrant sizes.

Now, images are reduced to a texture vector, and these can be compared to each other using the metric of choice. Images with different frame sizes will produce vectors of different lengths. If

Code 16.9 The **GetParts** function

```
# wavelet.py
def GetParts( wvlt ):
    V,H = wvlt.shape
    v,h = V//2,H//2
    parts = []
    ok = True
    while ok:
        parts.append( wvlt[:v,h:H] )
        parts.append( wvlt[v:V,:h] )
        parts.append( wvlt[v:V,h:H] )
        V,H = V//2, H//2
        v,h = V//2,H//2
        if v<2 or h<2:
            ok = False
    return parts
```

Code 16.10 The **WaveletEnergies** function

```
# wavelet.py
def WaveletEnergies( parts ):
    vec = np.zeros( len( parts ))
    for i in range(len(parts)):
        v,h = parts[i].shape
        vec[i] = (parts[i]**2).sum()/(v*h)
    return vec
```

the aspect ratio of two images differs then a comparison can still be performed on the rightmost elements in the texture vectors.

16.2.4 GRAY-LEVEL CO-OCCURRENCE MATRIX

In previous methods, the edge was defined as the intensity difference between neighboring pixels. The difference did not take into consideration the overall intensity, and thus, a bright region containing two pixels with values 100 and 200 would produce the same edge as a dark region with two pixels containing values of 0 and 100. Also, the intensity of the pixels lends another level of complexity which is captured in the *co-occurrence matrix*. For texture application, this matrix is named GLCM (gray-level co-occurrence matrix), and it defines the relationship between pairs of pixels sensitive to the intensity values. Traditionally, the GLCM is defined as

$$P_{\delta_x,\delta_y}[i,j] = \sum_{p=1}^{V}\sum_{q=1}^{H}\begin{cases}1 & \texttt{if } I[p,q]=i \texttt{ and } I[p+\delta_x,q+\delta y]=j \\ 0 & \texttt{Otherwise}\end{cases}, \quad (16.10)$$

where I represents the image with linear dimensions (V,H). This matrix considers the shift in just one direction, which is defined by (δ_x,δ_y). In operator notation, the elements of the matrix are represented by

$$p[i,j] = \sum_{\vec{x}}(\mathbf{a}[\vec{x}] \stackrel{?}{=} i) \times (D_{\vec{\delta}}\mathbf{a}[\vec{x}] \stackrel{?}{=} j). \quad (16.11)$$

Here, the $\stackrel{?}{=}$ operator returns a matrix that is the same size as $\mathbf{a}[\vec{x}]$ with boolean values. These are True if the pixels are equal to the value of i in the first term of Equation (16.11). The D is the shift operator. The indexes i and j span the gray levels and so the size of the matrix is $N \times N$, where N is the number of gray levels. There is a large computation savings if the number of gray levels is reduced. The user must decide if the texture is contained in images with reduced intensity resolution. Furthermore, a matrix $\mathbf{p}$ is computed for a single shift $\vec{\delta}$. This is not too different from wavelet decomposition, which computed values for three different shifts in each iteration.

The operator $\mathfrak{C}_{\vec{\delta}}$ is reserved for this computation as in,

$$\mathbf{p} = \mathfrak{C}_{\vec{\delta}}\mathbf{a}[\vec{x}]. \tag{16.12}$$

Consider a simple example for a very small 5×5 image with values of

$$\begin{bmatrix} 0 & 0 & 0 & 1 & 1 \\ 0 & 0 & 1 & 1 & 1 \\ 0 & 1 & 2 & 2 & 2 \\ 0 & 2 & 2 & 3 & 3 \\ 2 & 2 & 3 & 3 & 3 \end{bmatrix}.$$

For this example, the shift is $\vec{\delta} = (0,1)$ so this will consider values that are to the right. The number of pixels that have a value of 0 and the pixel to the right also has a value of 0 is 3. Thus, $p[0,0] = 3$, which is the first element in the matrix $\mathbf{p}$ in keeping with the Python convention which begins index numbering at 0. In this example, there are four gray levels, and thus, the matrix $\mathbf{p}$ is 4×4 and is

$$\mathbf{p} = \begin{pmatrix} 3 & 3 & 1 & 0 \\ 0 & 3 & 1 & 0 \\ 0 & 0 & 4 & 2 \\ 0 & 0 & 0 & 3 \end{pmatrix}. \tag{16.13}$$

Of course, other matrices can be computed for vertical shifts, diagonal shifts, or any shift that the user defines.

Equation (16.11) is easily employed in Python as depicted in the function **Cooccurrence** shown in Code 16.11. The inputs are the grayscale image, the desired shift, and the total number of possible gray levels as it is possible that the maximum gray value of an image is not the maximum possible gray level. Line 4 shifts the input by the prescribed amount, and line 7 computes the co-occurrence value for a single element in the matrix.

This program has four nested loops with two overtly presented in lines 5 and 6. Line 7 is performed over an array and so there are two nested loops in its execution but are contained within the NumPy modules which are very efficient. Still, the two loops in lines 5 and 6 can create a slow program if `N` is large. This is the number of gray levels and if `N` is 256, then line 7 will be called

Code 16.11 The **Cooccurrence** function

```
1 # texture.py
2 def Cooccurrence( mat, shift, N):
3     p = np.zeros((N,N))
4     b = nd.shift( mat, shift )
5     for i in range(N):
6         for j in range(N):
7             p[i,j] = ((mat==i)*(b==j)).sum()
8     return p
```

65,536 times. If the number of gray levels is reduced to 16 then the number of iterations is drastically reduced to 256. For the sake of speed, it is prudent to consider a reduction in the number of gray levels.

The co-occurrence matrix contains a lot of information and Haralick et al. [11] presented 14 metrics that relied on the GLCM. While all are reviewed here, only a few are realized in Python scripts.

16.2.4.1 Angular Second Moment

One of the simplest metrics is the *angular second moment*, which is defined as

$$f_1 = \sum_i \sum_j p_{i,j}^2, \tag{16.14}$$

where both i and j are indexes that span the number of gray levels.

The value $\sum_i \sum_j p_{i,j}$ is a constant as it counts the different transitions. This is the total number of transitions which is dependent on the size of the matrix. The lowest value possible for f_1 occurs when all of the values in **p** are the same. The highest value occurs when all of the values of **p** are 0 except for one. So, in a very simple case where **p** is 2×2 the minimum value of f_1 occurs when

$$\mathbf{p} = \begin{pmatrix} 1 & 1 \\ 1 & 1 \end{pmatrix}$$

and the maximum value occurs when all of the transitions are of a single type as in

$$\mathbf{p} = \begin{pmatrix} 0 & 0 \\ 0 & 4 \end{pmatrix}.$$

In the first case, $f_1 = 4$, and in the second case, $f_1 = 16$. For the computation of f_1, it does not matter which element in **p** is nonzero. The first case would occur if the data in the image were random and the second case would occur if the data in the image were all the same values, or in other words, the opposite of random. Thus, this metric is also called *homogeneity*.

Instituting this metric is Python, which is very easy as shown in Code 16.12. The function **HHomogeneity**.

16.2.4.2 Contrast

The *contrast* is defined as,

$$f_2 = \sum_{n=0}^{N_g-1} n^2 \left(\sum_{i=1}^{N_g} \sum_{\substack{j=1 \\ |i-j|=n}}^{N_g} p_{i,j} \right). \tag{16.15}$$

The term inside the parenthesis collects values according to the difference in gray-level values as defined by n. For example, for $n = 1$, the summation inside the parenthesis sums all values in **p** in which i and j differ by 1. These would be the number of pixels that have an intensity that differs

Code 16.12 The **HHomogeneity** function

```
# texture.py
def HHomogeneity( p ):
    return (p*p).sum()
```

by 1 to their selected neighbor (as defined by $\vec{\delta}$). This result is then multiplied by n^2, which would magnify the effect of pixel pairs that have a larger difference in intensity, which is a larger contrast.

An implementation is shown as the **HContrast** function shown in Code 16.13. The two loops started on lines 5 and 6 perform the summations that are inside the parenthesis in Equation (16.15). The result is a vector with N elements, which is multiplied by another vector containing the n^2 values in lines 9 and 10. The sum of this final vector is returned to the user.

16.2.4.3 Correlation

The *correlation* is computed by,

$$f_3 = \sum_{i=1}^{N_g} \sum_{j=1}^{N_g} \frac{(ij)p_{i,j} - \mu_x \mu_y}{\sigma_x \sigma_y}, \tag{16.16}$$

where μ_x and μ_y are the averages in the horizontal and vertical dimensions of the matrix. Likewise, σ_x and σ_y are the standard deviations in the horizontal and vertical dimensions. Subtracting the mean is essentially removing any bias from the data and the division by the standard deviation is a normalization process. The term $(ij)p_{i,j}$ has larger contributions from elements that are in the lower right portion of the **p** matrix. These correspond to higher intensity values in the original image. In the case above where there were only 4 gray levels, the element in the lower right corner of **p** corresponded to the case where both a pixel and its neighbor where at the maximum intensity level.

The discrete correlation of two signals $f[x]$ and $g[x]$ can be described as

$$c[n] = \sum_{m=-\infty}^{\infty} f^{\dagger}[m] g[m+n], \tag{16.17}$$

and for a single element n in the answer, the computation is an inner product of the signal $f[x]$ with a shifted version of $g[x]$. Large contributions to the computation of $c[n]$ occur when bright pixels in $f[x]$ align with bright pixels in the shifted version of $g[x]$. The computation of f_3 is quite similar in this manner.

The μ_x is the average of values in one direction of the **p** matrix and is easily computed in Python by summing values in that direction and then computing the average. Thus, μ_x is calculated in a single line 4 in Code 16.14. The values of μ_y, σ_x and σ_y are computed likewise in the following lines. Thus, the calculation of f_3 is relegated to a single line inside with the nested `for` loops. This completes the **HCorrelation** function, and the result is a single value that is the correlation value.

Code 16.13 The **HContrast** function

```
1  # texture.py
2  def HContrast( p ):
3      N = len(p)
4      temp = np.zeros( N )
5      for i in range( N ):
6          for j in range( N ):
7              ndx = abs(i-j)
8              temp[ndx] += p[i,j]
9      nvec = np.arange( N )**2
10     f2 = (nvec * temp).sum()
11     return f2
```

Code 16.14 The **HCorrelation** function

```
# texture.py
def HCorrelation( p ):
    N = len( p )
    mux = p.sum(1).mean()
    muy = p.sum(0).mean()
    sigx = p.sum(1).std()
    sigy = p.sum(0).std()
    f3 = 0
    for i in range( N ):
        for j in range( N ):
            f3 += ((i+1)*(j+1)*p[i,j] - mux*muy)/(sigx*sigy)
    return f3
```

16.2.4.4 Variance

The calculation of the *variance* is straightforward,

$$f_4 = \sum_{i=1}^{N_g} \sum_{j=1}^{N_g} (i - \mu)^2 p_{i,j}, \tag{16.18}$$

where μ is the average of the values in **p**.

This is implemented in the **HVariance** function shown in Code 16.15. Line 3 computes the average, and line 8 performs the computation for a single iteration.

16.2.4.5 Entropy

The Haralick metrics are presented slightly out of order from the original paper as the Python script for only a few of these metrics are presented here. The next metric uses the standard definition of entropy for digital data, and defines the *entropy* of the matrix as

$$f_9 = -\sum_{i=1}^{N_g} \sum_{j=1}^{N_g} p_{i,j} \log p_{i,j}. \tag{16.19}$$

This is also easy to implement in Python as shown in Code 16.16 with the function **HEntropy**. This is a single line of computation that adds a little value to **p** since the matrix can contain 0 values. This addition of a small value was recommended by Haralick et al.

Code 16.15 The **HVariance** function

```
1 # texture.py
2 def HVariance( p ):
3     mu = p.mean()
4     N = len(p)
5     f4 = 0
6     for i in range( N ):
7         for j in range( N ):
8             f4 += ((i-mu)**2)*p[i,j]
9     return f4
```

Code 16.16 The **HEntropy** function

```
# texture.py
def HEntropy( p ):
    return -( p * np.log(p+0.00001) ).sum()
```

16.2.4.6 The Remaining Haralick Metrics

The rest of the Haralick metrics are presented without supporting Python scripts or lengthy explanations. The measure of *inverse difference* is

$$f_5 = \sum_{i=1}^{N_g} \sum_{j=1}^{N_g} p_{i,j} \frac{1}{1+(i-j)^2} p_{i,j}. \tag{16.20}$$

The term $(i-j)$ is large for large differences in intensity values of pixel pairs. However, this is in the denominator so this function receives large contributions from pixel pairs that have the same intensity value.

The *sum average* is

$$f_6 = \sum_{i=2}^{2N_g} i p_{x+y}(i), \tag{16.21}$$

where the definition of the final term is

$$p_{x+y}(k) = \sum_{i=1}^{N_g} \sum_{\substack{j=1 \\ i+j=k}}^{N_g} p_{i,j}. \tag{16.22}$$

The *sum variance* and *sum entropy* are similar in nature for their respective manners of measurement. These are

$$f_7 = \sum_{i=2}^{2N_g} (i-f_8)^2 p_{x+y}(i) \tag{16.23}$$

and

$$f_8 = -\sum_{i=2}^{2N_g} p_{x+y}(i) \log p_{x+y}(i). \tag{16.24}$$

The *difference variance* computes f_{10} as the variance of $p_{x-y}(i)$, where

$$p_{x-y}(k) = \sum_{i=1}^{N_g} \sum_{\substack{j=1 \\ |i-j|=k}}^{N_g} p_{i,j}. \tag{16.25}$$

The *difference entropy* is computed likewise as

$$f_{11} = -\sum_{i=0}^{N_g-1} p_{x-y}(i) \log p_{x-y}(i). \tag{16.26}$$

There are two measure of information of correlation, which are

$$f_{12} = \frac{HXY - HXY1}{\max(HX, HY)}, \tag{16.27}$$

and

$$f_{13} = (1 - \exp[-2(HXY2 - HXY)])^2. \tag{16.28}$$

These equations use the following definitions:

$$HXY = -\sum_i \sum_j p_{i,j} \log p_{i,j},$$

$$HXY1 = -\sum_i \sum_j \log(p_x(i) p_y(j)),$$

$$HXY2 = -\sum_i \sum_j p_x(i) p_y(j) \log(p_x(i) p_y(j)),$$

$$p_x(i) = \sum_j p_{i,j},$$

and

$$p_y(j) = \sum_i p_{i,j},$$

The final metric is the *maximal correlation coefficient* which is defined as the square root of the second largest eigenvalue of the matrix $\mathbf{Q}$, where

$$Q_{i,j} = \sum_k \frac{p_{i,k} p_{j,k}}{p_x(i) p_y(j)}.$$

The ensuing test uses just 5 of the 14 Haralick metrics to demonstrate its use in texture similarity classification. Each of the metrics produces a scalar value, and these are placed in a vector of five elements. If all 14 metrics were to be used then the vector would be 14 elements long. The **Haralick** function shown in Code 16.17 shows the implementation of this subset of metrics. There are calls to each function and the returned values are placed in a vector. Thus, this function receives the GLCM and returns a vector of the Haralick metrics.

The vectors for all of the Brodatz images are computed in Code 16.18. The variable `names` is a list of strings of which each one is the path and file name from one of the Broadatz images. The second argument in line 1 is the shift used in computing the co-occurrence matrix, which in this case is a shift of one pixel to the right. The range of values for the Haralick metrics can be quite large and very different for each measure. So, line 2 is employed to normalize the values. Now, for each metric the minimum value is 0 and the maximum value is 1.0. Line 3 uses the method from above to compute the Euclidean distance from a target image and the rest of the images. Once again, the target is D100. Line 4 determines the sort order according to this distance measure. The best five results are shown, and of course, the first one is the target images. The other four are the best matches.

Code 16.17 The **Haralick** function

```
# texture.py
def Haralick(p):
    hvec = np.zeros( 5 )
    hvec[0] = HHomogeneity(p)
    hvec[1] = HContrast(p)
    hvec[2] = HCorrelation(p)
    hvec[3] = HVariance(p)
    hvec[4] = HEntropy(p)
    return hvec
```

These examples used the shift of only one pixel to the right. Certainly, shift of one pixel in the vertical direction should also be considered. It is also possible to consider shifts of much larger distances in an attempt to capture longer range features. This is accomplished by running Code 16.18 with a different shift in line 1. The best five matches using a shift of (20,0) were: D100, D4, D13, D2, and D31. Whereas the best five matches for a shift of (0,20) were D100, D4, D18, D2, and D31.

There are two issues of interest. The first is the differences between a shift of (0,1) and (0,20), and the second is the differences between a shift of (20,0) and (0,20). In the first comparison both trials found similarities to images D4 and D2. However, the small shift case preferred D83 and D13 while the large shift case preferred D18 and D31. The target image is shown in Figure 16.1(c) while the differed results are shown in Figure 16.9. The first two images are from the short shift test and the last two are from the long shift test. Clearly, D83 does not have the long-range structure that appears in the target image.

Recall that the (0,20) test considers pairs of pixels that are 20 units apart in the horizontal direction. In the target, there are contiguous regions that are much bigger than 20 pixels in size, and so

Code 16.18 Using the **Haralick** function

```
1 >>> hmat = txu.RunHaralick(names, (0,1))
2 >>> hmat = (hmat - hmat.min(0))/(hmat.max(0) - hmat.min(0))
3 >>> dists = np.sqrt( ((hmat-hmat[61])**2).sum(1))
4 >>> ag = dists.argsort()
5 >>> np.take( names, ag[:5] )
6 array(['../Brodatz/D100.gif', '../Brodatz/D4.gif', '../Brodatz/D83.gif',
7        '../Brodatz/D2.gif', '../Brodatz/D13.gif'],
8       dtype='<U19')
```

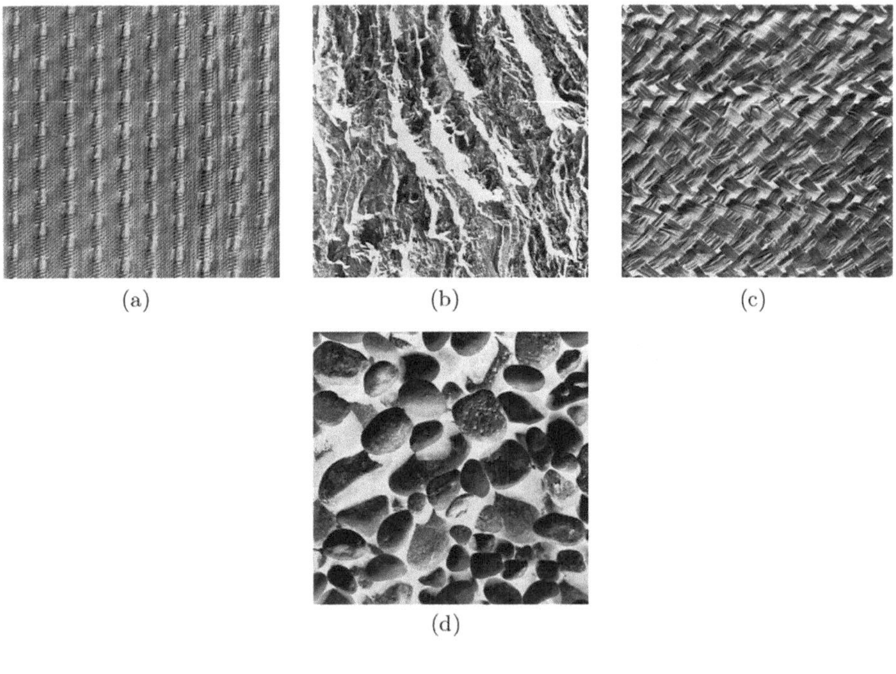

Figure 16.9 Brodatz images selected by the target. (a) D83, (b) D13, (c) D83 and (d) D31.

Figure 16.10 Brodatz images selected by the target. (a) D4 and (b) D2.

there are many pairs of pixels that are both dark or both bright. This type of feature must then appear in the lower two images as well. The dimension of these images are 640 × 640 and a rectangular thatching segment in Figure 16.9(d) is about 55 pixels horizontally from corner to corner. Thus, 20 pixels is a little less than half and as the thatching seems to have mostly uniform intensities on the left (or right) half of a rectangle the number of pixel pairs with similar intensities that are 20 pixels apart is actually quite large.

The second test compared the results from a shift of (20,0) to that of (0,20). The first case preferred D13 but the second case did not. Easily seen in this image is that there are longer stretches of similar pixel intensity in the vertical direction than there are in the horizontal direction. Since long stretches of similar intensities is a feature of the target image, D13 matched well for a long vertical shift.

This test considered only three of the many possible shifts. In those three tests both D4 and D2 were selected as good matches. These are shown in Figure 16.10. The similarity between D100 and D2 are strong. The similarity of D100 to D4 is less apparent. The granules in D4 are about 20 pixels wide, and so in this image, there are several pixel pairs that are 20 pixels apart, in both the vertical and horizontal dimensions, which are similar. While it is expected that D2 would match well with D100 for almost any shift, the image D4 would not match so well for shifts of other lengths.

These tests used only 5 of the 14 Haralick measures and considered only three of a large number of possible shifts. A more conclusive test would employ all of the metrics and many different shifts. This would also be an expensive computation to execute. However, that is the price to pay for match texture information across a wide range of distances and directions.

16.3 FILTER-BASED METHODS

Filter-based methods in texture classification employ a set of correlation filters that are used to extract pertinent information from the input image. Generally, the filters are small and so it is possible to use the **scipy.signal.correlate2d** function. Correlations are detailed in Chapter 12.

16.3.1 LAW'S FILTERS

The Law's filters are a set of 25 filters that were introduced in the early days of digital image processing [21]. The 25 filters are created from all possible pairings of five vectors shown in Code 16.19. The 25 filters are each a 5 × 5 matrix created by the outer product of two of the vectors.

Code 16.19 The five Law's vectors

```
[ 1, 4, 6, 4, 1]
[-1, -2, 0, 2, 1]
[-1, 0, 2, 0, -1]
[-1, 2, 0, -2, 1]
[1, -4, 6, 4, 1]
```

Code 16.20 The **BuildLawsFilters** function

```
1  # texture.py
2  from numpy import outer
3  def BuildLawsFilters( ):
4      a = np.array( [[1,4,6,4,1],[-1,-2,0,2,1],[-1,0,2,0,-1],\
5                  [-1,2,0,-2,1],[1,-4,6,-4,1]])
6      laws = []
7      for i in range( 5 ):
8          for j in range( 5 ):
9              laws.append( np.outer( a[i], a[j] ))
10     return laws
```

The **BuildLawsFilters** function shown in Code 16.20 creates the five vectors in lines 4 and 5. Line 9 considers all possible pairings and creates the 5×5 matrices using the **outer** function from *numpy*.

Let the set $\{\mathbf{f}[\vec{x}]\}$ represents the 25 Law's kernels. Consider a single input image, $\mathbf{a}[\vec{x}]$. Application of the Law's filters is

$$\mathbf{c}_i[\vec{x}] = \mathcal{S}_n\left(\frac{\mathcal{L}_L\mathbf{a}[\vec{x}]}{\bigvee\mathbf{a}[\vec{x}]} \otimes \mathbf{f}_i[\vec{x}]\right), \quad \forall i. \tag{16.29}$$

The normalized grayscale image is correlated with each filter, and each result is smoothed. The set $\{\mathbf{c}[\vec{x}]\}$ is 25 correlation images. The volume of data has increased by a factor of 25 and it is not necessary to process all of that data to achieve and end goal. Usually, this set of data is sampled to reduce the computational load.

A jet is a vector with the 25 correlation values from a single pixel location. In some cases, the pixel locations are specified by the type of data. For example, in face recognition, the locations are at the corners of features such as the eyes and mouth. In cases such as the texture recognition, the locations of the jets are randomly selected. Given a set of random pixel locations, $\vec{r}_j$, the jets are extracted by

$$v_{j,i} = \mathbf{c}_i[\vec{r}_j], \quad \forall i \forall j. \tag{16.30}$$

The user determines N, the number of jets to extract, and thus, texture information for a single image is reduced to a set of N jets each of length 25. The j corresponds to the vector number, and i to the element number.

The function **LawsJets** (Code 16.21) receives an input image, creates the Law's filters, performs the correlations, and then extracts random jets. The number of jets is controlled by the variable `NJ`.

Comparison of two images becomes the comparison of two sets of vectors. In the case of the face recognition, each jet has a specific location. For example, the first jet is associated with the right corner of the right eye. Comparison is easy as each vector is compared only to its counterpart. The first jet of one image is compared to only the first jet of the other image. In the case of the texture recognition, the process is complicated in that each jet of one set needs to be compared to every jet in the other set. The process becomes even more cumbersome, if the task is to compare texture of one image to textures of multiple images.

Code 16.21 The **LawsJets** function

```
# texture.py
def LawsJets( indata, NJ=100):
    filts = BuildLawsFilters()
    jets = np.zeros( (NJ,25) )
    corrs = list(map( lambda x: ss.correlate2d( indata, x ), filts ))
    for j in range( 25 ):
        corrs[j] = ss.cspline2d( abs(corrs[j]), 200 )
    corrs = np.array( corrs )
    V,H = indata.shape
    vs = np.arange( V )
    hs = np.arange( H )
    np.random.shuffle( vs ); np.random.shuffle( hs )
    for j in range( NJ ):
        jets[j] = corrs[:,vs[j], hs[j] ]+0
    return jets
```

16.4 SUMMARY

Texture is an ill-described quality of images. It refers to smoothness or roughness within localized regions of an image, but it also encompasses structure and patterns. Since it incorporates different properties, there are various algorithms that have been developed to encapsulate the properties of texture. Most methods convert the property of texture into pixel intensities. These can then be measured and compared with the same measure of other images to find similar texture qualities. This chapter reviews some of the popular methods and shows examples using a small data set.

PROBLEMS

Most of the problems in this chapter need a data set to analyze. Two Brodatz images (D1 and D2) have been cut up into five overlapping images and are stored in *data/texturehw* in the included data set. These ten images are named as the T-Data set.

1. For each image in the T-Data set compute,

$$f = \frac{\mathcal{M}\mathbf{a}[\vec{x}]}{\mathcal{T}\mathbf{a}[\vec{x}]}.$$

   Is there a single value of f that separates the images of D1 from the images of D2?
2. For each image in the T-Data set, compute the histogram by Code 16.2. Create a plot for all ten of these histograms. Do the histograms from the D1 images differ from the histograms of the D2 images?
3. For each image in the T-Data set, compute the wavelet energies. Create a plot for all ten of these histograms. Do the histograms from the D1 images differ from the histograms of the D2 images?
4. For each image in the T-Data set, compute the first five Haralick metrics. For the co-occurrence matrix, use the shift of (0,1). For each metric, determine if there is a threshold value that separates the images of D1 from the images of D2.
5. For each image in the T-Data set, compute the first five Haralick metrics. For the co-occurrence matrix, use the shift of (0,20), and reduce the input images to 16 levels of gray. For each metric, determine if there is a threshold value that separates the images of D1 from the images of D2.

6. Using the T-Data set, reduce each image to 16 gray levels and compute the co-occurrence matrix with a shift of (0,1). Compute the *inverse difference* for each image and determine if there is a threshold value that separates the values of the D1 images from the values of the D2 images.
7. Using the T-Data set, reduce each image to 16 gray levels and compute the co-occurrence matrix with a shift of (0,1). Compute the *sum average* for each image and determine if there is a threshold value that separates the values of the D1 images from the values of the D2 images.
8. Using the T-Data set, reduce each image to 16 gray levels and compute the co-occurrence matrix with a shift of (0,1). Compute the *sum entropy* for each image and determine if there is a threshold value that separates the values of the D1 images from the values of the D2 images.
9. Using the T-Data set, reduce each image to 16 gray levels and compute the co-occurrence matrix with a shift of (0,1). Compute the *difference variance* for each image and determine if there is a threshold value that separates the values of the D1 images from the values of the D2 images.
10. Using the T-Data set, reduce each image to 16 gray levels and compute the co-occurrence matrix with a shift of (0,1). Compute the *difference entropy* for each image and determine if there is a threshold value that separates the values of the D1 images from the values of the D2 images.
11. Using the T-Data set, reduce each image to 16 gray levels and compute the co-occurrence matrix with a shift of (0,1). Compute the *maximal correlation coefficient* for each image and determine if there is a threshold value that separates the values of the D1 images from the values of the D2 images.

17 Gabor Filtering

Edges and corners contain signature attributes of describable shapes. Thus, many target recognition algorithms rely on the detection of these features. However, corners and edges are not restricted to be only on the perimeter of the object. For example, many face recognition algorithms detect the corners and edges of the eyes, mouth, nose, chin perimeter and other items that are not on the perimeter. Complicating the detection is that edges are not always sharp as the transition from bright to dark may be over a span of several pixels. Furthermore, edges can be of any length and any orientation.

One approach of detecting edges of any orientation, clarity, and length is to analyze the correlation responses of the image with a set of Gabor filters. This chapter will review the construction of these filters and then present a few examples of implementation.

17.1 GABOR FILTERING

Gabor filters are a set of correlation filters designed to respond to edges of various orientations, scale, and length. A family of filters have a few parameters that the user can define. Some Gabor filters are shown in Figure 17.1. These exhibit alterations in frequency, orientation, and extent. Sensitivity to the width of an edge is controlled by the frequency in the filter. The orientation in the filter matches the orientation of its sensitivity. The spatial extent of the filter defines the response to lengths of edges in the input image. A properly designed suite of Gabor filters can then respond to all edges in the image or set of images.

Formally, the Gabor filter is defined as

$$G(x,y;f,\theta) = \exp\left\{-\frac{1}{2}\left[\frac{x'^2}{\delta_x^2} + \frac{y'^2}{\delta_y^2}\right]\right\}\cos\left(2\pi f x'\right), \tag{17.1}$$

where

$$\begin{aligned} x' &= x\cos(\theta) + y\sin(\theta) \\ y' &= -x\sin(\theta) + y\cos(\theta) \end{aligned} \tag{17.2}$$

The construction of the filter is controlled by several parameters. The pair δ_x and δ_y control the extent of the ridges in the filter. The f controls the frequency, and θ controls the angle. Commonly, δ_x and δ_y are held constant throughout a set of filters.

Code 17.1 shows the **GaborCos** function which creates the filters shown in Figure 17.1. The input `VH` is the frame size, the `f` controls the frequency, the `theta` controls the angle, and the two parameters, `deltax` and `deltay`, control the perimeter shape. It should be noted that the filter is also sensitive to the initial frame size. In some cases, it may be better to compute the filters in small frames and then place them into larger blank frames before using them. Lines 3 and 4 create two matrices from a specified frame size. These contain the coordinates x and y for all pixels. Lines 5 thought 13 compute Equation (17.1) for all points.

Each Gabor filter is defined by a state vector, $\vec{\xi}$, which contains four elements describing the filter, $\{\vec{\xi} : (f, \theta, \delta_x, \delta_y)\}$. The set Ξ contains all variations of $\vec{\xi}$ needed for a particular application. The operator notation to create a set of filters is

$$\{\mathbf{f}[\vec{x}]\} = \mathfrak{G}_{\Xi}.$$

The curly braces represent a set of images as in $\{\mathbf{f}[\vec{x}]\} = \mathbf{f}_i[\vec{x}], i = 1, \ldots, N$.

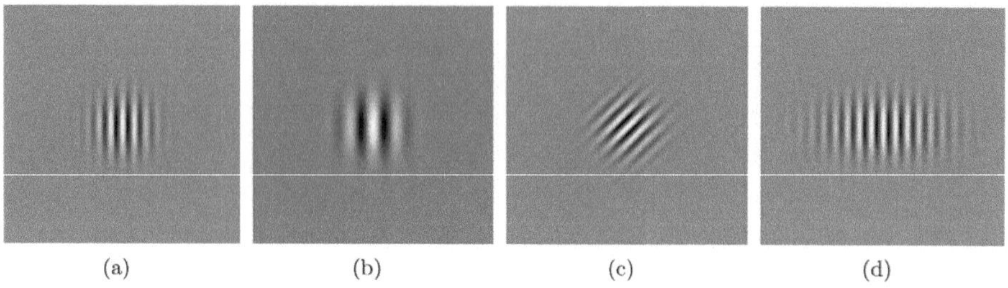

Figure 17.1 Some Gabor filters with different frequencies, orientation, and extent.

Code 17.1 The **GaborCos** function

```
# gabor.py
def GaborCos( VH, f, theta, deltax, deltay ):
  hindex = np.multiply.outer( np.ones(VH[0]), np.arange(VH[1]))- (VH[1]/2)
  vindex = np.multiply.outer( np.arange(VH[0]), np.ones(VH[1]))- (VH[0]/2)
  ct,st = np.cos(theta), np.sin(theta)
  xp = ct * hindex + st * vindex
  yp = -st * hindex + ct * vindex
  cg = np.cos( f * xp )
  t1 = xp*xp/(2. * deltax * deltax)
  t2 = yp*yp/(2.*deltay*deltay)
  t12 = -t1-t2
  while t12.min()<-100: t12 = t12 / 10.
  ans = cg * np.exp(t12)
  return ans
```

Usually, a set of filters is generated by increments of the parameters f and θ. The function **Filts** in Code 17.2 creates a set of filters for a user-defined range of f and eight equally spaced values of θ ranging from 0 to π. The first input is VH is the frame size of the filter, which is often much smaller than then image, for example, 128×128. The second input, `frange`, is the list of f values. Line 5 starts an inner loop which uses eight different rotations, and the return is a list of filters.

Each of these filters are correlated with the input $\mathbf{a}[\vec{x}]$, which produces a set of correlation surfaces,

$$\{\mathbf{c}[\vec{x}]\} = \{\mathbf{a}[\vec{x}] \otimes \mathbf{f}[\vec{x}]\}. \tag{17.3}$$

The notation $\{\mathbf{a}[\vec{x}] \otimes \mathbf{f}_i[\vec{x}]\}$ places the correlation inside the braces, which indicates that correlation is applied to each image. Code 17.3 shows the function **ManyCorrelations**, which performs N

Code 17.2 The **Filts** function

```
1 # gabor.py
2 def Filts( VH, frange ):
3     filts = []
4     for f in frange:
5         for q in range( 8 ):
6             theta = q*np.pi/8.
7             filts.append( GaborCos( VH, f, theta, 1, 1 ))
8     return filts
```

Code 17.3 The **ManyCorrelations** function

```
1 # gabor.py
2 def ManyCorrelations( indata, filts ):
3     VH = indata.shape
4     corrs = []
5     N = len( filts )
6     for i in range( N ):
7         f = mgc.Plop( filts[i], VH )
8         corrs.append( crr.Correlate2D( indata, f ) )
9     return corrs
```

correlations. The inputs are the image and the list of filters. The output is the list of correlation images. The function **Correlate2D** requires that both inputs have the same frame size, and thus, the filters are placed in a larger frame in line 7.

The output is a set of correlation filters, which are then analyzed to pursue the desired result. There are various algorithms that can be used to perform this analysis, and the selection is problem dependent. Examples, of the application of Gabor filters are presented in the following sections.

17.2 EDGE RESPONSE

Before demonstrating the response of the Gabor filters to a complicated photographic image, a much simpler example is presented. In this case, the input has straight lines with different sharpness. Two filters are constructed to demonstrate the response to edge clarity and orientation.

Figure 17.2(a) shows an input image that is a square annulus with the edges on the right side being blurred. This image is created from two images, a sharp-edged annulus $\mathbf{a}[\vec{x}]$ and a fuzzy-edged annulus $\mathbf{b}[\vec{x}]$. Each image is 512×512, and by defining $\vec{v}_1 = (0,0)$ and $\vec{v}_2 = (512,256)$, the operation $\square_{\vec{v}_1,\vec{v}_2}\mathbf{a}[\vec{x}]$ selects the left half of the image. The right half of the other image is selected using $\vec{v}_3 = (0,512)$ and $\vec{v}_4 = (512,512)$. They are combined with the concatenation operator as in,

$$\mathbf{c}[\vec{x}] = \mathcal{C}_1\left(\square_{\vec{v}_1,\vec{v}_2}\mathbf{a}[\vec{x}], \square_{\vec{v}_3,\vec{v}_4}\mathbf{b}[\vec{x}]\right), \tag{17.4}$$

with the subscript in $\mathcal{C}_1$ representing the argument used in the **concatenate** function. The image is created in lines 5 through 7 in Code 17.4 and the negative is shown in Figure 17.2(a).

Line 8 creates 24 Gabor filters with three different f values and eight different θ values; two of these, `filts[0]` and `filts[8]`, are shown in Figure 17.2(b) and (c). In order to see the details, the filter images are shown at twice the magnification as the original image. The correlations are performed in line 9, and the absolute values of the two from the shown filters are displayed in Figure 17.2(d) and (e).

The Gabor filter shown in Figure 17.2(b) is sensitive to thick vertical lines, and the corresponding correlation surface shows an equal response for the vertical lines in the input. This filter is reacting to the thick lines in the input and not on the edges of those thick lines. The Gabor filter in Figure 17.2(c) has a higher frequency value and therefore is more sensitive to thinner lines such as the edge of the annulus. In this case, the response is stronger for the sharper edges, which are on the left side of the original image. Both of these filters are, of course, sensitive to only vertical lines, and thus, the horizontal lines in the input provide no responses in the outputs. A complete family of Gabor filters would include filters at various orientations to respond to all edges in the input.

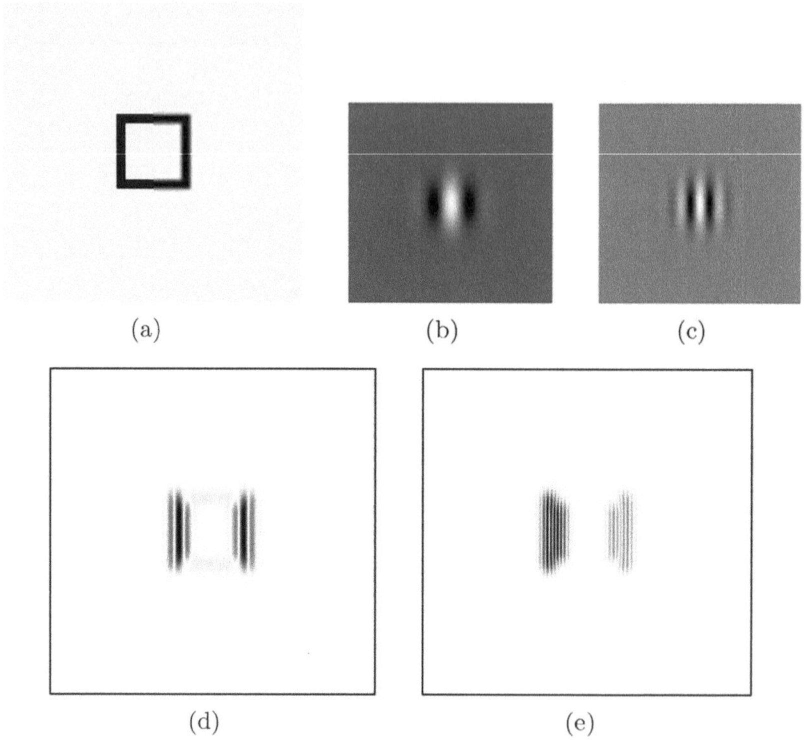

Figure 17.2 (a) An input image, (b) and (c) are two Gabor filters shown at twice the magnification as the original image, and (d) and (e) are the absolute values of the respective correlations.

Code 17.4 Complete steps to create an image, Gabor filters, and the correlations

```
>>> import numpy as np
>>> import imageio
>>> import mgcreate as mgc
>>> import gabor as gbr
>>> adata = mgc.SquareAnnulus()
>>> bdata = mgc.SoftSquareAnnulus()
>>> cdata = np.concatenate( (adata[:,:256], bdata[:,256:]) ,1)
>>> filts = gbr.Filts((128,128),(0.25,0.5,1) )
>>> corrs = gbr.ManyCorrelations( cdata, filts )
```

17.3 TEXTURE EXTRACTION WITH GABOR FILTERS

Correlating the input with several Gabor filters expands the information space. The original image had the dimensions of $V \times H$ and the set of correlation surfaces has the dimensions of $N \times V \times H$. Quite often, the pertinent information does not require that many elements. One approach, then, is to extract a few vectors at specified pixel locations. These vectors are called *jets*, and each have a length N. The selection of pixels depends on the application. Applications such as face recognition have specific geometries, and the points of interest, named fiducial points, are often at features such as the corners of the eyes, the eyebrows, the nostrils, and the mouth. In this application, all of the images should have a majority of these features. Jets are extracted from these fiducial points and can be compared to similar vectors from other faces. In other words, the jets for the left corner of the left eye can be compared to jets at that same location in the other images.

In the case of texture classification fiducial points do not exist. Furthermore, the location of the features of the texture are in arbitrary locations. Unlike the face application, it is not meaningful to compare the first jet from one image to only the first jet of another image. Instead, every jet from one image must be compared to every jet in the other images.

Consider a case in which there are multiple images, $\mathbf{a}_i[\vec{x}]$. Each image is correlated by the set of Gabor filters, $\mathbf{b}_k[\vec{x}]$. Thus, for a single image, several correlations are constructed:

$$\mathbf{c}_{jk}[\vec{x}] = \mathcal{S}\left(\mathbf{a}_i[\vec{x}] \otimes \mathbf{b}_k[\vec{x}]\right). \tag{17.5}$$

From a set of correlation surfaces, jets are extracted from random pixel locations, $\vec{r}_i$. So for image $\mathbf{a}_j[\vec{x}]$, the set of jets are

$$v_{ijk} = \Box_{\vec{r}_i} \mathbf{c}_{jk}[\vec{x}]. \tag{17.6}$$

The v_{ijk} is the k-th element of an N length vector, where N is the number of Gabor filters, for the j-th image at the i-th pixel location. The set of jets for one image is $\{\vec{v}_j\}$. Similarity of two images is reduced to the similarity of two sets of vectors.

Implementation of this process is shown in Code 17.5. The function **RandomJets** selects the jets from random locations. The inputs is the data cube that contains the correlation images in the format $N \times V \times H$ and the number of desired jets `NJ`. Line 6 creates two random numbers and these are scaled to the image size, and in line 9, a single jet is extracted. The result is a matrix in which each row is a jet.

Comparison of a set of jets from one image to a set of jets from other images can be performed through many different methods. The one selected here is PCA (see Chapter 8), which creates a new mapping space for the set of jets from an entire set of images. In this case, 100 jets from random locations were selected from each image. The vectors in PCA space are defined by

$$\vec{y}_m = \mathfrak{P}_n \vec{v}_m \quad \forall m, \tag{17.7}$$

where n is the selection of dimensions that the user decides to keep. An alternative notation is $\{\vec{y}\} = \mathfrak{P}_n\{\vec{v}\}$.

Code 17.6 shows the entire process given a list, `names`, of image names. The variable `VH` in line 1 is the frame size of the Gabor filters, which can be much smaller than the frame size of the images. Line 7 loads an image as a gray scale, and line 8 performs the correlations with all of the filters created in line 4. These are smoothed and random jets are extracted in line 11, completing Equation (17.6). Line 14 maps this data to PCA space, completing Equation (17.7).

Each of the Brodatz images is correlated with the set of filters, and a set of 100 jets is extracted from each correlation set. The data now consists of 111 sets of 100 jets. The PCA space is constructed from all of these jets. However, a plot of all of these points is very confusing, and so the

Code 17.5 The **RandomJets** function

```
1  # gabor.py
2  def RandomJets( corrs, NJ ):
3      N,V,H = corrs.shape
4      jets = np.zeros( (NJ,N), corrs.dtype )
5      for i in range( NJ ):
6          vh = np.random.rand(2)
7          v = int( vh[0] * V )
8          h = int( vh[1] * H )
9          jets[i] = corrs[:,v,h] + 0
10     return jets
```

Code 17.6 The entire process of gathering correlations, extracting jets, and mapping in PCA space

```
>>> import scipy.signal as ss
>>> import gabor as gbr
>>> import pca
>>> filts = gbr.Filts( VH, [0.25,0.5,1,2] )
>>> alljets = []
>>> for nm in names:
        orig = imageio.imread( nm, as_gray=True )
        corrs = gbr.ManyCorrelations( orig, filts )
        for c in range(len(corrs)):
            corrs[c] = ss.cspline2d(corrs[c].real, 10 )
        jets = gbr.RandomJets( corrs.real, 100 )
        alljets.extend( jets )
    alljets = np.array( alljets )
    ndata = pca.Project( alljets )
```

result will show the mapping from only four of the input images. The PCA space, though, was constructed from the jets from all of the images. The four selected images are shown in Figures 17.3(a) through 17.3(d), and the mapping of these points in the PCA space is shown in Figure 17.3(e).

Even though the data clouds overlap, each have a definable cloud shape. This application is different than the previous face recognition system. In the face system, each image was mapped to a single point in the PCA space, and the desire was to have all of the images from a single person collect in a very small region of the map. Each image mapped to a single point. In this

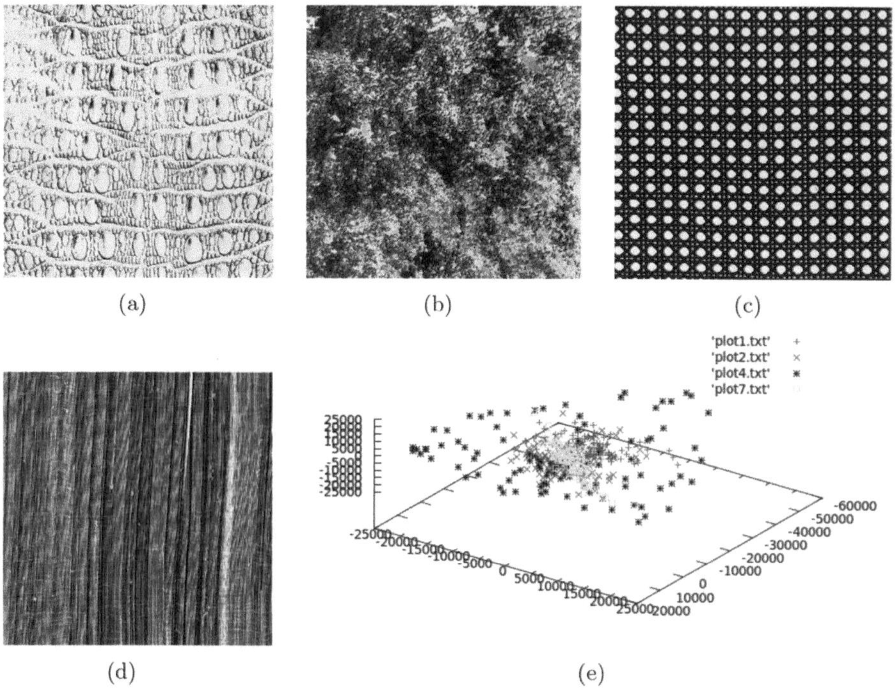

Figure 17.3 (a) Image D10, (b) image D100, (c) image D102, (d) image D105, and (e) the PCA space mapping.

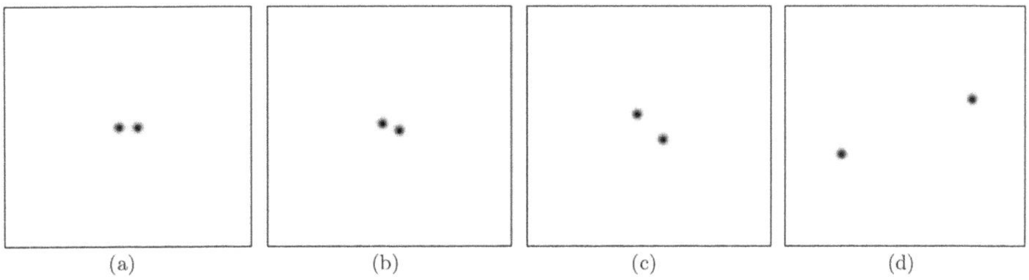

Figure 17.4 The Fourier transforms of four Gabor filters.

texture application, each original image is eventually mapped to 100 points in the PCA space. Each image defines a cloud in the PCA space. The shape of the cloud is important, but its location is not. Therefore, the fact that the clouds overlap is of no concern. Success of the system is achieved if each texture in the input space creates a signature cloud in the output space.

The red crosses in Figure 17.3(e) correspond to Figure 17.3(a). The jets create a small region in PCA space perhaps because the image is washed out and many of the jets are similar. The green crosses correspond to Figure 17.3(b), which has no regular pattern. The jets in the plot form a cloud that is mostly homogeneous. The blue asterisks correspond to Figure 17.3(c), and as seen, these jets had a much wider range of values. This figure has very sharp contrast, and thus, the jets are vastly different from each other. The final markers correspond to Figure 17.3(d). This cloud is long and quite thin, forming a cylinder with tapered ends. The regularity formed jets with similarities. Thus, each input image contained definable qualities that lead to characteristic shapes in the PCA shape. Images with similar qualities should produce clouds in the PCA space with similar images.

There are alterations to this experiment that were not considered. Changing the size of the frames of the Gabor filters does alter the results as the filters can span larger swaths of the original image. Only 100 jets were extracted but since these images are 640×640, there are almost 410,000 jets that could have been used. A larger number of jets would provide more definition to the cloud shape.

17.4 GABOR FILTERS IN FOURIER SPACE

Each Gabor filter contains a unique frequency range and thus has a concise signature in Fourier space. Figure 17.4 shows the Fourier transforms of four different Gabor filters, using ($f = 0.25$, $\theta = 0$), ($f = 0.25$, $\theta = 22.5°$), ($f = 0.5$, $\theta = 45°$), and ($f = 2.0$, $\theta = 157.5°$). Each image isolates opposing regions in the Fourier space. The locations of the regions rotate with the angle θ and expand outwards from the center as f increases. However, the size of the spots do not change.

A correlation with a Gabor filter is equivalent to isolating a small set of frequencies, which is related to the distance from the DC term. These frequencies are only for a small rotation angle, which is related to the rotation of the spots in the image. Therefore, the Gabor filters are basically localized frequencies in Fourier space.

17.5 SUMMARY

Gabor filters are another approach used to extract edge information from an image. The filers are often grouped as a set with varying frequencies, orientations, and extents. A few examples, displayed the use of Gabor filters for extracting texture information.

PROBLEMS

1. Generate filters for four different values of θ (0, $\pi/6$, $\pi/3$, $\pi/2$), with constant values $f = 1$, $\delta_x = 1$, and $\delta_y = 1$. The frame size is $\vec{w} = (128, 128)$.

2. Compute the Fourier transforms for all filters generated in Problem 1. Display the Fourier transforms according to the display Equation (10.23).
3. Generate filters for three different values of f (0.5, 1, 1.5), with constant values $\theta = 0$, $\delta_x = 1$, and $\delta_y = 1$. Use a frame size of $\vec{w} = (128, 128)$.
4. Compute the Fourier transforms for all filters generated in Problem 3. Display the Fourier transforms according to the display Equation (10.23).
5. Generate filters for three different values of δ_x (0.5, 1, 1.5), with constant values $f = 1$, $\theta = 0$, and $\delta_y = 1$. The frame size should be $\vec{w} = (128, 128)$.
6. Compute the Fourier transforms for all filters generated in Problem 5. Display the Fourier transforms according to the display Equation (10.23).
7. Using Code 17.4, correlate the input with all filters. Determine which of the filters provided the largest response (largest values in the correlation surface) for the right side of the annulus.
8. Create a Gabor filter using $f = 0.5$, $\theta = 0$, $\delta_x = 1$, and $\delta_y = 1$. Correlate this filter will all of the Brodatz images. Which image provided the largest response and why. Note: The frame size of the Brodatz images are not powers of 2. Just use the first 512 rows and 512 columns of the Brodatz image to get the best performance.
9. Extract the row of data from the negative of Figure 17.4(a) through the center of the points. Are the profiles of these signals Gaussian?
10. Create a new image similar to the negative of Figure 17.4(a), except that the signals should be made twice as large. Compute the inverse Fourier transform of this new image. Does it look like a Gabor filter?

18 Describing Shape

Often objects in an image have a defined shape. However, the presentation of that shape may not be complete. For example, a two-dimensional view of a three-dimensional shape does not capture all of the information of that shape. While it is possible to capture the entire shape of a two-dimensional object in an image, it is still difficult to mathematically describe the shape.

Consider the attempt to define the shape of a square. One method may be to claim that it has equally spaced parallel lines positioned at right angles. In this manner, the description of the shape is based on its perimeter. Other shapes, such as the silhouette of a dog, do not have any definitive description, and therefore defining shape by its perimeter is too restrictive. Other methods propose to describe a shape by its interior mass. These consider the relationship between larger contiguous regions. A third method is to describe the structure of the object. The important features of a dog would then be a large middle mass, with five thin extensions (legs and tail), and one thicker extension with a protrusion (neck and head).

The three major categories of shape description methods reviewed here are the contour, interior regions, and structure. The following lists some of the methods used in these categories. These are described in more detail in later sections and in some cases example Python scripts are provided.

- Contour Methods
  - Perimeter
  - Polygon Estimation
  - Metrics
  - Fourier Descriptors
  - Wavelets
  - Elastic Matching
- Region Methods
  - Metrics
  - Quadtrees
  - Pyramids
  - Euler Numbers
  - Geometric Moments
  - Zernike Moments
- Structure Methods
  - Convex Hull
  - Medial Axis

18.1 CONTOUR METHODS

Contour methods describe the shape by qualitative measures of the perimeter. Some of these methods only work well for simple shapes, and many have limitations on their performance.

18.1.1 CHAIN CODE

The *chain code* method snakes around the perimeter of a shape and encodes the direction of the snake by integers. Figure 18.1 shows a simple hexagon on the left which is the shape to be described. On the right is the orientation key which considers movement only in 45° increments. The hexagon is described by the vector (0,0,1,1,2,2,3,3,4,4,5,5,6,6,7,7,8,8). In this manner, each shape can then be converted to a string of integers which can thn be used for comparisons with other shapes.

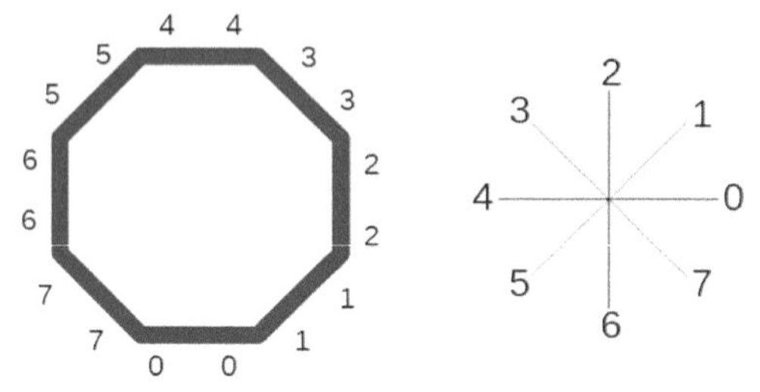

Figure 18.1 The chain code example.

This method is simple to employ but does have its drawbacks. First, it does not include any interior holes in the shape. For example, the hexagon is not filled, and there is no description of the interior. Even if the interior is coded, there is no measured relationship to the exterior perimeter. The thickness of the line separating the interior from the exterior is not described by these vectors.

This method is sensitive to scale and somewhat sensitive to the starting location of the chain. It also only considers the localized portions of the perimeter. Consider two images of a dog, and in the second photo, the tail is at a different angle to the body. The tail's code would be different because the perimeter is rotated with respect to the rest of the body, but the images are of the same object.

18.1.2 THE POLYGON METHOD

The polygon method estimates the shape of the perimeter with a polygon. Figure 18.2(a) shows an arbitrary shape, which is currently poorly described with in inscribed triangle. There are regions that have huge error, which indicates that more sides need to be added to the polygon. For example, the gap between the perimeter and the line AB is quite large, and so a new node would be places on the perimeterand named D as shown in Figure 18.2(b). The line AB would be replaced with lines AD and DB. The process continues until all of the errors are below a defined threshold.

This method, like its predecessor, does not consider any interior holes and is susceptible to changes of scale and rotation.

18.1.3 METRICS USED TO DESCRIBE SHAPE

There are several metrics that have been proposed to measure qualities of shape. Often the results of these are used in concert to provide a more accurate definition of shape. Rarely will any single metric be sufficient. The ones reviewed here are as follows:

- Area,

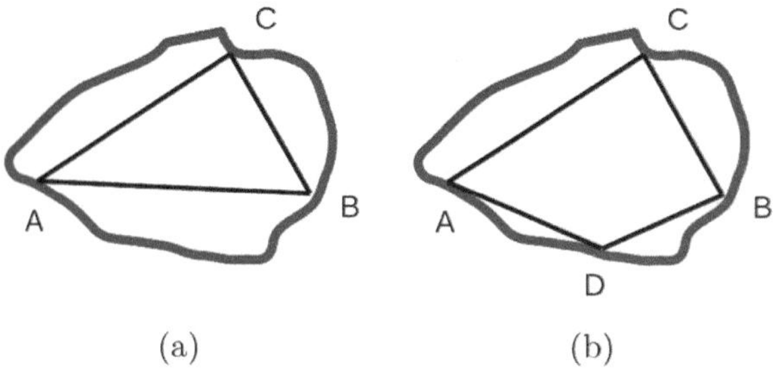

Figure 18.2 (a) The shape with the first three points, and (b) the shape with a four point added to the definition.

- Perimeter,
- Curvature,
- Bending Energy, and
- Circularity

The area is a very simple metric that is just the number of pixels that are on the object. Since the images are binary, the area is also the number of pixels and so the first metric is easily computed as

$$A = \sum_{\vec{x}} \mathbf{a}[\vec{x}]. \tag{18.1}$$

The length of the perimeter is also easy to measure as it is the number of pixels that are along the edge of the image. The perimeter is calculated by subtracted an eroded image from the original. The eroded image is one pixel smaller along the perimeter, and so the perimeter is the difference between the two images as in

$$\mathbf{b}[\vec{x}] = \mathbf{a}[\vec{x}] - \triangleright_1 \mathbf{a}[\vec{x}], \tag{18.2}$$

where the operator $\triangleright_n$ is the erosion over n iterations. The length of the perimeter is the number of pixels in $\mathbf{b}[\vec{x}]$. The perimeter length is thus

$$T = \sum_{\vec{x}} \mathbf{b}[\vec{x}]. \tag{18.3}$$

Other measures require that the perimeter points be ordered so that consecutive points follow along the edge of the shape. The operator notation is

$$\vec{v} = \mathbb{P}\mathbf{a}[\vec{x}], \tag{18.4}$$

which returns a set of pixel locations.

Extracting points on the perimeter is easy, but putting them in order is more involved. One crude method is presented in the function **PerimeterPoints**, which is shown in Code 18.1. This function sequentially finds closest points. It starts with a seed point, finds the closest one to it, and then replaces the seed with the new point. The process continues until all points have been considered. The image is converted to binary values in line 3, and the perimeter is isolated in lines 4 and 5. These are converted to a matrix in line 6. The matrix `mat` contains the locations of the perimeter points. Line 7 uses the **scipy.spatial.distance.cdist** function to compute the Euclidean distance between every pair of points. The value `cd[i,j]` is the Euclidean distance between points `mat[i]` and `mat[j]`. If the number of points is N, then the matrix `cd` is $N \times N$. The location of the minimum value in row `cd[k]` indicates which point is closest to point `mat[k]`. Of course, this search needs two exclusions. First, the closest point to `mat[k]` is itself, and thus it needs to be excluded from consideration. Second, each point can be used only once. The value `big` in line 8 is bigger than any other value in `cd`. When a pixel is used, the corresponding value in `cd` is replaced by `big`, so that it will not be considered again. The list `pmpts`, initialized in line 9 will contain the answer.

The extraction of points occurs inside the loop starting in line 13. Line 14 prevents a point from being its own closest point. Line 15 identifies the closest point to the seed, and line 16 places this in the answer list. Line 17 removes the current point form further consideration. The call to the function occurs in line 23. The previous two lines load the image and converts it so the background is black and the target is white.

Displaying the result is performed by the **ShowPerimPoints** function shown in Code 18.2. Basically, the pixel at each point is converted to a gray value that increases as the process iterates. Thus, the order of the points from **PerimterPoints** is shown in an increasing gray scale. The result is shown in Figure 18.3.

The length of the perimeter is the sum of the Euclidean distances between the points. The function **ChainLength** computes this distance excepting the distance from the last to the first point.

Code 18.1 The **PerimeterPoints** function

```
# shape.py
def PerimeterPoints( adata, pick=0 ):
    temp = adata > 0.5
    bdata = temp - nd.binary_erosion( temp )
    v,h = bdata.nonzero()
    mat = np.vstack((v,h)).T
    cd = sp.distance.cdist(mat,mat)
    big = cd.max()*2
    pmpts = [pick]
    k = pick
    N = len( cd )
    cd[:,k] = big
    for i in range(N-1):
        cd[k,k] = big
        x = cd[k].argmin()
        pmpts.append( x )
        cd[:,x] = big
        k = x
    return pmpts, mat

>>> adata = imageio.imread('data/bear1.png', as_gray=True )
>>> adata = adata.max() - adata
>>> pmpts, mat = shape.PerimeterPoints( adata )
```

Code 18.2 The **ShowPerimPoints** function

```
# shape.py
def ShowPerimPoints( vh, mat, pmpts ):
    showme = np.zeros( vh )
    for i in range(len(pmpts)):
        v,h = mat[pmpts[i]]
        showme[v,h] = i
    return showme

>>> showme = shape.ShowPerimPoints( adata.shape, mat, pmpts )
```

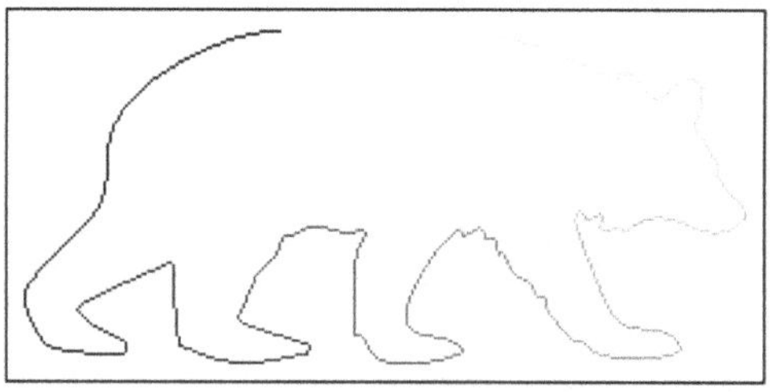

Figure 18.3 The perimeter points are extracted in order of progression as evidenced by the increasing gray value.

Code 18.3 The **ChainLength** function

```
# shape.py
def ChainLength( perimpts ):
    a = perimpts[1:] - perimpts[:-1]
    dist = np.sqrt((a**2).sum())
    return dist
```

Code 18.4 The **Curvature** function

```
1 # shape.py
2 def Curvature( perimpts ):
3     a,b = np.gradient( perimpts )
4     c,d = np.gradient( a )
5     kt2 = c[:,0]**2 + c[:,1]**2
6     return kt2
```

Code 18.3 shows this function in which the input is a matrix of which each row is a point in the space and that these points are in the correct order.

With the tools of ordering the perimeter pixels in hand, it is possible to proceed with the third metric. Curvature is defined as

$$|k(t)|^2 = \left(\frac{d^2x}{dt^2}\right)^2 + \left(\frac{d^2y}{dt^2}\right)^2, \tag{18.5}$$

where t is the variable that follows the contour. Basically, this is the square of the Euclidean distance of the second derivative of the data points. The derivative of values in a vector is computed by the **numpy.gradient** function. For a matrix, this returns the gradients in the vertical and horizontal dimensions, respectively. In this computation, only the gradients in the vertical direction of the matrix are desired and so only variables a from line 3 and c from line 4 are needed. Line 4 is the second derivative, and the Euclidean distance is computed in line 5. The return is a vector that describes the curvature of the perimeter.

The fourth metric, bending energy, is described as,

$$E = \frac{1}{N}\sum_i |k(t)|^2, \tag{18.6}$$

which is merely the sum of the values in the vector returned by **Curvature**.

The final metric reviewed in this section is circularity which is,

$$\gamma = \frac{T^2}{4\pi A}. \tag{18.7}$$

A similar attempt at measuring geometric qualities of a shape is geometric moments. The set of moments is defined as

$$M_{ij} = \sum_x \sum_y x^i y^j I(x,y). \tag{18.8}$$

The function $I(x,y)$ defines the points on the perimeter and i and j are a series of small integers.

18.1.4 FOURIER DESCRIPTORS

One of the issues with the previous method is that it does not capture relative positions of the shapes. For example, in an image of a standing bear, the bear's head is at the top of the frame. It would be an odd image to see the bear's head at the bottom of the frame. A particular object has expected relationships among its components. The process of *Fourier descriptors* is a method that attempts to capture the relative shape of an object. In this regard, it is insensitive to size, location, or even rotation.

Consider a point that is located at the center of mass of the object. Each point on the perimeter can be described as an (x,y) point with reference to this center point. Figure 18.4 shows the concept with the vector pointing to a single point on the perimeter. For each angle, θ a (x,y) point is collected and stored in a complex format as $x+\imath y$. A vector is created from these points for a given set of θ values.

To gather the Fourier descriptors, the first step is to define a center point. One possible method is to just find the center of mass (assuming that each pixel has a mass of 1). Given a binary-valued image $\mathbf{a}[\vec{x}]$, the process is

$$\vec{w} = \boxtimes \mathbf{a}[\vec{x}]. \tag{18.9}$$

The **center_of_mass** function from the *scipy.ndimage* module performs this calculation.

If the intent is to compare descriptor vectors to each other, then it is necessary that the vectors have the same length. The **RPolar** function used in Code 6.11 in Chapter 6.7 uses the horizontal dimension of the image frame to determine the number of angles used in the transformation. Therefore, it is important that all images have the same frame size before computing the Fourier descriptors. The vector-scaling vector is then

$$\vec{z} = (1, 360/H), \tag{18.10}$$

and the transformation is

$$\mathbf{b}[\vec{\rho}] = S_{\vec{z}} \mathcal{P}_{\vec{v}} \mathbf{a}[\vec{x}], \tag{18.11}$$

where S is the Scaling operator, $\mathcal{P}$ is the Radial-polar operator, and $\vec{v}$ is the center of the radial-polar transformation.

At this juncture, the shape has been scaled and converted to the radial-polar representation. Thus, radial lines from the center point in $\mathbf{a}[\vec{x}]$ are vertical lines starting at the top of the frame in $\mathbf{b}[\vec{x}]$. The distance from the top of the frame to the bright pixels in the new frame is the radial lengths in the original. Furthermore, the width of the new image is 360, and so each column in the image represents a radial sample at 1° increments from the original.

The next step in the process is to measure the distances from top of the frame to the bright pixels. This is an iterative process denoted by

$$y_i = \mathcal{N} \square_i \mathbf{b}[\vec{\rho}], \quad \forall i, \tag{18.12}$$

where the $\square_i$ operator retrieves a subimage, and in this case is a vertical sample of width 1 at column i. The $\mathcal{N}$ is the Nonzero operator, which returns a vector of the index (or location) of the nonzero elements in the input. The last entry in this vector is the distance from the center of mass

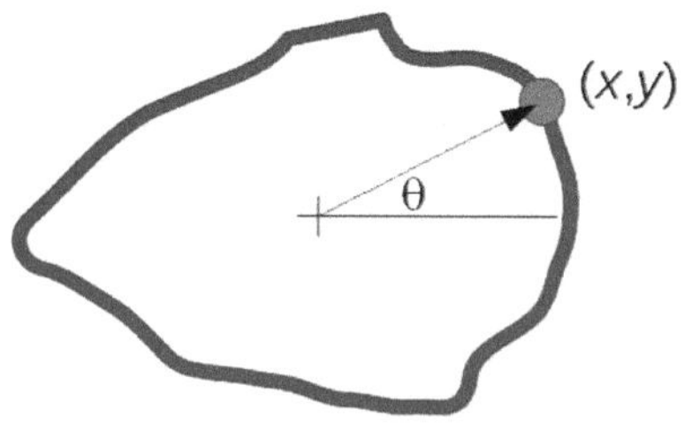

Figure 18.4 A single point on the perimeter of the shape is converted to complex notation and stored as an element in a vector.

to the perimeter of the shape along a radial in the original image. So, these values are collected in a vector $\vec{v}$:

$$v_i = y_i[-1], \quad \forall i. \tag{18.13}$$

This vector has 360 values, which are the distances from the center to the perimeter along the radials.

Finally, the Fourier transform of these elements are computed.

$$\vec{\xi} = \mathfrak{F}\vec{v}. \tag{18.14}$$

The Fourier Descriptor operator is $\mathcal{F}$, and so the whole process is encapsulated by

$$\vec{\xi} = \mathcal{F}\mathbf{a}[\vec{x}]. \tag{18.15}$$

Recall from Section 12.2.1 that a multiplication in Fourier space is equivalent to a correlation in the input space. The correlation of two descriptor vectors as computed in Equation (18.13) is the comparison of the two vectors for all possible relative shifts. In this case, the vectors represent data taken at angular increments. So the correlation of two such vectors is equivalent to compare the two shapes for all possible rotations. This information is available by multiplying the Fourier descriptor vector ($\vec{\xi}$) of two shapes and then computing the inverse Fourier transform of the result. This completes the correlation operation. A large spike in this result indicates that there is a rotation in which the Fourier descriptor vectors are similar. This can be interpreted as an indication that the two shapes are equivalent.

Code 18.5 displays the **FourierDescriptors** function, which computes the Fourier descriptor vector. The input `orig` is a binary-valued image with the target pixels set to 1, and the background set to 0. Line 3 computes the center of mass, which is used as the center point for rotation, and line 5 creates the radial-polar image.

Usually, the Fourier descriptors are used to compare images, and therefore, it is necessary that all images have the same width. Lines 6 through 8 convert the radial-polar image so that the width is 360 pixels. The `for` loop starting in line 11 samples each vertical column of this image and finds the location of the last pixel with a value of 1. This distance is the radial distance from the center of

Code 18.5 The **FourierDescriptors** function

```
1  # shape.py
2  def FourierDescriptors( orig ):
3      cofm = nd.center_of_mass( orig )
4      vv,hh =int(cofm[0]), int(cofm[1])
5      rp = rpolar.RPolar( orig, (vv,hh))
6      V,H = rp.shape
7      scale = 360/H
8      rp = nd.zoom( rp,(1,scale))
9      H = 360
10     vec = np.zeros( H, complex )
11     for i in range(H ):
12         nz = rp[:,i].nonzero()[0]
13         if len(nz)==0:
14             nz = old
15         theta = np.radians( i*360./H )
16         vec[i] = nz[-1]*np.cos( theta ) + 1j*nz[-1]*np.sin(theta)
17         old = nz
18     return vec
```

mass to the perimeter. There are 360 columns and so the Fourier descriptors are collected for every 1° increment for θ. Line 16 converts the information about the radius and angle back to rectilinear coordinates and stores the information in a vector. The output is a complex-valued vector with 360 elements. This completes the process, and the shape has been converted into a complex-valued vector.

There are two major disadvantages to the use of Fourier descriptors. The first is that the process is heavily reliant on the choice of the center point from which the radial lines begin. A small change in the location of the center point will change all of the lengths of the radial lines and likewise all of the values of $\vec{\xi}$. One method of locating the center point is to use the center of mass, but if part of the shape is changed, perhaps due to an occlusion, then the center of mass changes and therefore does also all of the (x, y) points.

A second issue is that each radial line has only one (x, y) point that is defined, but more complicated shapes will have the perimeter of the shape intersect with a single radial line in multiple locations. Thus, this method should be applied only to convex shapes.

18.1.5 WAVELETS

In this application, the perimeter of the image is the input to the wavelet decomposition. The elements of a vector are the energies of the individual quadrants of the decomposition. Each image creates a shape description vector, and vectors for different images can be compared by employing the user's desired metric.

The latter is described in three steps, where the first converts the input binary-valued image to a perimeter by

$$\mathbf{b}[\vec{x}] = \mathbf{a}[\vec{x}] - \rhd_1 \mathbf{a}[\vec{x}]. \tag{18.16}$$

The wavelet decomposition is applied as in

$$\mathbf{c}[\vec{x}] = \mathcal{W}\mathbf{b}[\vec{x}], \tag{18.17}$$

and the energy vector over the subimages is collected,

$$v_i = \mathcal{E}\square_i \mathbf{c}[\vec{x}], \quad i = 1, \ldots N, \tag{18.18}$$

where $\square_i$ represents the windows needed to extract the information from each of the subimages in $\mathbf{c}[\vec{x}]$. The value N is the number of subimages generated by the wavelet process, and the final result is a vector $\vec{v}$ that contains the energy values at different resolutions.

This vector is not sensitive to location or orientation of the shape. Small occlusions of the shape will result in only small changes in $\vec{v}$. However, this process is sensitive to the size of the shape. The length of the vector is related to the size of the image frame, but that is easily controlled by the user limiting the number of iterations in the wavelet process.

18.1.6 ELASTIC MATCHING

Elastic matching is the energy necessary to warp one image into another. Consider two images, $\mathbf{a}[\vec{x}]$ and $\mathbf{b}[\vec{x}]$, which are of similar objects such as faces. Elastic matching is reduced to determining the energy needed to convert one grid into another. More simply, it is the sum of the distances between fiducial points of the same index. In the case of faces with 45 fiducial points, there are 45 distances from a fiducial point in one image to the same fiducial point in the other image. These distances are then combined to create a single energy value for the differences in the grids. This process does not actually need the image or the warping process from Section 7.1. It needs only the fiducial grids.

However, the raw grids are not sufficient for the comparison. Three issues that first need to be considered. The first of these is shift. If the faces are not in the same location in their respective frames, then there will be added distances between the fiducial points. So, one of the grids needs to be shifted so that they center point aligns with the other. The second issue is rotation. It is quite possible that the face in one of the images is slightly tilted which would again infuse differences in distances that are not related to the actual grid. So, one of the grids may need to be rotated to align with the other. The third issue is scale. It is possible that the camera was closer to the face when collecting the data. Different scales will infuse errors in comparison that also need to be removed.

None of these issues is particularly difficult to resolve, and the following codes are provided to remove these invasive errors. Consider a very simple case of three face images, $\mathbf{a}[\vec{x}]$, $\mathbf{b}[\vec{x}]$, and $\mathbf{c}[\vec{x}]$. The first two are different photographs of the same person, and the last one is a different person. The fiducial grids were obtained manually (see Section 7.1.1). The dancer program stores the data for each fiducial point in a row of a text file. Each row has the format `(x,y) r g b`, where `(x,y)` is the location and the other values are the color values which are not important. The function **ReadFiducial** shown in Code 18.6 reads in this file and isolates the `x` and `y` values. The function returns a 45×2 matrix.

This process is applied to each image. The notation for the i-th fiducial grid is $\vec{g}_i$. The goal of the process is then to compare all of the vectors in one grid to the partner vectors in the other grid. The comparison metric is not critical, but for this case is chosen to be the Euclidean distance. Thus, the measure of similarity between two grids is

$$u = \frac{1}{N} \sum_i \left(\mathfrak{D} \left(\vec{g}_i, \vec{h}_i \right) \right)^2, \tag{18.19}$$

where $\mathfrak{D}$ is the Distance operator, and $\vec{g}_i$ and $\vec{h}_i$ are the fidicual grids for the images after the biases are removed. In this example, there are $N = 45$ fiducial points in each grid. This operation will compute the 45 Euclidean distances between the similar fiducial points of each grid. The value of u is the energy required to warp one grid into the other.

The first bias is the shift. Both grids need to be centered in the same frame. The second grid is shifted so that its center is in the same location as the first grid. The shift for the i-th image is determined by

$$\Delta \vec{v}_i = \vec{v}_1 - \vec{v}_i, \tag{18.20}$$

where $\vec{v}_1$ is the center of the first grid and $\vec{v}_i$ is the center of the i-th grid. The vectors are determined by

$$\vec{v}_i = \mathcal{M}\{\vec{g}_i\}. \tag{18.21}$$

Code 18.6 The **ReadFiducial** function

```
# elastic.py
def ReadFiducial( inname ):
    a = open(inname).read()
    a = a.split('\n')
    N = len( a )
    if len(a[-1])<10:
        N -= 1
    mat = np.zeros( (N,2) )
    for i in range( N ):
        b = a[i].split()
        b1 = int( b[0][1:-1])
        b2 = int( b[1][:-1])
        mat[i] = b1,b2
    return mat
```

Removing the rotation bias is a similar process in which the angles from the center to each fiducial point are collected. The average angle is the average over these 45 values. The second grid is rotated so that it has the same average angle as the first grid. The rotation angle is determined by

$$\Delta\theta_i = \theta_1 - \theta_i, \tag{18.22}$$

where

$$\theta_i = \tan^{-1}\left(\frac{g_{i;y}}{g_{i;x}}\right), \tag{18.23}$$

and $g_{i;y}$ is the y-th coordinate value of the g_i grid.

The scaling bias is adjusted through a scaling factor:

$$s_i = \frac{d_1}{d_i}, \tag{18.24}$$

where d_i is a parameter for the i-th grid as determined by

$$d_i = \mathcal{M}\mathfrak{D}\left(\vec{g}_i, 0\right), \quad \forall i, \tag{18.25}$$

with $\mathfrak{D}$ representing the Distance operator.

Now that all of the parameters are in place, the biases in a grid $\{\vec{g}_i\}$ are removed via,

$$\vec{h}_i = S_{s_i}\mathcal{R}_{\Delta\theta_i}D_{\Delta\vec{v}_i}\vec{g}_i, \quad \forall i. \tag{18.26}$$

With the biases removed, it is possible to accurately measure the differences as in Equation (18.19).

The following implementation in Python scripts solves each of the three biases separately. The first is to center the grid, which uses the **RemoveCenterBias** function shown in Code 18.7. The input is `mats`, which is a list of grid matrices. Line 3 computes the center of the first grid, and line 5 computes the center of the subsequent grids. Line 6 computes the shift, which is applied in line 7 for the i-th grid. The data inside the list is changed, and so it is not necessary to have a `return` statement.

The second process removes the angular bias, which is accomplished in the **RemoveRotateBias** function shown in Code 18.8. The average angle for a grid is the average of the angles from a center point to each fiducial point. This is computed in line 7 for each grid. The rotation of the i-th grid is then the difference between the average angle for the i-th grid and the first grid as computed in line 9. The distance from the center point to each fiducial point is computed in line 12, and thus combined with the earlier computation of the angle; each point has been converted to its polar coordinates. The rotation is applied in lines 13 and 14. Once again, the contents of the list `mats` are being changed, and so it is not necessary to have a `return` statement.

The final bias to be removed is a scale difference. Once again, the data is considered in polar coordinates and the distances from the center point to each fiducial point are collected in line 5 of

Code 18.7 The **RemoveCenterBias** function

```
1 # elastic.py
2 def RemoveCenterBias( mats ):
3     ctr1 = mats[0].mean(0) # center of first grid
4     for i in range( 1, len(mats)):
5         ctr = mats[i].mean(0)
6         shift = ctr1 - ctr
7         mats[i] += shift
```

Code 18.9 which shows the **RemoveScaleBias** function. The change in scale is computed in line 10 as the ratio of the average distance values of the first grid and the i-th grid.

Now, all of the parts on in place to convert a grid to the correct location, orientation, and scale. These three functions are applied to all of the grids. Each receives the grid from all of the images, since the biases are converting the data to a common location, rotation, and scale. Since each grid consumes only a few hundred bytes, these functions can process a large number of grids.

The final step is to compare the grids. This can be performed in many different manners as suits the user. The example here chooses to use the average of the Euclidean differences between the like fiducial points. The Euclidean differences are computed in the **GridDifference** function shown in Code 18.10.

The matrix `matA` is the fiducial grid for $\mathbf{a}[\vec{x}]$ after the biases are removed. Likewise, `matB` and `matC` are associated with $\mathbf{b}[\vec{x}]$ and $\mathbf{c}[\vec{x}]$ respectively. Images $\mathbf{a}[\vec{x}]$ and $\mathbf{b}[\vec{x}]$ are different views of the same person, while $\mathbf{c}[\vec{x}]$ is an image of a different person. Line 8 shows the score of the two images of the same person was 4.8 and for the images of two different persons was 7.5. A lower score is a better match, and so it is seen that in this small example, elastic matching could distinguish the different people. Of course, this is a very small example and scaling up to a larger number of images

Code 18.8 The **RemoveRotateBias** function

```
# elastic.py
def RemoveRotateBias( mats ):
    N = len( mats )
    D = len( mats[0] )
    T = np.zeros( (N,D) )
    for i in range( N ):
        T[i] = np.arctan2( mats[i][:,1], mats[i][:,0] )
    for i in range( 1, N ):
        diff = T[0] - T[i]
        avgdiff = diff.mean()
        T[i] += avgdiff
        r = np.hypot( mats[i][:,0], mats[i][:,1] )
        mats[i][:,0] = r * np.cos(T[i])
        mats[i][:,1] = r * np.sin(T[i])
```

Code 18.9 The **RemoveScaleBias** function

```
1  # elastic.py
2  def RemoveScaleBias( mats ):
3      N = len( mats )
4      D = len( mats[0] )
5      R = np.hypot( mats[0][:,0],  mats[0][:,1] )
6      avg = R.mean()
7      for i in range( 1, N ):
8          R = np.hypot( mats[i][:,0],  mats[i][:,1] )
9          ravg = R.mean()
10         scale = avg/ravg
11         R *= scale
12         T = np.arctan2( mats[i][:,0], mats[i][:,1] )
13         mats[i][:,0] = R * np.sin( T )
14         mats[i][:,1] = R * np.cos( T )
```

Code 18.10 The **GridDifference** function

```
# elastic.py
def GridDifference( grid1, grid2 ):
    diff = (grid1 - grid2)**2
    diff = np.sqrt(diff.sum(1))
    return diff

>>> GridDifference( matsA, matsB ).mean()
4.8520996692580685
>>> GridDifference( matsA, matsC ).mean()
7.4959247663264224
```

may provide different performance results. However, in terms of face recognition, other algorithms can be used to assist in the recognition process. For example, results from eigenface comparisons could compliment the evidence provided by elastic matching to correctly classify faces.

18.2 REGION METHODS

Many of the previous methods analyzed the perimeter of the shape in the attempt to define the shape. The region methods in this section create a description of the shape by analyzing the interior.

18.2.1 EIGENVECTORS AND EIGENVALUES

The simplest measure of the interior is the area. As often is the case, the very simplest method is also the poorest performing metric. Certainly it is possible to have a star and a circle with the same area but their shapes are very different. So the description of shape by its region must be more extensive than the measurement of area.

Eigenvectors provide an insight to an objects overall shape including cases in which the perimeter is poorly defined. The following examples use multivariate distributions, which define interior data points but poorly define the boundaries of the data. Figure 18.5 displays an object with a definable shape but a fuzzy perimeter. This data was created by the **numpy.random.multivariate_normal** function in **Shape1** shown in Code 18.11. The function generates 1,000 points randomly distributed according to a multivariate distribution. Line 4 computes the covariance matrix which is require by the **multivariate_normal** function.

The function **Shape1** creates a distribution of data points, but the measurement of these points must be performed without knowledge of how they were generated. Thus, the covariance matrix used in the creation of the points is not available to the function that will measure properties of these points. A proper measurement of the data points should produce a covariance matrix that is similar to the matrix that was used to generate the points. Measurements of the data are performed in the **ExtractStats** function shown in Code 18.12. The covariance matrix computed in line 4 is based on just the data and so it should be similar but not exactly the same as the covariance matrix used in the **Shape1** function to generate the data. The output is the computed covariance matrix, the eigenvalues, and the eigenvectors (columns in the last matrix). The first eigenvector aligns with the long extent of the data cloud, and the eigenvalues are related to the eccentricity of the distribution.

Table 18.1 shows the matrices and vectors produced by **ExtractStats** for the three different cases. The first is the original data, which has generated using the covariance matrix

$$\begin{pmatrix} 10 & 0 \\ 0 & 1 \end{pmatrix}.$$

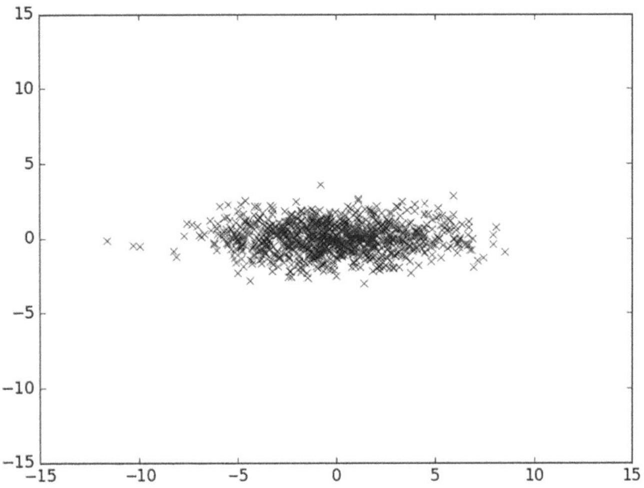

Figure 18.5 The output of **Shape1**.

Code 18.11 The **Shape1** function

```
# eshape.py
def Shape1():
    mean = np.array((0,0))
    cov = np.array(((10,0),(0,1)))
    x,y = np.random.multivariate_normal(mean,cov,1000).T
    return x,y
```

Code 18.12 The **ExtractStats** function

```
1  # eshape.py
2  def ExtractStats( x,y ):
3      mat = np.array( (x,y))
4      cv = np.cov( mat )
5      evals, evecs = np.linalg.eig( cv )
6      return cv, evals, evecs
7
8  (array([[ 9.75458434,  0.04500644],
9         [ 0.04500644,  1.00827913]]),
10  array([ 9.75481593,  1.00804754]),
11  array([[ 0.99998676, -0.00514556],
12        [ 0.00514556,  0.99998676]]))
```

As seen in lines 8 and 9, the calculated covariance matrix is nearly the same. The eigenvalues (line 10) are very close the diagonal values of the original covariance matrix, and the eigenvectors (lines 11 and 12) are columns of the original covariance matrix, each scaled to the length of 1.

Shapes 2 and 3 are shown in Figure 18.6. Shape 2 starts with the same covariance, but the distribution is rotated by 45°. Thus, the eigenvalues are nearly identical with Shape 1, and the eigenvectors differ by the same rotation. Shape 3 is at the same orientation as Shape 2 but the distribution is

Table 18.1
Statistics from the Three Shapes

	Covariance	Eigenvalues	Eigenvectors
Shape 1	$\begin{pmatrix} 9.75 & 0.05 \\ 0.045 & 1.01 \end{pmatrix}$	$9.75, 1.01$	$(1.00, 0.01), (-0.01, 1.00)$
Shape 2	$\begin{pmatrix} 5.33 & 4.17 \\ 4.17 & 4.99 \end{pmatrix}$	$9.33, 0.98$	$(0.72, 0.69), (-0.69, 0.72)$
Shape 3	$\begin{pmatrix} 25.27 & 24.78 \\ 24.78 & 26.37 \end{pmatrix}$	$50.61, 1.033$	$(-0.71, 0.70), (-0.70, -0.71)$

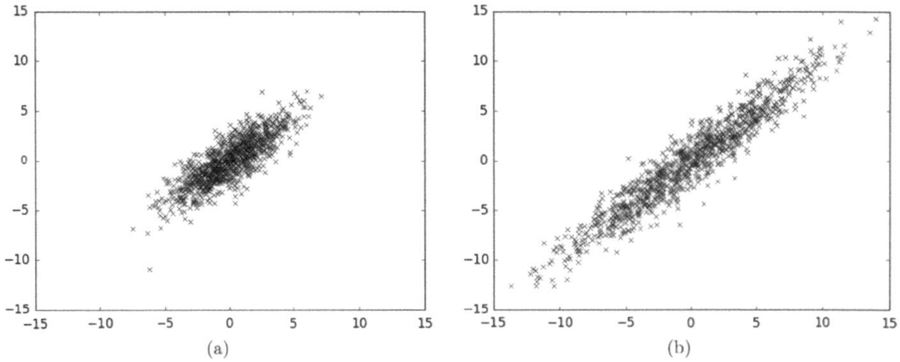

Figure 18.6 (a) Shape 2 and (b) Shape 3.

$$\begin{pmatrix} 50 & 0 \\ 0 & 1 \end{pmatrix}.$$

Thus, there is a change in the eigenvalues but not the eigenvectors. As seen in this simple example, the eigenvectors are indicative of the shape and the eigenvectors are indicative of the orientation.

Two-dimensional problems produce only two eigenvalues. While there was success on the previous test, it is not hard to develop examples that foil this attempt to define shape. Consider the shapes shown in Figure 18.7 and their accompanying results shown in Table 18.2.

Shape 4 and Shape 3 have nearly identical statistics. This indicates that this method is not capable of distinguishing between the two of them. Shape 5 has slightly more distinguishing values because

Table 18.2
Statistics from Figure 18.7

	Covariance	Eigenvalues	Eigenvectors
Shape 4	$\begin{pmatrix} 26.46 & 24.78 \\ 24.78 & 26.37 \end{pmatrix}$	$1.03, 50.61$	$(-0.72, 0.70), (-0.70, -0.71)$
Shape 5	$\begin{pmatrix} 22.70 & 20.64 \\ 20.64 & 22.86 \end{pmatrix}$	$2.14, 43.42$	$(-0.71, 0.71), (-0.71, 0.71)$
Shape 6	$\begin{pmatrix} 19.86 & 17.69 \\ 17.69 & 19.57 \end{pmatrix}$	$37.40, 2.02$	$(0.71, 0.70), (-0.70, 0.71)$

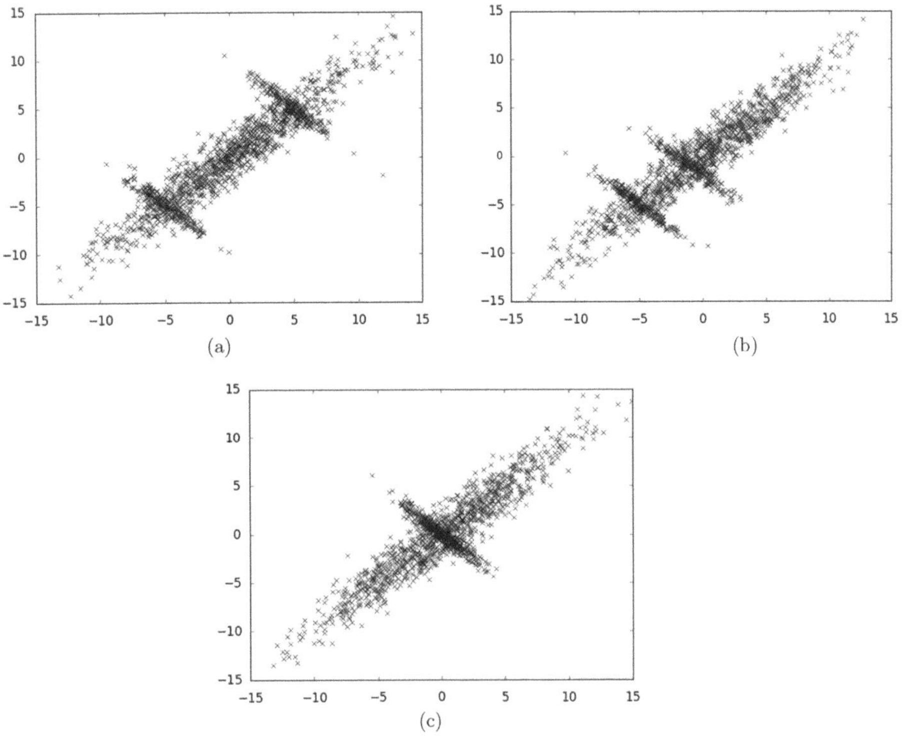

Figure 18.7 Shapes 4, 5, and 6.

one of the small branches has moved. Shape 6 has the same number of points in its small branch as does Shape 4 in its two small branches. It does show a difference in its eigenvalues. Since Shape 5 is the only one that is not symmetric the conclusion is that the symmetry cannot be determined from these values. Perhaps the change in eigenvalues is related to the distance between the small branches, but with the absent symmetry quality, the conclusion is that this method may be good at distinguishing the distance between the branches but not the location of the branches.

18.2.2 SHAPE METRICS

Eigenvalues, however, should not be discarded as they can still provide information for a shape metric. Another group of four shape metrics uses the eigenvalues in one of the definition.

The first metric in this group is area

$$m_1 = \sum_{\vec{x}} \Gamma_{>\gamma} \mathbf{a}[\vec{x}], \tag{18.27}$$

where γ is set so that $\Gamma \mathbf{a}[\vec{x}]$ becomes a binary-valued representation of the shape. All of the pixels inside the shape are set to 1, and the rest are set to 0. The second metric is the perimeter length

$$m_2 = \mathbf{a}[\vec{x}] - \rhd_1 \mathbf{a}[\vec{x}]. \tag{18.28}$$

The third metric is the eccentricity, which is the ratio of the first two eigenvalues

$$m_3 = \frac{\mu_1}{\mu_2}. \tag{18.29}$$

Recall that the eigenvalues are related to the elongation of the space. A perfect circle creates identical eigenvalues and thus $m_3 = 1$. An oblong shape produces $\mu_1 \gg \mu_2$ and thus $m_3 \gg 1$. This definition works well for 2D shapes, but the shapes in higher dimensions produce more than two eigenvalues.

The fourth metric is compactness, which is the ratio of area to perimeter. These values are already at hand, so the expression is

$$m_4 = \frac{m_1}{m_2}. \tag{18.30}$$

A perfect circle is the most compact shape as it has the smallest perimeter for a given area. An oblong shape with the same area will have a larger perimeter, and thus, a smaller value of m_4.

Only a few lines of Python script are required to determine the values of these four metrics, as shown in Code 18.13 with the function **Metrics**. The area is computed in line 3. Line 5 computes the perimeter, and line 6 sums up these values to get m_2. The computation of m_3 begins on line 7 in which the locations of all of the on-target pixels are collected. These are placed into a matrix `mat`, which is then used to compute the 2×2 covariance matrix, and then the eigenvectors and eigenvalues. Line 12 insures that the largest eigenvalue is in the numerator. The final metric is computed in line 13.

Consider the case shown in Figure 18.8 of six binary shapes. Three of these are bears and three are birds. A successful description of shape should produce a vector bear images that are self-similar, but different than the vector from the bird image.

Code 18.14 calls **Metrics** and computes the four values for a single image. Line 1 loads the image and line 2 inverts the black and white values. The images as shown have a dark target and a white background. If this image was used, then the area would be the background and not the target. So, line 2 inverts the image and makes sure that the values are floats. The function is called in line 3, and it returns four values for the area, perimeter length, eccentricity, and compactness.

Code 18.13 Computing four geometric values

```
1  # eshape.py
2  def Metrics( orig ):
3      area = (orig > 0.5).sum()
4      a = nd.binary_erosion( orig )
5      perim = orig-a
6      perimsum = perim.sum()
7      vs,hs = a.nonzero()
8      mat = np.zeros( (2,len(vs)) )
9      mat[0] = vs; mat[1] = hs
10     evls, evcs = np.linalg.eig( np.cov( mat ) )
11     eccen = evls[0]/evls[1]
12     if eccen < 1: eccen = 1/ eccen
13     compact = perimsum**2 / area
14     return area, perimsum, eccen, compact
```

(a) (b) (c) (d) (e) (f)

Figure 18.8 Six silhouette shapes.

Code 18.14 The metrics for the six shapes

```
1  >>> data = imageio.imread( 'data/shapes/bear1.png', as_gray=True)
2  >>> data = (data < 100).astype( float )
3  >>> Metrics( data )
4  (19497.0, 952.0, 4.710, 46.484)
5  ...
6  (15864.0, 647.0, 7.570, 26.387)
7  (23619.0, 775.0, 3.279, 25.429)
8  (29962.0, 766.0, 3.461, 19.583)
9  (30277.0, 782.0, 3.296, 20.197)
10 (22003.0, 696.0, 2.878, 22.015)
```

Lines 6 and beyond shows the same computation for the images in the order shown in Figure 18.8. The area and perimeter measures are not good discriminators between the two types of objects. The third column is the eccentricity, and as seen, the first two bears have larger values. That is due to the targets being longer along one axis. The fourth column is the compactness, and again, there is a quantitative difference between the two classes.

The success here is very limited. For example, the bears have their limbs extended in the images whereas the birds do not. That drastically affects the values of m_3 and m_4. Since it is quite possible to have images of birds with wings expanded and bears in a more compact sitting position, the metrics used here are doomed to failure. Fortunately, there are many more methods in which shape can be described.

18.3 DESCRIBING STRUCTURE

The final type of shape descriptors that will be reviewed here attempts to describe the structure of the shape. The convex hull method is based on a series of level sets and attempts to describe a shape as the evolution of its perimeter. The medial axis, on the other hand, attempts to describe the structure via the spine of the shape.

18.3.1 CURVATURE FLOW

In many cases, the recognition of shape must be insensitive to small changes and focus more on the overall structure rather than the details at the perimeter. One approach is to use curvature flow, which considers the localized curvatures of the shape. The flow is the action of changing that shape such that the border of the shape move toward the local center of curvature. This idea is shown in Figure 18.9, which shows a solid shape, and the arrows point in the direction in which the border will move. The length of the arrows shows the speed. Regions that have a smaller radius of curvature change at a faster rate.

Small perturbations in the perimeter are removed in the first iterations of this process. Then, the progression is based on the structure of the object rather than the details of the perimeter. Eventually, all shapes (in two dimensions) become a circle and then collapse to a point. Before it reaches that circle, the process contains information about the structure of the shape. This can be captured as the sum of the perimeter or the shape itself. As time progresses, the perimeter evolves and thus the chosen summation changes. This information is captured in a vector and then used to compare to other shapes.

Computing a single iteration of the evolution of the shape is relatively easy. The shape is represented by binary-valued pixels in which the interior pixels are 1, and the exterior pixels are 0. When a smoothing operator is applied to this shape, then the values of the pixels near the perimeter

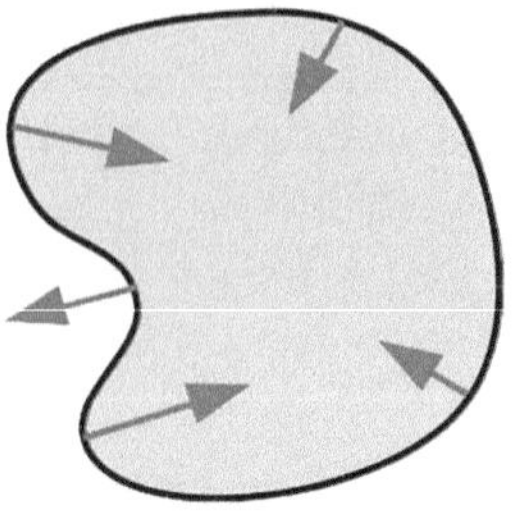

Figure 18.9 The directions of curvature flow.

change. Exterior pixels near the perimeter increase in value, and this increase is dependent on the local curvature of the shape. Likewise, interior pixels near the perimeter of the shape are reduced in value, and this reduction is sensitive to the local curvature. Thus, the pixels that need to be set to 1 are those that are exterior with a large value after the smoothing process. The pixels that need to be set to 0 are those interior pixels that have a small value after smoothing.

Given an image with a solid shape, $\mathbf{a}[\vec{x}]$. The selection of pixels to turn on is

$$\mathbf{b}[\vec{x}] = \mathbf{a}[\vec{x}] + \Gamma_{>(1-\gamma)}\left(\mathcal{S}\mathbf{a}[\vec{x}] \times (1 - \mathbf{a}[\vec{x}])\right), \tag{18.31}$$

where γ controls the speed of the progression and is a value between 0 and 1. The value of $\gamma = 0.75$ works well on most shapes. The smoothed image is multiplied by $1 - \mathbf{a}[\vec{x}]$, which are the pixels outside the shape. Thus, the only pixels that survive the multiplication are exterior pixels that have been raised in value by the smoothing. These are compared to a threshold $1 - \gamma$ which keeps only the selected pixels that will grow the perimeter. These are added to the original image, and thus, the pixels turned on in $\mathbf{b}[\vec{x}]$ are from the original shape and the growth.

The shrinking process is described as,

$$\mathbf{c}[\vec{x}] = \mathbf{b}[\vec{x}] - \left(\Gamma_{<\gamma}\mathcal{S}\mathbf{b}[\vec{x}]\right) \times \mathbf{b}[\vec{x}]. \tag{18.32}$$

The result from the previous computation is smoothed and compared to a threshold. This keeps all pixels below a threshold and this is multiplied by the shape to remove from consideration of any external pixels. The surviving pixels are interior and below a threshold. These are subtracted from the shape, which moves portions of the shape inwards.

There are only two equations that perform a single iteration of the curvature flow, and so the Python script is simple to assemble. The **CurveFlow** function shown in Code 18.15 has only three lines of instructions. Line 3 ensures that the incoming data is treated as floats. Line 4 performs Equation (18.31), and line 5 performs Equation (18.32). The output is the new shape after a single iteration.

Many iterations of the **CurveFlow** will morph the shape to a circle and then eventually to collapse to a point. The user can extract any desired information from the output of the function after each

Code 18.15 The **CurveFlow** function

```
1 # shape.py
2 def CurveFlow( indata, gamma=0.75 ):
3     adata = indata + 0.0
4     bdata = adata + (ss.cspline2d(adata,70)*(1-adata)) > (1-gamma)
5     cdata = bdata - (ss.cspline2d(bdata,70)<gamma)*bdata
6     return cdata
```

iteration. An example is shown in Code 18.16, which starts with a binary image as variable a. The loops perform 10 iterations of a process that itself includes 20 iterations of **CurveFlow** and then the capturing of the results. These are shown in Figure 18.10. The original shape is seen, and then the progression of the boundary is overlain on it. As seen, the shape morphs toward a circle. The user can design an experiment to change the speed of the flow by adjusting γ and to collect the information after desired iterations.

18.3.2 MEDIAL AXIS

The final approach to be reviewed is a medial axis, which is basically the spine of a shape. Consider a case in which the image is a front view of a human with the arms out to the side. A second image is of the human with the arms raised up higher. This is the same object, but is it the same shape? Certainly, the skeleton is the same with the only difference being the angle of two joints. The medial axis approach attempts to garner this type of information.

One method of finding the medial axis of a shape is shown in Figure 18.11. Inside this shape are four of many inscribed circles that have arcs, which are tangent to the perimeter in two locations. The centers of these, and other circles not shown, are connected by the dashed line. This is the medial axis.

Computation of the axis in this manner can be tedious and slow. A faster method starts with the recognition that the radius of one of these circles is also the minimum distance from the center

Code 18.16 Running iterations of curvature flow

```
>>> answ = np.zeros( a.shape )
>>> orig = a + 0
>>> for i in range (10):
        for i in range( 20):
            a = shape.CurveFlow( a, 0.75 )
        b = nd.binary_erosion( a > 0.01)
        answ += (a-b/2)
>>> imageio.imsave('figure.png', orig-answ)
```

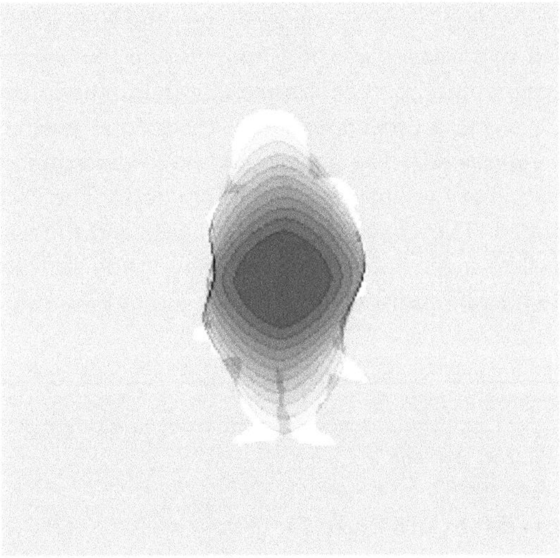

Figure 18.10 Demonstration of the curvature flow.

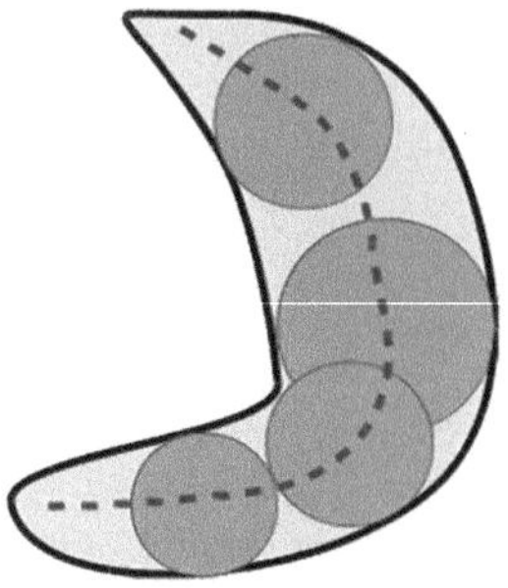

Figure 18.11 A view inscribed circles which have arcs tangent to the perimeter are used to define the medial axis.

of the circle to a point in the background. The computation of the distancesfrom all points in the frame to the nearest background point is performed by the **scipy.ndimage.distance_transform_edt** function. This function is expressed by the Euclidean distance transform operator $\mathcal{D}_E$ and so the transformation is

$$\mathbf{b}[\vec{x}] = \mathcal{D}_E \mathbf{a}[\vec{x}]. \tag{18.33}$$

In Code 18.17, the image is placed in a larger frame in line 3 so that the shape is not near the edge of the frame. Equation (18.33) is executed in line 4. The input and result of this line are shown in Figure 18.12(a) and (b). The remaining black background of these two images are converted to gray for the purpose of printing. The black pixels in Figure 18.12(b) are those that are computed from the function.

The spine of the image is the bright ridge through the middle of the object. However, the intensity of this object is not uniform. Pixels on the spine near the edge are dimmer than the off-spine pixels in the center. Thus, a simple threshold is not sufficient to delineate the medial axis.

The goal is to find the contour of the ridge. Pixels on both sides of the ridge are dimmer, and thus, the ridge is also an edge. Since the gradient of the intensities in Figure 18.12(b) is continuous a simple edge detection algorithm is sufficient. One possible approach is to use the **scipy.ndimage.laplace** function which is called in line 5 of Code 18.17. The output of this function (without alteration to the background) is shown in Figure 18.12(c). The dark line, which is easily detected by a threshold is the medial axis. Secondary axes (the ribs) are also seen.

Consider a case in which two images are of a human with the arms at a different position and the third is of a dog. The goal would be to determine that the first two images have similar objects. The analysis of the structure is one approach, and thus, the medial axes are considered. Comparison of medial axes have many approaches. The goal would be to determine that the only difference in the first two images is the angle at two intersections (shoulders). The medial axes (the major bones of the skeleton) are unchanged. They have the same lengths and the same connections. Whereas, the third image would produce a medial structure that is vastly different in relative lengths and connections of the axes. Such a comparison is not trivial and an area of ongoing research.

Code 18.17 Computing the medial axis

```
1 >>> import scipy.ndimage as nd
2 >>> import mgcreate as mgc
3 >>> adata = mgc.Plop( data, (512,512) )
4 >>> bdata = nd.distance_transform_edt( adata )
5 >>> cdata = nd.laplace(bdata )
```

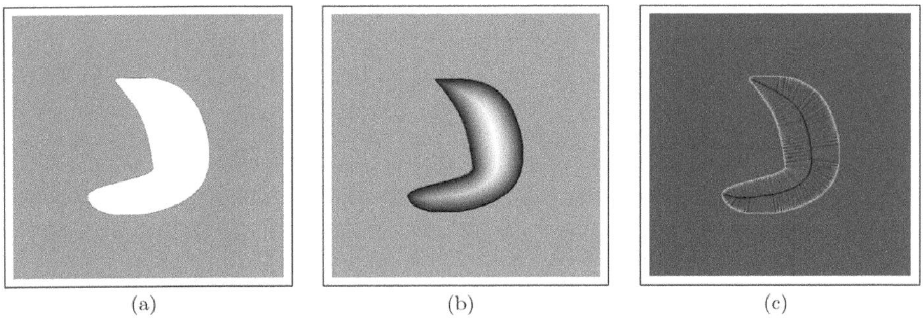

Figure 18.12 (a) The original image, (b) the intermediate image, and (c) the medial axis.

18.4 PROBLEMS

For these problems, use a silhouette database. One possible set is the mythological creatures 2D database v 1.0 at `http://tosca.cs.technion.ac.il/book/resources_data.html` [3].

1. Compute the area and perimeter length of one of the images from the database.
2. Write a Python script that computes the area of all images in the database. Determine the sort order from smallest to largest. Did this method sort the images by shape?
3. Repeat Problem 2, except use the perimeter length as the metric instead of the area.
4. Repeat Problem 2, except use the bending energy (Equation (18.6)) as the metric instead of the area.
5. Repeat Problem 2, except use the circularity (Equation (18.7)) as the metric instead of the area.
6. Compute the Fourier descriptors of the first image in the database *centaur1.bmp*. Print out the first three components of this vector.
7. Using the *centaur1.bmp* image, compute covariance matrix for the data points. Compute the eigenvectors and eigenvalues. Draw the eigenvectors on top of the shape.
8. Using the *centaur1.bmp* image, compute covariance matrix for the data points. Compute the eigenvectors and eigenvalues. Explain how the magnitude of the eigenvalues is related to the shape.
9. Create $\mathbf{a}[\vec{x}] = \mathbf{o}_{\vec{w};\vec{v},r}$ where $\vec{w} = (256,256)$, $\vec{v} = (128,128)$, and $r = 50$. Show that the eigenvalues of this shape are equivalent.
10. Create $\mathbf{a}[\vec{x}] = \mathbf{p}_{\vec{w};\vec{v},r_1,r_2}$ where $\vec{w} = (256,256)$, $\vec{v} = (128,128)$, $r_1 = 50$, and $r_2 = 25$. What is the ratio of the two eigenvalues of this shape?
11. Create $\mathbf{a}[\vec{x}] = \mathbf{p}_{\vec{w};\vec{v},r_1,r_2}$ where $\vec{w} = (256,256)$, $\vec{v} = (128,128)$, $r_1 = 75$, and $r_2 = 25$. What is the ratio of the two eigenvalues of this shape?
12. Create $\mathbf{a}[\vec{x}] = \mathbf{p}_{\vec{w};\vec{v},r_1,r_2}$ where $\vec{w} = (256,256)$, $\vec{v} = (128,128)$, $r_1 = 50$, and $r_2 = 25$. Compute the four moments m_1, m_2, m_3, and m_4. (See Equations (18.27) through (18.30).)
13. Compute the four moments m_1, m_2, m_3, and m_4 (Equations (18.27) through (18.30)) for all shapes in the database. For each image create a vector (m_1, m_2, m_3, m_4). Compute the Euclidean distance from each vector to the vector for the first image (*centaur1.bmp*). Sort the images according to this distance.
14. Compute the medial axis image for *centaur1.bmp*.

Part V

Basis

19 Basis Sets

The previous chapters have presented several algorithms for specific purposes. Even though the intent and mathematics of these algorithms differ greatly, they still have one major feature in common. The first part of the operation creates multiple images, and the second part extracts the decision from these intermediate images. For example, the application of Gabor filters creates a correlation of the input with each filter. If there are N filters, then the intermediate stage is N times larger than the input in terms of number of pixels. In this case, the Gabor filters create a *basis set*. This set expands the information space as shown in Figure 19.1. The input is processed by a set of filters (or equivalent) producing several intermediate images. These are then analyzed to form the final image or decision.

The purpose of the expansion is to reduce the complexity of the information. This is accomplished if it is much easier to extract the answer from the intermediate images, then it is to extract it from the original input. These intermediate images can be obtained through a set of filters, layers in neural networks, or other processes.

There are many different algorithms that fit in this class of algorithms, some of which have been reviewed in previous chapters. More algorithms will be reviewed in this and the subsequent chapter. The algorithms reviewed in this text are:

1. Wavelets (Section 18.1.5),
2. Discrete Cosine Transform (Section 19.1),
3. Gabor filters (Section 17.1),
4. Zernike filters (Section 19.2),
5. Fourier filters (Chapter 11),
6. Fractional power filters (Section 12.5.1),
7. Pulse-coupled neural networks (Chapter 20), and
8. Empirical Mode Decomposition (Section 19.3).

Each operation is sensitive to different features within the image. The selection of the algorithm depends on the task at hand. For example, the wavelet transformation creates new images that are sensitive to lines in the image including a sensitivity to the line thickness. In this case, the generated images are smaller than the input image and by design this particular transformation does not generate more pixels than the input. This transformation, however, still flows as shown in Figure 19.1 since there are multiple outputs.

The discrete cosine transform (DCT) is sensitive to frequencies within the image. Multiple cosine functions are applied, thus producing several outputs that depict the strength of each frequency inherent in image. This is detailed in Section 19.1.

Gabor filters are sensitive to edges and orientation within an image. In this respect, this type of filters consider the same information as does the Wavelet transforms. In an application, a family of Gabor filters are generated, and each is correlated with the input thus producing a set of correlation output surfaces. Once again, the information in the image is expanded into a larger space with the intent of providing easier access to information concerning edges in the image.

Zernike polynomials are sensitive to radial symmetries. Thus, they are more suitable to images that have somewhat circular objects. An example shown in Section 19.2 considers the application of these polynomials to images of cells. These objects have a center and are somewhat circular in nature.

Fourier filtering is a generic process as the user defines the filters that are needed for their particular application. Section 11.2 built wedge filters that were sensitive direction, and Section 11.1.3 created filters that were sensitive to object size. Chapter 12 demonstrated that Fourier filters can be

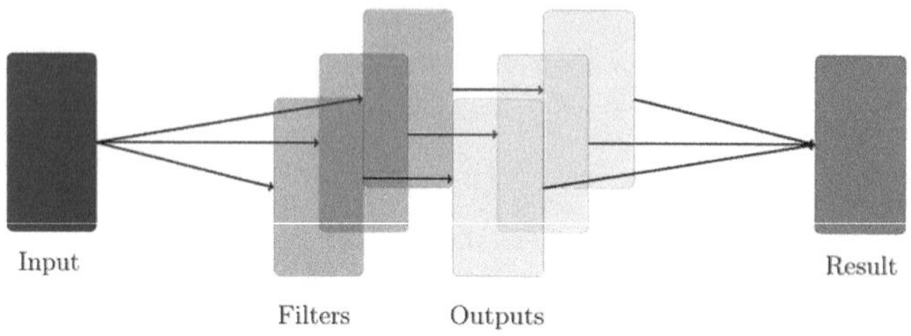

Figure 19.1 Simple schematic of a basis set.

defined by the user for many different purposes. Section 12.5.1 presented Fractional Power Filters that create filters based on multiple inputs, but an application may use several FPFs such as shown in Section 12.5.1.5.

The Empirical Mode Decomposition is an alternative method of extracting frequency information. This is reviewed in Section 19.3.

The final presentation is the pulse-coupled neural network (Chapter 20), which generates several output images that have binary value pixels. This system is based on models of the visual cortex and is not designed to be sensitive to any particular feature of the image. Instead, different outputs provide different features from the image.

19.1 DISCRETE COSINE TRANSFORM

The DCT (DCT) decomposes the signal into a set of cosine functions. There are, in fact, several variants of this transform, but only one is reviewed here. The Python function used in the 1D transform does have a switch to select the other variants.

Given a vector $\vec{x}$ that is of length N, the DCT is,

$$y[k] = 2\sum_{n-0}^{N-1} \cos\left(\frac{\pi k(2n+1)}{2N}\right) \quad 0 \leq k < N. \tag{19.1}$$

This is actually the type II DCT and is the default transform use in the Python function. This function multiplies the input signal by a set of orthogonal cosines. The output is a vector $\vec{y}$, and each element corresponds to the magnitude of each of the cosine functions. The value `y[k]` is the magnitude of the function $\cos(\pi k/2N)$ that exist in the original data.

The *scipy.fftpack* module contains the function **dct**, which performs this transformation. This is shown in lines 1 and 2 in Code 19.1, where the vector `orig` is the original input data.

The inverse transform is defined as,

$$y[k] = x[0] + 2\sum_{n=1}^{N-1} x[n]\cos\left(\frac{\pi(k+0.5)n}{N}\right). \tag{19.2}$$

The function **idct** performs this function and this is actually a type III forward DCT. The call is shown in line 3 in Code 19.1, and the output `outdata` is the same as `orig` except for a scaling factor that is dependent solely on the number of elements in the array.

Code 19.1 Using the 1D discrete cosine transform

```
1 >>> import scipy.fftpack as fft
2 >>> b = fft.dct( orig )
3 >>> outdata = fft.idct( b )
```

Code 19.2 shows a simple example. Line 3 defines the input space, while lines 4 through 6 create the input signal. This is constructed from two cosine signals, and thus it is expected that the transform should have just two large values.

Partial results of the transform is shown in Figure 19.2. The rest of the transform is to the right, and all of the values are very close to 0 and therefore not shown. The two peaks are at $x = 8$ and $x = 16$, and the length of the input vector is 512 elements. The values of 8/512 and 16/512 match the scaling factors 1/32 and 1/64 in lines 4 and 5 in Code 19.2.

The operator symbol $\mathcal{J}$ is used to represent this transform and thus a forward transform is

$$\mathbf{b}[\vec{\omega}] = \mathcal{J}\mathbf{a}[\vec{x}], \tag{19.3}$$

and the reverse transform is

$$\mathbf{c}[\vec{x}] = \mathcal{J}^{-1}\mathbf{b}[\vec{\omega}]. \tag{19.4}$$

The 2D DCT is defined as

$$X_{k_1,k_2} = \sum_{n_1}\sum_{n_2} x_{n_1,n_2} \cos\left[\frac{\pi}{N_1}\left(n_1 + \frac{1}{2}\right)k_1\right] \cos\left[\frac{\pi}{N_2}\left(n_2 + \frac{1}{2}\right)k_2\right]. \tag{19.5}$$

The *fftpack* module contains transforms for only the 1D case, and thus, a new function must be written for the 2D case. There are two approaches to consider. The first would be to write the function as shown. This would result in four-nested loops in Python script, which should be avoided for large arrays as it will be slow. The second approach realizes that the 2D transform is the same

Code 19.2 An example of a 1D DCT

```
1  >>> import numpy as np
2  >>> import scipy.fftpack as fft
3  >>> x = np.arange( 512 )
4  >>> y1 = np.cos( x/32 * np.pi )
5  >>> y2 = np.cos( x/64 * np.pi )
6  >>> y = y1+ y2
7  >>> a = fft.dct( y )
8  >>> b = fft.idct( a )
9  >>> b[0],y[0]
10 (2048.0, 2.0)
```

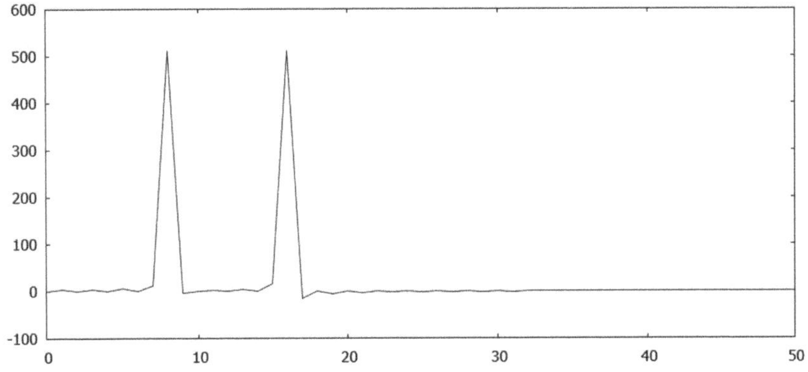

Figure 19.2 The DCT of a simple signal.

as computing the DCT of all of the rows and then columns of an array. This will utilize the function from the *fftpack* module, which should enhance the speed.

Code 19.3 shows the **dct2d** function that performs the 2D DCT. The first `for` loop computes the 1D transforms of all of the rows, and the second **for** loop computes the 1D transforms of all of the columns. The inverse 2D transform is shown in Code 19.4 with the function **idct2d** function. The same logic is followed in that the inverse transform of the rows are computed and then the inverse transform of the columns is computed.

Perhaps the most popular use of a 2D DCT is in the JPEG compression algorithm. After JPEG has converted the YIQ color format (see Equation (5.12)), it downsamples the two chroma channels. The next step computes the DCT over small blocks of the image, and these results are quantized and then compressed. The downsampling separates the data in one of the YIQ channels into 8×8 blocks. Each of these blocks are transformed by a 2D DCT. These extract the various frequency component in both the vertical and horizontal direction.

Figure 19.3 shows an 8×8 array of the various DCTs for this process. The first one (upper left) is pure white, which indicates that it multiplies the input pixels by 1 and sums them. To its right is the second transform which is a cosine function in the horizontal direction. This is a full wave and is sensitive to similar content in the input. The images to the right increase the frequency of the cosine function. The images down the left side show the increasing sensitivity to the vertical waves in the input. The rest shows sensitivities in both directions and different frequencies. Each of these depict the type of information that is being extracted in the transform. The upper left portion corresponds to the lower frequencies, and the lower right corresponds to the higher frequencies. This set of 8×8 images creates a complete, orthonormal set. This is another example of a basis set. The 2D DCT is basically the inner product of the 8×8 input which each of the images in the basis set.

Code 19.3 The **dct2d** function

```
# dct2d.py
def dct2d( mat ):
    V,H = mat.shape
    temp = np.zeros((V,H))
    for i in range( V ):
        temp[i] = fft.dct( mat[i] )
    answ=np.zeros((V,H))
    for j in range( H ):
        answ[:,j] = fft.dct( temp[:,j] )
    return answ
```

Code 19.4 The **idct2d** function

```
# dct2d.py
def idct2d( mat ):
    V,H = mat.shape
    temp = np.zeros((V,H))
    for i in range( V ):
        temp[i] = fft.idct( mat[i] )
    answ=np.zeros((V,H))
    for j in range( H ):
        answ[:,j] = fft.idct( temp[:,j] )
    return answ
```

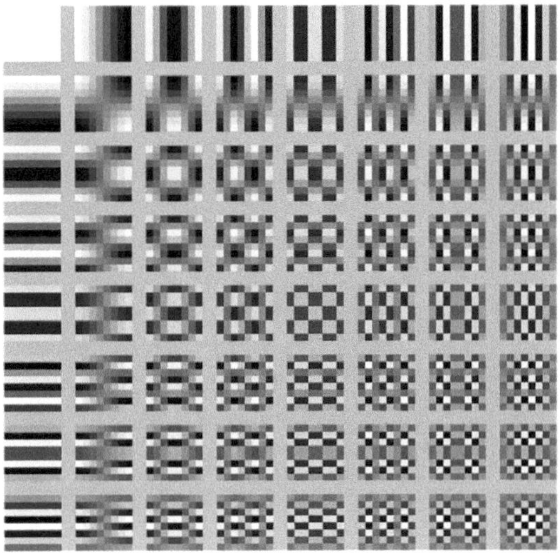

Figure 19.3 The 8×8 DCTs.

A very similar set of images was shown in Figure 9.6 with the computation of the natural eigenimages. This too was a set of orthonormal images that were created in a very different fashion. However, there are several images that are the same, or nearly the same, in both sets. Both of these are orthonormal basis sets and have very similar properties. This leads to the pending question of: Is there a performance difference in the two basis sets? This answer is pursued in Section 19.4.

19.2 ZERNIKE POLYNOMIALS

The DCT responds to frequencies and orientations that are inherent in the image. However, these filters do not necessarily extract pertinent information in the most efficient manner. The Zernike polynomials also create a basis set but these are more sensitive to radial and polar changes.

Generating a Zernike filter begins with the definition two integers m and n. The values of n start at 0 and increment to a number defined by the user. The values of m go from $-n$ to $+n$ increment by a value of 2. So, the first filter is $(n=0,\ m=0)$, the next two filters are $(n=1,\ m=-1)$ and $(n=1,\ m=1)$ and so on. The generating equations are:

$$Z_n^m(\rho,\phi) = R_n^m(\rho)\cos(m\phi), \tag{19.6}$$

or

$$Z_n^{-m}(\rho,\phi) = R_n^m(\rho)\sin(m\phi), \tag{19.7}$$

where

$$R_n^m(\rho) = \sum_{k=0}^{(n-m)/2} \frac{(-1)^k (n-k)!}{k!\left(\frac{n+m}{2}-k\right)!\left(\frac{n-m}{2}-k\right)!}\rho^{n-2k}. \tag{19.8}$$

The initial filter images are shown as circular regions in Figure 19.4. Each row corresponds to an increasing value of n and each column corresponds to an increasing value of m. These filters have radial and polar regularities and are thus more sensitive to images with radial and polar features.

The $\mathcal{Z}$ operator is used to generate a single Zernike filter as in

$$\mathbf{b}[\vec{x}] = \mathcal{Z}_{\vec{w};r,m,n}[\vec{x}], \tag{19.9}$$

where $\vec{w}$ is the frame size, r is the radius of the image in the frame, and m and n are the integers used in the equations. The Python function to generate these polynomials is lengthy and so it is

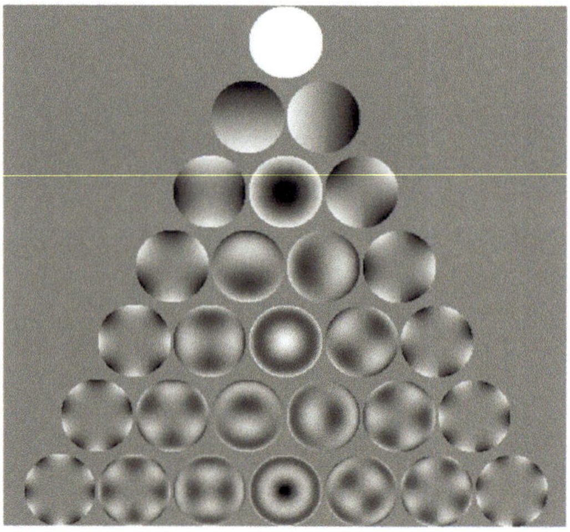

Figure 19.4 Several Zernike filters.

placed in Code C.2. Most of the code establishes parameters based on the integers m and n. The code then generates the sine or cosine pattern in radial-polar space. The final step is to apply an inverse radial-polar transformation to convert the image to the representation that is shown.

Consider the image shown in Figure 19.5, which is a Pap smear slide that has several abnormal cells and one normal cell which is at the 3:00 position in the cluster. The small red regions are white blood cells, which need to be ignored. The abnormal cells have no particular orientation, location or shape, which renders useless some of the previous algorithms. There is little benefit in defining the perimeter of the cells since the perimeters are quite different and susceptible to major alterations such as cytoplasm folding.

Features that are useful are the size of the nucleus and the texture of the cytoplasm. Some previous texture algorithms relied on the comparison of the image to a set of filters which had ridges with various frequencies. Zernike polynomials also define a set of varying frequencies but with circular symmetry. Since the nature of both the nucleus and cytoplasm are circular, Zernike filters are the

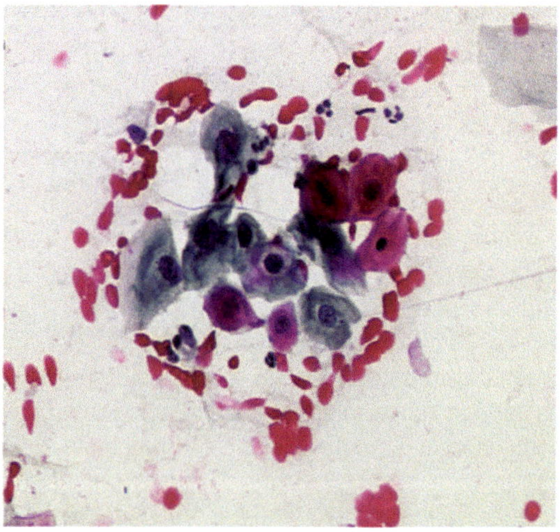

Figure 19.5 An original image.

chosen tool. Each of the filters in the set $\{\mathbf{b}[\vec{x}]\}$ is put into a frame that is the same size, $\vec{w}$, as the original image, $\mathbf{a}[\vec{x}]$. The correlation of each filter with the input is computed as

$$\mathbf{c}_i[\vec{x}] = \mathbf{a}[\vec{x}] \otimes U_{\vec{w}}\mathbf{b}_i[\vec{x}] \quad \forall i. \tag{19.10}$$

The next step of the process prefers that the correlations be stacked into a data cube. So, the process uses the Concatenate operator to collect the correlations as in

$$\mathbf{c}_i[\vec{x}] = \mathcal{C}\left(\mathbf{a}[\vec{x}] \otimes U_{\vec{w}}\mathbf{b}_i[\vec{x}]\right) \quad \forall i. \tag{19.11}$$

The i-th pixel location in $\mathbf{c}[\vec{x}]$ is defined as $\vec{r}_i$. The location at the center of the target cell is denoted as $\vec{r}_t$. The jet at the center of the target cell is extracted by

$$\vec{v}_t = \square_{\vec{r}_t}\mathbf{c}[\vec{x}]. \tag{19.12}$$

The jets for all locations are defined as

$$\vec{v}_i = \square_{\vec{r}_i}\mathbf{c}[\vec{x}], \quad \forall i. \tag{19.13}$$

The first test is to determine if the jet at the center of the target is vastly different from the other jets. Strong similarities in non-target regions would indicate that this is a poor method of image analysis. The distance from the target jet to all jets is computed as

$$\mathbf{d}[\vec{x}] = \mathcal{D}(\vec{v}_t, \vec{v}_{\vec{x}}), \tag{19.14}$$

where $\vec{v}_{\vec{x}}$ is the jet at each location $\vec{x}$. The negative of this image is shown in Figure 19.6. The darkest regions indicate the best matches to the target. As seen, there is only one dark region and it is on target. Thus, there are no other jets that are extremely similar to the target. This is sufficient reason to proceed with further testing.

The next test rotates the input image and repeats the process. However, the final test compares the target jet from the non-rotated image to the jets created from the rotated image. The entire test is expressed as

$$\mathbf{d}[\vec{x}] = \mathcal{D}(\vec{v}_t, \square_{\vec{r}_{\vec{x}}}(\mathcal{C}_j\left((\mathcal{R}_\theta\mathbf{a}[\vec{x}]) \otimes U_{\vec{w}}\mathbf{b}_j[\vec{x}]\right))). \tag{19.15}$$

The result is shown in Figure 19.7. In this case, the normal cell is now located at 6:00 with respect to the cluster. There, a dark region is seen, and it is three times less than any other region. Thus, the Zernike polynomials showed a sensitive to information with only a small sensitivity to rotation.

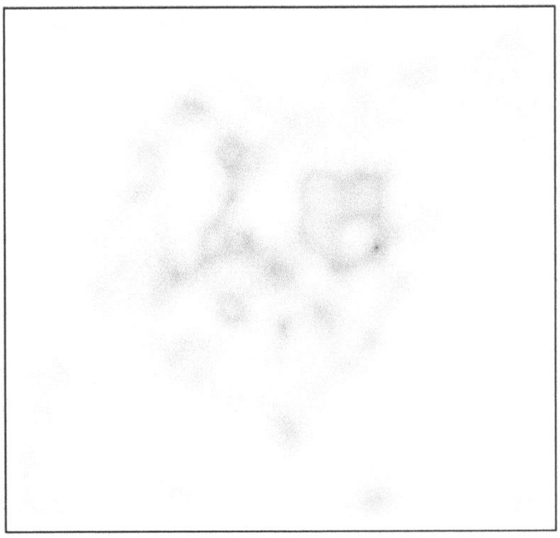

Figure 19.6 The Euclidean distances to the normal jet.

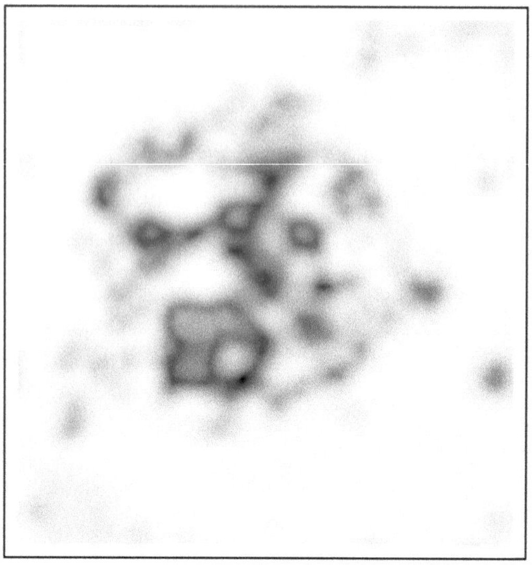

Figure 19.7 The Euclidean distances to the normal jet applied to a rotated input.

19.3 EMPIRICAL MODE DECOMPOSITION

Empirical mode decomposition (EMD)[12] is a method by which a signal is decomposed into a set of IMFs (intrinsic mode functions), which are sensitive to different frequency bands of the original signal. An example for a 1D signal is shown in Figure 19.8, which displays the original signal in part A. Part B shows several IMFs, and part C shows the residual signal. The sum of the IMFs, and residual is the original signal.

As seen, each IMF contains a certain frequency range for the signal. The first IMF contains the highest frequencies, and subsequent signals have lower frequencies. The original process begins with a signal $\vec{w}$, and the EMD process iteratively extracts a series of IMFs. The IMF process is itself an iterative process, which begins with creating a temporary signal $\vec{v} = \vec{w}$. The upper envelope, $\vec{u}$, is computed by spline fitting the peaks of $\vec{v}$, and the lower envelope $\vec{l}$ is computed by spline fitting the valleys of $\vec{v}$. The average of the envelopes is $\vec{a} = (\vec{l} + \vec{u})/2$, and this is subtracted from the temporary signal $\vec{t} = \vec{v} - \vec{a}$. The iteration stops when the standard deviation of $\vec{t}$ falls below a user-defined threshold. If this stop condition is not met, then $\vec{v} = \vec{t}$ and the process repeats. Eventually, the stop condition is met, and the IMF is $\vec{v}$.

Since there are several IMFs to be computed, they are denoted as $\vec{m}_i$ and each is the $\vec{v}$ returned from the IMF process. To delineate the process, the original signal is denoted as $\vec{w}_0$, and when the first IMF is computed this signal is modified $\vec{w}_1 = \vec{w}_0 - \vec{m}_1$ and the next IMF, $\vec{m}_2$, is extracted from $\vec{w}_1$. This process repeats with the user deciding how many IMFs to extract. In the case of N iterations, the residual is $\vec{w}_{N+1}$.

The original signal can be exactly reconstructed by

$$\vec{w}' = \vec{w}_{N+1} + \sum_{i=1}^{N} \vec{w}_N. \tag{19.16}$$

Using the EMD process for suppression of noise can be as simple as reconstructing the signal without some of the first IMFs [37]. Moving the EMD to a two-dimensional image [24] requires a method of creating the estimates of the upper and lower envelopes.

There are some concerns that accompany the EMD process, which also need to be addressed before it can become an automated tool. An early concern was that the tails of the signal (in the

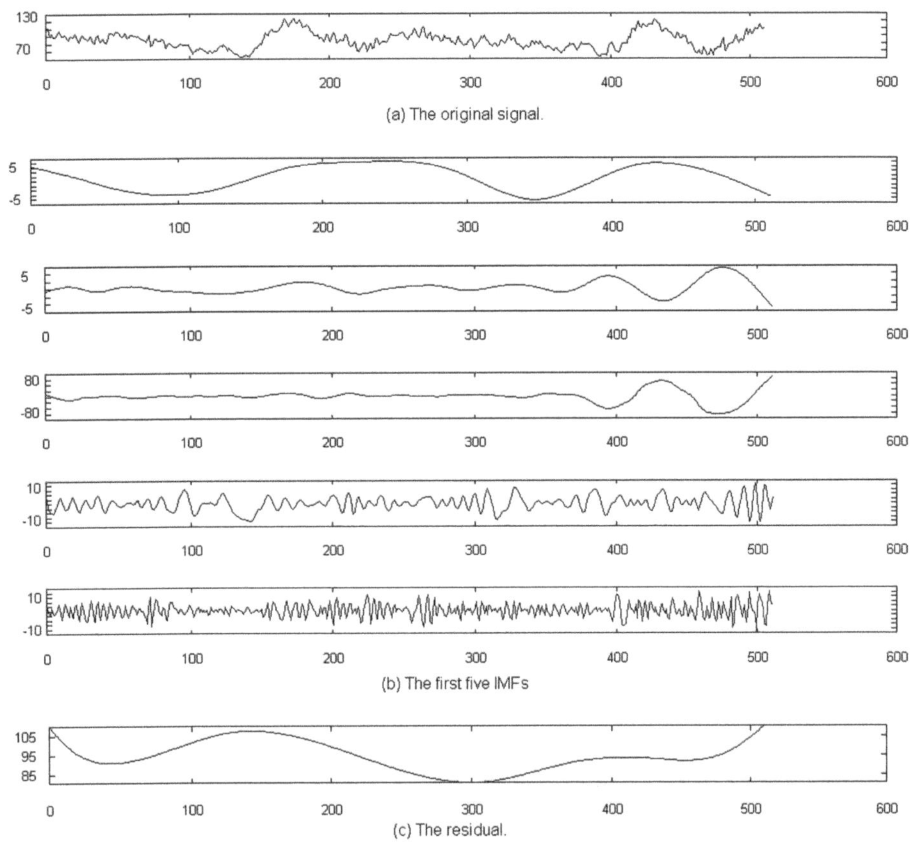

Figure 19.8 An exampled of EMD.

1D case) would become very large in the IMFs. There have been multiple solutions processed of which the most popular is to consider attach mirror images of the signal to each end of the original signal. The problematic ends occur in the attached signals, and the process behaves nicely over the region of the original signal. Another problem arises if a portion of the signal is highly oscillatory and another portion of the signal is quite smooth. In this case, one part of a signal IMF has higher frequencies, while the other portion has lower frequencies, and thus, a single IMF has different frequency bands at different locations in the signal. A third concern is that events can shift and be located in slightly altered locations from the original event. A fourth concern is on signals that are circular in nature such as data extracted from a sampling ring from an image. In this case, the signal begins and ends at a user selected angle and the sampled ring is converted to a vector. It is expected that the IMFs should be the same excepting a shift independent of the angle that the user selects for cutting the sampling ring. This is not the case, and therefore, the original EMD is sensitive to the angle selection when it should not be.

These four concerns are alleviated by using a different approach in calculating the IMFs. In the original algorithm, the upper and lower envelopes were used to calculate an average. It was from this average that further calculations extended. Thus, for a 2D case, the average calculation is replaced here with a spline fit, which is a built-in function in the *scipy.signal* module.

Code 19.5 contains the routines for the modified EMD. Starting on Line 20 is the function **EMD2d**, which is the main driver. It receives the input matrix representing grayscale pixel values, and N which is the number of IMFs to generate. The optional `lam` argument controls the differences between consecutive IMFs. Increasing this number increases the differences between

Code 19.5 Modified 2D EMD

```
1  # emdspl.py
2  def IMFr2d( mat, lam ):
3      # lam is the coeff for cplsine2d
4      temp = mat + 0
5      ok = 1
6      i = 0
7      while ok:
8          avg = cspline2d(temp, lam )
9          old = temp + 0
10         temp -= avg
11         sd = (old-temp).std()
12         if sd<0.3:
13             ok = 0
14         if i>=40:
15             ok = 0
16             print '40 Iterations'
17         i += 1
18     return temp
19
20 def EMD2d( mat, N, lam=1 ):
21     # N = niters
22     w = mat + 0.0 # make sure it's float
23     imfs = []
24     for i in range( N ):
25         sig = IMFr2d( w, lam )
26         imfs.append( sig )
27         w -= sig
28         lam *= 10
29     return imfs, w # resid
30
31 >>> imfs, resid = emdspl.EMD2d( mat, 5 )
32 >>> imfs, resid = emdspl.EMD2d( mats, 6, 5 )
```

the IMFs. This function repeatedly calls **IMFr2d**, which computes a single IMF. Instead of calculating the mean through the spline of the upper and lower envelopes this function merely computes the mean through a spline of the values in Line 8. This is not exactly the EMD process but for image processing applications it is faster with similar results.

Some of the IMFs of Line 32 are shown in Figure 19.9. As seen, the earlier IMFs contain the higher frequency information. Furthermore, Equation (19.16) still holds, which is critical.

High frequency noise removal again is simply the reconstruction without the earliest IMFs. The reconstruction without the first IMF is shown in Code 19.6, and the result is shown in Figure 19.10. The high frequency noise is significantly suppressed and as with the other methods, there is damage to the edges of the image.

EMD offers some of the same advantages as do other algorithms. The original image is expanded according to the inherent frequencies. From this expansion, metrics can be applied to extract pertinent information for a specific application.

(a) (b) (c) (d)

Figure 19.9 (a) The original image, (b) the first IMF, (c) the third IMF, and (d) the fifth IMF.

Code 19.6 Reconstruction

```
>>> recon = imfs[1] + imfs[2] + imfs[3] + imfs[4] + imfs[5] + resid
>>> sophia.a2i( recon ).show()
```

Figure 19.10 Reconstruction without the first IMF.

19.4 IMAGE ANALYSIS WITH BASIS SETS

Consider again the analysis of an image using Gabor filters (some of which are shown in Figure 17.1). The correlations of the filters $\{\mathbf{f}[\vec{x}]\}$ was described in Equation (17.3). In general, these types of analysis follow the schema shown in Figure 19.11. The input image is correlated with several filters thus producing several correlation surfaces. Information from these correlations are extracted leading then to the final decision as to the nature of the input image. In many cases, the input image is processed before the correlations to remove noise and undesired artifacts or to present

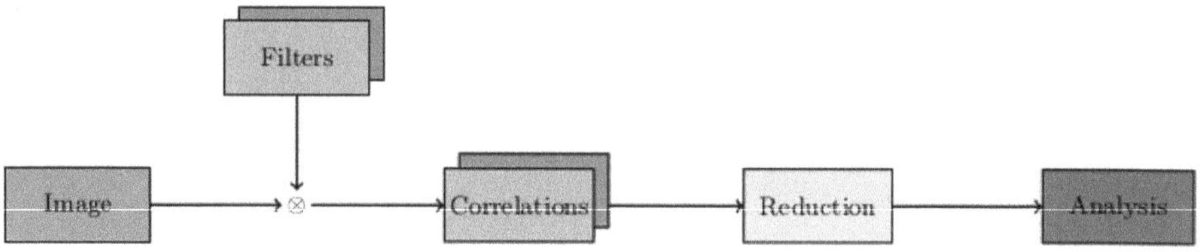

Figure 19.11 The generic schema.

the image in a different manner. This is a generic protocol that has been used in several applications in this text including the Gabor filters and Zernike filters. The differences between these systems is the set of filters used and the method of reduction and analysis. For now, the concentration of the discussion is on the generic scheme.

This is a large volume of data. If the expansion protocol were effective then the complexity of the data would also be reduced, thus it should not be necessary to process the entire data cube in order to reach a decision. Consider an image with the dimensions $V \times H$, where V and H are the number of pixels in the vertical and horizontal directions. There are N filters, and thus, there are a total of $N \times V \times H$ pixels in the correlation surfaces. The amount of space required to hold the information has been expanded by a factor of N. In the face of facial recognition, this step was to extract the jets at specific, fiducial locations. The amount of information at this stage is usually much less that $V \times H$. In this process, the information from the input image was first expanded and then contracted. The intent of the expansion is to reduce the complexity of the important information and the intent of the contraction is to isolate the important information. In the case of applying Gabor filters to a face image, the expansion separated the edge and orientation information in a larger space, and the reduction was to gather that information at only the important locations.

To explain this intent in more detail, consider the incredibly simple XOR (exclusive or) problem. The data is shown in Table 19.1. In this case, there are only two input elements, x_1 and x_2, and a single output element y. The XOR problem is solvable by a neural network shown in Figure 19.12. The two inputs are on the right and the output on the left. Each neuron sums their total inputs and emits a 1.0 if the total is above the threshold value 0.5. The hidden neuron is activated only when both inputs are high.

Table 19.1
Exclusive Or.

x_1	x_2	y
0	0	0
0	1	1
1	0	1
1	1	0

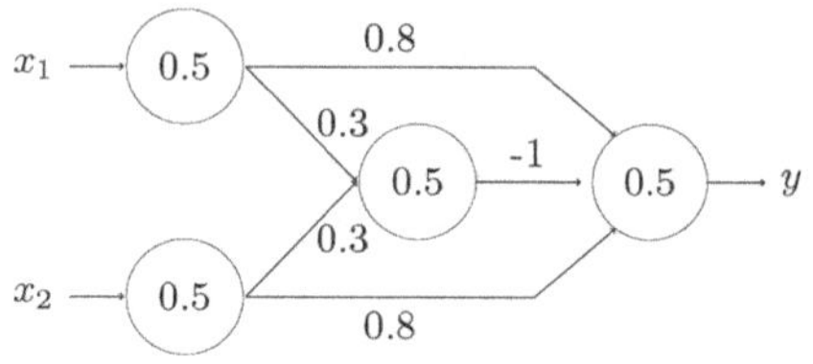

Figure 19.12 The XOR problem mapped to a neural network with a single hidden neuron.

The hidden neuron is required because this is a second-order problem. The XOR problem can also be described by

$$y = x_1 + x_2 - 2x_1x_2, \tag{19.17}$$

which is also a second-order equation, and is the simplest polynomial that can solve this problem.

The output neuron receives three input lines even though the original data had only two lines. The output neuron does not distinguish between lines from the input or hidden neuron. Its only concern is to receive three inputs and make a decision. The data entering into the last neuron is shown in Table 19.2 where z is the value coming out of the hidden neuron.

The output neuron receives three inputs and makes its decision based on those lines. The decision is,

$$y = x_1 + x_2 - 2z, \tag{19.18}$$

where, according to the last neuron, is merely a first-order equation. The neural network has expanded the space from two lines to three lines and in doing so reduced the complexity of the decision from second-order to first-order. It has expanded the information space and then reduced the space to make a final decision.

This very simple example demonstrates the same philosophy as in Figure 19.11. The input information is expanded in an attempt to reduce the complexity of the important information, and then the information space is collapsed to make a decision. This expansion is highly dependent on the information that is important in the application. Consider again the task of face recognition using Gabor filters (Section 17.3). In this case, the images were correlated with several Gabor filters in an attempt to reduce the complexity of the important information, which was determined to be the behavior of the edges. This expanded the information space. The reduction occurred by selecting jets at specific locations. The decision was then based on these jets.

While this process followed the scheme in Figure 19.11, the question arises: Was the Gabor filters the best set of filters to use?

Recall that that Gabor filters are sensitive to edge thickness and orientation. The jets were extracted from locations that were only at edges and corners of specific features. It follows then that the logic of this system is that faces can be recognized by features of the edges. This algorithm does not use skin color, skin texture, or even the shape of the face in making its decision. It relies merely on the characteristics of edges.

The process of developing a protocol, though, should use the reverse logic. The first step is to determine what is important in the input data, and that should lead to the selection of the set of filters or basis set. Of course, it is highly possible that the designed algorithm is more complicated than that shown in Figure 19.11. If the user decides that other features are important then it will be necessary to have several steps that follow this design such, as the outline shown in Figure 19.13. Here the user has decided to attack four different features and each of the blocks would then be designed in the style of Figure 19.11. However, the overall schema follows the same pattern. The information of the original input image is expanded by multiple analytic tools and then collapsed to form a decision.

Table 19.2
Data of the Last Neuron

x_1	x_1	z	y
0	0	0	0
0	1	0	1
1	0	0	1
1	1	1	0

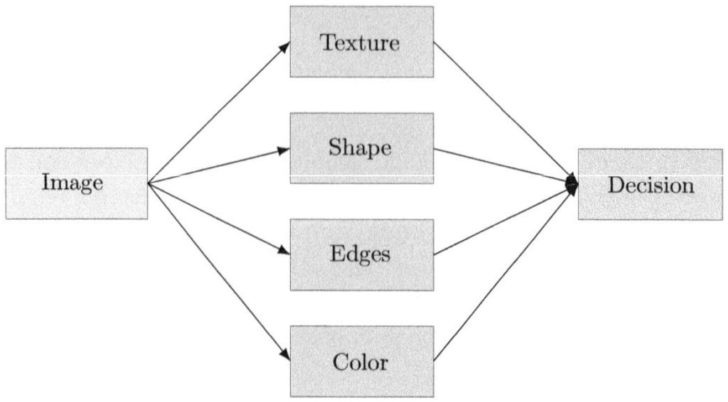

Figure 19.13 A larger schema extracting multiple features.

It is important, then, to understand the real purpose of each of the tools presented in earlier chapter, as well as those that are not presented in this book. As just reviewed, the Gabor filters are sensitive to edge thickness and orientation. So also are the filters created in the DCT process, wavelet decomposition, Law's filters, and so on. Not all filters are sensitive to edges and orientation. For example, the Zernike polynomials are more sensitive to radial and polar changes. Fourier filtering is more pliable and has been used for multiple purposes. Recall that the Gabor filters were also isolated frequency responses which were demonstrated to be isolated regions in Fourier space (see Figure 17.4). Thus, it is also possible to create a set of Fourier filters that are sensitive to edge thickness and orientation. However, other types of Fourier filters have also been demonstrated. The wedge filters (Section 11.2) were sensitive to orientation and the bandpass filters (Section 11.1.3) were sensitive to size.

So far, the filters discussed are sensitive to well-defined features such as edges, texture, orientation, shape, etc. Systems have also been developed that create a basis set for features that are now easily described in words or understood the concept. Consider again the process of creating eigenimages. These were created from a set of images rather than from a description of a feature. Some of these are shown in Figure 9.4, and in these images it is possible to describe some of the features that were evident in the images. For example, one eigenimage showed a sensitive to facial hair, another to illumination glint, and so on. However, there are quite possibly other features captured by the eigenimages but are not described in words.

In the case of the eigenimages, the location of a data point in the PCA space was performed by the inner product of the input with the eigenimages. The inner product is a two-step process in which all of the elements are multiplied by elements in a filter, and then results are summed. Consider a slightly different manner of performing the same computation for an input $\mathbf{a}[\vec{x}]$ and a set of eigenimages $\mathbf{b}_i[\vec{x}]$,

$$v_i = \sum_{\vec{x}} (\mathbf{a}[\vec{x}] \times \mathbf{b}_i[\vec{x}]), \quad \forall i. \tag{19.19}$$

The multiple computations performed inside the parenthesis is the expansion, and the summation is the contraction.

The advantage of the eigenimage approach is that it can create filters that are sensitive to information that the user may not be able to mathematically describe. The disadvantage is that such an algorithm cannot distinguish between important information and noise. Consider a set of of images of several people in which each person had a unique background. Eigenimages attempts to find differences in the image set, and since all of the image have faces, the major differences would be the background objects. The eigenimages would have large magnitudes for pixels from the background and small magnitudes for pixels on the faces. In the previous implementation of the eigenfaces, the backgrounds were removed for this very reason. The point is that the eigenimages were being trained on the data in an uncontrolled manner. The user provided no assistance to determine which pixels are on target and which are noise.

Fourier filtering can also be used to capture information that is difficult to describe. This was the intent of the fractional power filters (see Section 12.5.1). This filter (or set of filters) was created from training on real images in an attempt to capture the important discriminating features in the image set. Again, the resultant filters were based solely on the data with no aid from the user. In such a case, noise pixels and target pixels are treated with the same importance.

Each of these filter systems, and many that have not been mentioned here, rely on the creation of a basis set. This is the filter set or equivalent. In the case of the eigenimages and Gabor filters, the basis set is orthonormal but in the case of the FPF suite they are not.

A basis set, though, does not have to be a set of correlation filters. Consider the objective of converting an RGB image to the YIQ image space. This was accomplished with a matrix conversion as shown in Equation (5.12). This equation considers the pixels individually, but it is also possible to consider the conversion as multiplication of each image plane as in

$$\mathbf{Y} = a\mathbf{R} + b\mathbf{G} + c\mathbf{B}, \tag{19.20}$$

where a, b, and c are constants obtained from the matrix. In a sense, the right-hand side is a space that is expanded by a factor of 3, and the additions reduce the space back to the original size. The coefficients could also be represented as matrices, each with a single value for all elements, and the multiplication of a matrix a with $\mathbf{R}$ is an elemental multiplication just as performed in inner products and correlations. In that sense, these coefficients are also a basis set.

So the proper design of a classification system begins with the user defining the features that are important or recognizing that such a definition is impossible to precisely define. Then the next step is to define the basis set that best extracts the desired information.

Consider again the system in Figure 19.13 which is designed for a face recognition system. For each block, the user will need to define the best basis set to employ. Face texture has already been accomplished via the eigenface process and thus it would be an algorithm to use. Shape information is available by the warping process as shape could be defined as the distances that each fiducial point moved in order to warp to the standard grid. Edge information has already been extracted via the Gabor filters. Color comparisons could be as simple as converting the images to a space such as HSV (Section 5.2) or YIQ and computing the statistical moments (average, variance, skew, kurtosis, etc.) of the distribution of skin tone values. Each process expands the information and then condenses it. The final decision would collect all of the condensed arrays and use a voting system to make the final decision.

Incorrect choices of a basis set can be devastating to performance. This is demonstrated in an example of recognition of a fingerprint image from a library that contains only two fingers. In this case, there are only three images that are shown in Figure 19.14. The first two images are from the same finger, and the third is from a different finger. Thus, a recognition should be able to classify the first image as being similar to the second but not the third. Furthermore, the first two images are scaled and rotated so that the core and deltas are the same distance and angle apart, thus alleviating these factors.

Common acceptance is that the important features of fingerprints are the broad classifications (whorl, arch, loop, etc.), the number of ridges between the core and delta, and the location and nature of the minutia. Algorithms that fail to extract important information such as these features are doomed to failure. Consider the application of Fourier filtering (or FPF) in which the filter is constructed from the second image. The correlation of the input with the filter attempts to find the optimal alignment of the two images. This is already known due to the pre-processing of these images. The aligned image is shown in Figure 19.15, where one image is coded in red and the other in green. It is easy to see one color in between the ridges of the other. The reason for this is that when a fingerprint is obtained the finger is pressed against an item such as paper or glass. This applies

Figure 19.14 Fingerprint images $\mathbf{a}[\vec{x}]$, $\mathbf{b}[\vec{x}]$, and $\mathbf{c}[\vec{x}]$.

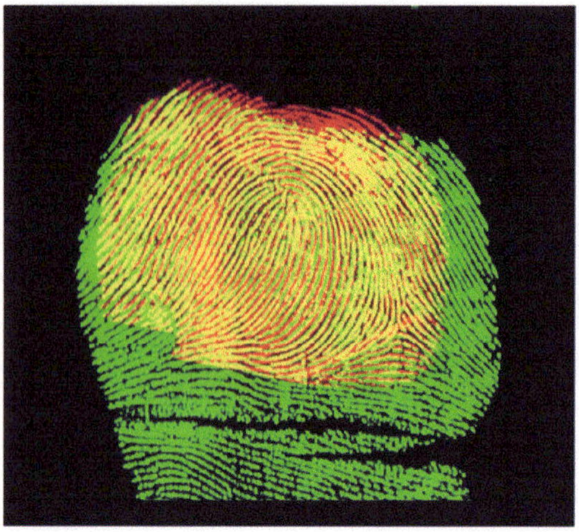

Figure 19.15 A color coding of the first two images.

pressure to the finger and each roll will be different due to the different deformations of the finger. Thus, two different images of the same finger do not align the ridges as shown.

The correlation is a first order operation in that it matches pixel intensities. If there is no shift that aligns a large number of pixels then there is no large correlation response. The two correlations are computed by

$$\mathbf{d}[\vec{x}] = \mathbf{a}[\vec{x}] \otimes \mathbf{b}[\vec{x}], \tag{19.21}$$

and

$$\mathbf{f}[\vec{x}] = \mathbf{a}[\vec{x}] \otimes \mathbf{c}[\vec{x}]. \tag{19.22}$$

The row of data from $\mathbf{d}[\vec{x}]$ that includes the correlation peak is extracted and is shown in Figure 19.16. Likewise, the row with the peak value from $\mathbf{f}[\vec{x}]$ is also extracted and is shown on the same graph. If the system could recognition the first image as being similar to the second, then it is expected that the correlation surface would have a large spike or at least a large value. While the correlation of the first two images did provide a larger value, it is not significantly larger than the other correlation. Furthermore, there is no peak. Basically, this method failed because the selected method is unable to extract important information.

Each application requires that the user to define or at least recognize the important features and then to select the correct basis set or equivalent. Consider the case of recognizing a pencil that can be at any orientation between $-\theta$ and $+\theta$. The pencil would display two long parallel, finite lines

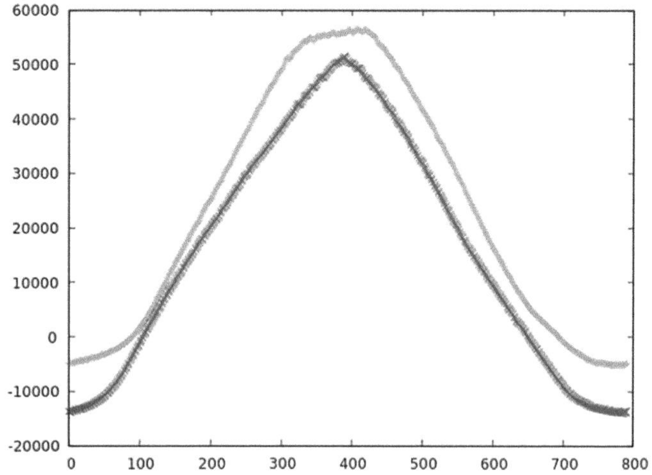

Figure 19.16 Each correlation created a 2D surface and these plots show a horizontal sample through along the row through the peak value.

due to the edges. Thus, an edge sensitive method (Hough, Gabor, DCT, etc.) would be appropriate to consider.

More difficult problems would require multiple schemes. Consider the task of finding Waldo from the Where's Waldo images. Waldo has a certain size, blue pants, white and red horizontally striped shirt, and so on. These are well-defined features and the userwould be well suited to find items according to color , size, equally spaced horizontal stripes and so on. Methods for extracting such information have been presented in previous chapters and in hundreds of publications. The features are defined and algorithms chosen. Only then should the user begin building a computer program to process the data. In this case, after construction and testing the user can then claim to be able to find Waldo, which by the way, is a task that most three-year-olds can accomplish. Such is the nature of image analysis applications.

PROBLEMS

1. Through an example, determine if the vectors generated by $\vec{v} = T\mathbf{q}_{\vec{w}}[\vec{x}]$ are orthonormal. Use $\vec{w} = (5,5)$.
2. Through an example, determine if the vectors generated by $\vec{v} = T\mathbf{q}_{\vec{w}}[\vec{x}]$ are orthonormal. In this case, the random matrix has complex values. Use $\vec{w} = (5,5)$.
3. Create three Gabor filters with $f \in (0.2, 0.3, 0.4)$ and $\theta = 0$. Determine if these filters are orthonormal.
4. Create Gabor filters, $\{\mathbf{a}[\vec{x}]\} = \mathfrak{G}_{\vec{w},f,\theta}[\vec{x}]$, $f \in (0.2, 0.3, 0.4)$ and $\theta \in (0, \pi/4, \pi/2)$. Compute the eigenimages of these filters.
5. Load the images from the Myth database (`http://tosca.cs.technion.ac.il/book/resources\_data.html`) as $\{\mathbf{a}[\vec{x}]\}$. Create a set of Gabor filters $\{\mathbf{b}[\vec{x}]\} = \mathfrak{G}_{\vec{w},f,\theta}[\vec{x}]$, $f \in (0.2, 0.3, 0.4)$, $\theta \in (0, \pi/8, \pi/4, 3\pi/8, \ldots, 7\pi/8)$, and $\vec{w}$ is the same size as one of the myth images. Create a matrix $\mathbf{M}$ such that $M_{i,j} = \mathbf{a}_i[\vec{x}] \cdot \mathbf{b}_j[\vec{x}]$. Compute the Euclidean distance of the vector in the first row to all rows. The Myth database has three different types of figures. The first row in $\mathbf{M}$ is associated with the first image. Vectors will lower Euclidean distances are supposedly more similar to the first image. Do the images of the same type of character have the lowest Euclidean distances?

6. Repeat Problem 5, but replace vectors generated by the Gabor filters with the wavelet decomposition energy vectors.
7. Repeat Problem 5, but replace the Gabor filters with Zernike images.
8. Repeat Problem 5, but replace vectors generated by the Gabor filters with the discrete cosine images (see Part V, Section 19.1).
9. Repeat Problem 5, but replace vectors generated by the Gabor filters with images with random pixel values. Does the performance of random images differ greatly from the results in Problem 6.
10. Create and image $\mathbf{a}[\vec{x}]$ which is 64×64. Each row is identical and created by $\sin((x - 32)\pi/6.4)$. Create a set of Gabor filters $\{\mathbf{b}[\vec{x}]\} = \mathfrak{G}_{\vec{w},f,\theta}[\vec{x}]$, $f \in (0.2, 0.5, 1)$, $\theta = 0$, and $\vec{w} = (64, 64)$. Compute the inner product of each filter with the image. Provide an explanation as to why the values are so unusual.
11. In Problem 10 the filters are obviously ill suited for the image. Create a new function that is a modification of **gabor.GaborCos**. This new function is **GaborSin** and replaces the **cos** function with the **sin** function. Using this function repeat Problem 10.
12. In the Gabor filter chapter, it was shown that Gabor filters have localized energy in the Fourier plane. Is the same true for Zernike filters?
13. In the Gabor filter chapter, it was shown that Gabor filters have localized energy in the Fourier plane. Is the same true for DCT filters?

20 Pulse Images and Autowaves

In the 1990s, a new process of analyzing images emerged which modeled pulsing activity of neurons from the mammalian visual cortex. In an image processing application pulse patterns of the neurons were used to extract shapes, textures, motion, and a variety of other features in the image. This chapter will review the theory of the pulse image generators and present a few of the applications.

20.1 PULSE-COUPLED NEURAL NETWORK

The pulse-coupled neural network (PCNN) [15] is a digital model based upon the visual cortex. When applied to images, each pixel is assigned to a neuron, and each neuron is connected to local neighbors. Each neuron can pulse and so an output at a specified time is a collection of pulsing neurons. The synchronization of these pulses extracts pertinent image information.

20.1.1 MAMMALIAN VISUAL CORTEX

There have been several models of the mammalian visual cortex starting with the HodgkinHuxley model of the 1950s and the Fitshugh-Nagumo model in the early 1960s. These models described neural behavior as leaky integrators and mathematically through coupled differential equations that linked neural potentials.

These models were followed by the Eckhorn model which measured the spiking activity in the cat visual cortex and created a model to describe the behavior. The potential of neuron k was described as

$$U_{m,k}(t) = F_k(t)\left[1 + L_k(t)\right], \tag{20.1}$$

where F is the *feeding* potential and *linking* potential. They are described as

$$F_k(t) = \sum_{i=1}^{N}\left[w_{ki}^{f} Y_i(t) + S_k(t) + N_k(t)\right] \otimes I(V^a, \tau^a, t), \tag{20.2}$$

and

$$L_k(t) = \sum_{i=1}^{N}\left[w_{ki}^{l} Y_i(t) + N_k(t)\right] \otimes I(V^l, \tau^l, t). \tag{20.3}$$

The output is described as

$$Y_k(t) = \begin{cases} 1 & U_{m,k}(t) > \Theta_k(t) \\ 0 & \texttt{Otherwise} \end{cases}. \tag{20.4}$$

The number of neurons is N, and the array $\mathbf{w}$ describes the neuron connections. The array $\mathbf{S}$ is the external stimulation and the ranges for the constants are $tau^a = [10, 15]$, $\tau^l = [0.1, 1.0]$, $\tau^s = [5, 7]$, $V^a = 0.5$, $V^l = [5, 30]$, $V^s = [50, 70]$, $\Theta_0 = [0.5, 1.8]$.

20.1.2 PCNN

The PCNN is a digital version of the Eckhorn model [14]. The major modification is that that time is now discrete which actually alters the behavior of the network. However, the main goal here is to find a useful image processing tool and not to attempt to replicate the functionality of a portion of the mammalian brain. Therefore, the digitization of the algorithm is acceptable.

20.1.2.1 Theory

Similar to the Eckhorn, each neuron in the PCNN has a potential that is comprised of a *feeding* and *linking* component. These are compared to a dynamic threshold to create the neuron's output. Neurons are locally connected. Given in input stimulus $\mathbf{a}[\vec{x}]$, the feeding potential is computed by

$$\mathbf{f}_{n+1}[\vec{x}] = \alpha \mathbf{f}_n[\vec{x}] + \mathbf{a}_n[\vec{x}] + \mathbf{m}[\vec{x}] \otimes \mathbf{y}_n[\vec{x}], \tag{20.5}$$

where α is a decay constant, n is the iteration index, V_F is the feeding constant which is less than 1, $\mathbf{m}[\vec{x}]$ is a small correlation kernel that describes local Gaussian connections, and $\mathbf{y}[\vec{x}]$ is the array of output values which are described in a subsequent equation.

The linking compartment is computed by

$$\mathbf{l}_n[\vec{x}] = \alpha \mathbf{l}_{n-1}[\vec{x}] + V_L \mathbf{w}[\vec{x}] \otimes \mathbf{y}_{n-1}[\vec{x}], \tag{20.6}$$

where V_L is a linking constant less than 1 and $\mathbf{w}[\vec{x}]$ describes local connections and in many applications $\mathbf{w}[\vec{x}] = \mathbf{m}[\vec{x}]$. The *internal potential* is determined by

$$\mathbf{u}_n[\vec{x}] = \mathbf{f}_n[\vec{x}]\left(1 + \beta \mathbf{l}_n[\vec{x}]\right), \tag{20.7}$$

where β is a linking constant and is usually much smaller than 1.

The output is computed by

$$\mathbf{y}_n[\vec{x}] = \mathbf{u}_n[\vec{x}] > \mathbf{t}_n[\vec{x}], \tag{20.8}$$

and the dynamic threshold is computed by

$$\mathbf{t}_n[\vec{x}] = g \mathbf{t}_{n-1}[\vec{x}] + V_\Theta \mathbf{y}_n[\vec{x}], \tag{20.9}$$

where V_Θ is a large constant commonly in the range of 20–50.

In Johnson's original work, the PCNN was applied to images with isolated shapes, and it was demonstrated that that the sum of the pulse activity became a repeating pattern sensitive to the input shape.

20.1.2.2 Pulse Streams

The Python module *pcnn.py* contains an object class for the PCNN part of which is shown in Code 20.1. Line 3 defines the constants, and Line 5 creates the constructor, which basically establishes all of the matrices. The variables correspond to the theory with the exception that the `self.T` represents Θ from Equation (20.9). A single iteration of the PCNN is defined in the function starting on line 10. The matrix `work` defined in line 12 or 14 is the correlation of the Gaussian connections with the output. This corresponds to the last term in Equations (20.5) and (20.6). The equations are executed in lines 1519.

Code 20.2 shows a typical run of the PCNN. In this case, all of the variables will be archived in lists for examination. Line 2 creates an instance of the PCNN class indicating that the array size is 256×256. Line 3 alters one of the constants. Line 4 begins the iteration with the actual computations being called in line 5. The input for this case was a 256×256 matrix named `mat`. The only caveat is that the pixel values in `mat` must be between 0 and 1.

The first example uses an input which is a solid circle centered in the frame with a radius of 20. Code 20.2 runs the experiment and collects the values for all parameters for the neuron located at (128,148). These values are shown with respect to iteration index in Figure 20.1.

The pulse pattern `Y[20]` is shown in Figure 20.2(a). The original input is a small circle in the middle of the frame and as seen the PCNN creates waves that expand from the circle. Once a pattern of waves is established then the repetitive nature of the signals shown in the latter part of Figure 20.1 is created. The large spiking pattern is the dynamic threshold. The small spikes at the bottom are the

Code 20.1 The original PCNN Python class

```
# pcnn.py
class PCNN:
    f,l,t1,t2, beta = 0.9,0.8,0.8,50.0, 0.2
    # constructor
    def __init__ (self,dim):
        self.F = zeros( dim,float)
        self.L = zeros( dim, float)
        self.Y = zeros( dim,float)
        self.T = zeros( dim,float) + 0.0001
    def Iterate (self,stim):
        if self.Y.sum() > 0:
            work = cspline2d(self.Y.astype(float),90)
        else:
            work = zeros(self.Y.shape,float)
        self.F = self.f * self.F + stim + 8*work
        self.L = self.l * self.L + 8*work
        U = self.F * (1 + self.beta * self.L )
        self.Y = U > self.T
        self.T = self.t1 * self.T + self.t2 * self.Y + 0.1
```

Code 20.2 Typical execution of the PCNN

```
>>> F,Y,L,T = [],[],[],[]
>>> net = pcnn.PCNN( (256,256))
>>> net.t2 = 100
>>> for i in range( 150 ):
        net.Iterate(mat)
        F.append( net.F + 0 )
        L.append( net.L + 0 )
        T.append( net.T + 0 )
        Y.append( net.Y + 0 )
```

output pulses. Of the remaining two signals, the upper one is the feeding and the lower one is the linking values.

However, the presence of other objects upset these waves. In a second test, a second circle is added to the input as shown in Figure 20.2(b). The PCNN is reset and executed as before. The output at `Y[20]` is shown in Figure 20.2(c), and as seen, the wave from the first circle is altered by the presence of the second circle. In terms of analyzing images, this is not a trivial matter. It will be discussed further in Section 20.2.1.

20.1.2.3 Applications

The PCNN has been applied to a variety of image processing applications. One of the early and most often used type of applications is to segment images [14, 8, 31]. Recently, PCNN activity is more dedicated to processing the image before the segmentation or recognition stages [36, 25]. Other applications have included converting shapes into sounds [29], logical computing [16], handwriting recognition [30], image fusion [20], and foveation [17].

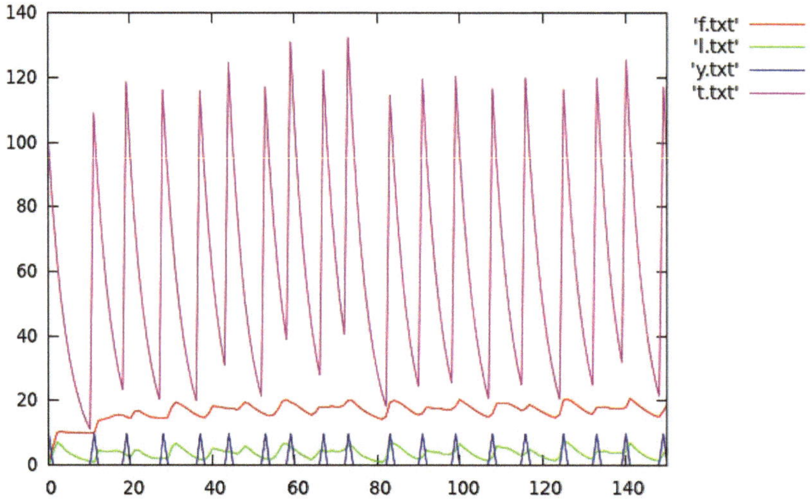

Figure 20.1 Values of the variables for a single neuron.

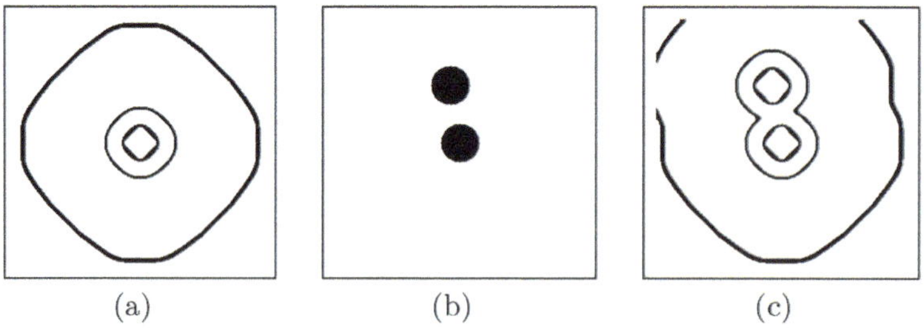

Figure 20.2 (a) The pulse image for $n = 20$, (b) A new input with two spots, and (c) the pulse image for $n = 20$ for the modified input.

There has been a recent resurgence of applications using the PCNN or its predecessors. Applications are seen in image segmentation again and edge detection [9]. This latter is the impetus for one of the exercises at the end of the chapter.

20.1.2.4 Operator Notation

The operator J_n is used to describe the PCNN process for n iterations. The output is a set of pulse images as in

$$\{\mathbf{y}[\vec{x}]\} = J_n \mathbf{a}[\vec{x}]. \tag{20.10}$$

20.2 INTERSECTING CORTICAL MODEL

Multiple cortical models have been proposed and they have common features with roots back to the HodgkinHuxley Model. The intersecting cortical model (ICM) is an artificial model designed for the purpose of analyzing images. While it borrows features from the common cortical models, it is not an attempt to replicate any biological system. The ICM does include one major deviation from biological models in developing inter-neural connections. The new connections are based on autowaves, which are shown in Figure 20.2(c).

20.2.1 CENTRIPETAL AUTOWAVES

Autowaves are waves the propagate but do not reflect or refract. They have been witnessed in several physical phenomena [22, 4], including the wave propagation inside the PCNN. In Figure 20.2(c), the waves from the two objects collide and the wavefronts of both are annihilated along the front. Furthemore, the waves do not bounce off of the walls or other objects.

One of the main drawbacks the PCNN is the interference of expanding autowaves [19]. This is shown in Figure 20.2(c) in which the two autowaves collide and annihilate each other. The presence of one object interferes with the waves of the other object. For image analysis, this is a detrimental feature of the network.

In the original PCNN, the autowave created by a solid object propagated outwards, and the wave would eventually become a circle in a non-pixelated environment. The behavior of the wavefront becoming a circle is similar to the behavior of level sets (see Section 20.2.1). The introduction of *centripetal autowaves* to the ICM model created waves with similar evolutionary behavior except that they did not expand away from the object. This behavior is very similar to that of level sets (Section 18.3). Such behavior would mitigate the interference and therefore the ICM adopted level sets as the manner in which the neurons are connected. Unlike the biological model, the ICM connections between the neurons are dependent upon the firing pattern of those neurons.

20.2.2 ICM

The ICM operates through P iterations of three equations. The first generates the states for all neurons (or pixels), the second generates the output, and the third updates the dynamic threshold values. The neural states are updated by

$$\mathbf{f}_{n+1}[\vec{x}] = \alpha \mathbf{f}_n[\vec{x}] + \mathbf{a}_n[\vec{x}] + \mathbf{w}[\vec{x}] \otimes \mathbf{y}_n[\vec{x}], \tag{20.11}$$

where $\mathbf{f}_n[\vec{x}]$ represents the state of all neurons at iteration n. The iteration index is $\{n = 1, \ldots, P\}$, where P is the number of iterations selected by the user. The coefficient α is the decay constant and is commonly near 0.9. The image $\mathbf{y}_n[\vec{x}]$ represents the output states at iteration n and initially all of these values are 0. The operator $\mathbf{w}[\vec{x}]$ represents the neural connections in is detailed subsequently.

The output is generated by

$$\mathbf{y}_{n+1}[\vec{x}] = \mathbf{f}_n[\vec{x}] > \mathbf{t}_n[\vec{x}], \tag{20.12}$$

where the output values of the individual neurons are set to 1 if their neural states exceed their thresholds which are represented by $\mathbf{t}_n[\vec{x}]$. These thresholds are computed by,

$$\mathbf{t}_{n+1}[\vec{x}] = \beta \mathbf{t}_n[\vec{x}] + \gamma \mathbf{y}_{n+1}[\vec{x}], \tag{20.13}$$

where $\beta < \alpha$ and $\gamma \gg 1$. Initially, all threshold values are set to a value slightly greater than 1.

As the ICM iterates, values of the threshold decay and quickly values in $\mathbf{f}[\vec{x}]$ become larger than the values in $\mathbf{t}[\vec{x}]$, thus causing the neurons to pulse which in turns greatly increases their threshold values. Each iteration produces an binary-valued output image $\mathbf{y}[\vec{x}]$. Johnson [15] proposed the conversion of the output pulse image set to a set of signatures, $\vec{g}$, by summing over the image space for each iteration

$$g_n = \sum_{\vec{x}} \mathbf{y}_n[\vec{x}], \quad \forall n. \tag{20.14}$$

One of the issues with the ICM is that P is determined by the user without much justification.

The kernel $\mathbf{w}[\vec{x}]$ that controls the neural communications is computed in two steps:

$$\mathbf{b}_n[\vec{x}] = \mathbf{y}_n[\vec{x}] + \Gamma_{>0.5}\left(\mathcal{S}\Gamma_{>0}\mathbf{y}_n[\vec{x}]\right)\left(\Gamma_{\leq 0}\mathbf{y}_n[\vec{x}]\right) \tag{20.15}$$

and

$$\mathbf{w}[\vec{x}] \otimes \mathbf{y}_n[\vec{x}] = \mathbf{b}_n[\vec{x}] - \Gamma_{<0.5}\left[\left(\mathcal{S}\mathbf{c}_n[\vec{x}]\right)\left(\mathbf{c}_n[\vec{x}]\right) + \left(1 - \mathbf{c}_n[\vec{x}]\right)\right], \tag{20.16}$$

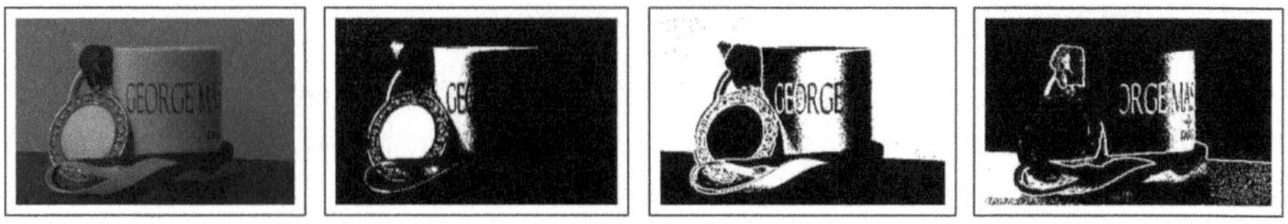

(a) (b) (c) (d)

Figure 20.3 (a)The original input and (b–d) three typical pulse images.

where

$$\mathbf{c}_n[\vec{x}] = \Gamma_{>0}\mathbf{b}_n[\vec{x}]. \tag{20.17}$$

The Γ_c is a threshold operator that sets pixel values to 1 if the condition c is true and to 0 otherwise. The $\mathcal{S}$ operator is the the smoothing operator.

The operator I_n represents the ICM process for n iterations, and the output is a set of pulse images as in

$$\{\mathbf{y}[\vec{x}]\} = I_n\mathbf{a}[\vec{x}]. \tag{20.18}$$

Consider an output $\mathbf{y}_n[\vec{x}]$, which contains a contiguous region of ON pixels. The connections $\mathbf{w}[\vec{x}] \otimes \mathbf{y}_n[\vec{x}]$ will be ON, where the perimeter of the contiguous region flows toward the local centers of curvature. If the system were not continually pumped with $\mathbf{a}[\vec{x}]$, then this contiguous region would evolve into a solid circle and then collapse to a point.

A typical example is shown in Figure 20.3, which shows an original image and three significant pulse images. As seen, the neurons still pulse in synchronized regions, and therefore, provide isolated segments in the image. Similar to the PCNN, as the iterations progress the synchronization decays according to the textures of the segments.

20.3 TEXTURE CLASSIFICATION WITH THE ICM

An example of texture classification of the ICM is applied to lung CAT scan images of which a typical example is shown in Figure 20.4. This patient has been diagnosed with idiopathic pulmonary fibrosis (IPF), which is unfortunately a fatal disease. One slice of the CAT scan is shown in Figure 20.4. One of the visual features of IPF is the *honeycomb* presentation, which is evident in the lower portions of both lungs and in the upper portion of the lung on the right. The texture of this portion of the image is different than the healthier portions of the lung which tend to be smoother.

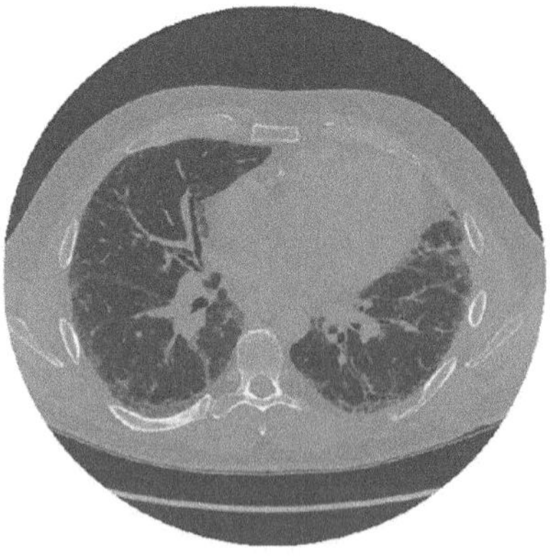

Figure 20.4 Original image of a patient with IPF [1].

The parameters were set, so that the pulse cycles did not overlap. All neurons pulse once before any neuron pulses a second time. The first cycle of pulses is shown in Figure 20.5, and the synchronized behavior of the neurons is evident.

The second cycle is shown in Figure 20.6. The synchronized behavior is beginning to decay but it does so with a sensitivity to the textures that are inherent in the image. The third cycle is shown in Figure 20.7, and the decay is more evident. Typically, the fourth cycle is so dispersed that it no longer provides information about the texture.

The output of the ICM is a set of pulse images, $\{\mathbf{y}[\vec{x}]\}$ shown in Figure 20.8. A sample of data from all iterations at a single pixel, $\vec{v}$, is $\vec{z} = \Box_{\vec{v}}\{\mathbf{y}[\vec{x}]\}$. In the experiment, several such vectors were extracted from different regions in the lung. Through an associative memory, these vectors were classified as belonging to healthy lung (white), not healthy or not inside the lung (gray), or unknown (black). Figure 20.9(a) shows the pixels classified as healthy, and 20.9(b) shows the pixels classified as fibrotic tissue. Applying the classification to all images in a set led to the volume measures of fibrotic tissue and to is density for lung strata.

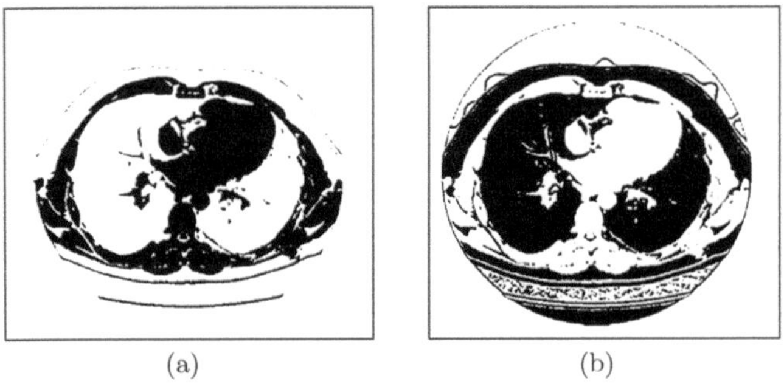

Figure 20.5 First cycle pulses for (a) $n = 1$ and (b) $n = 25$.

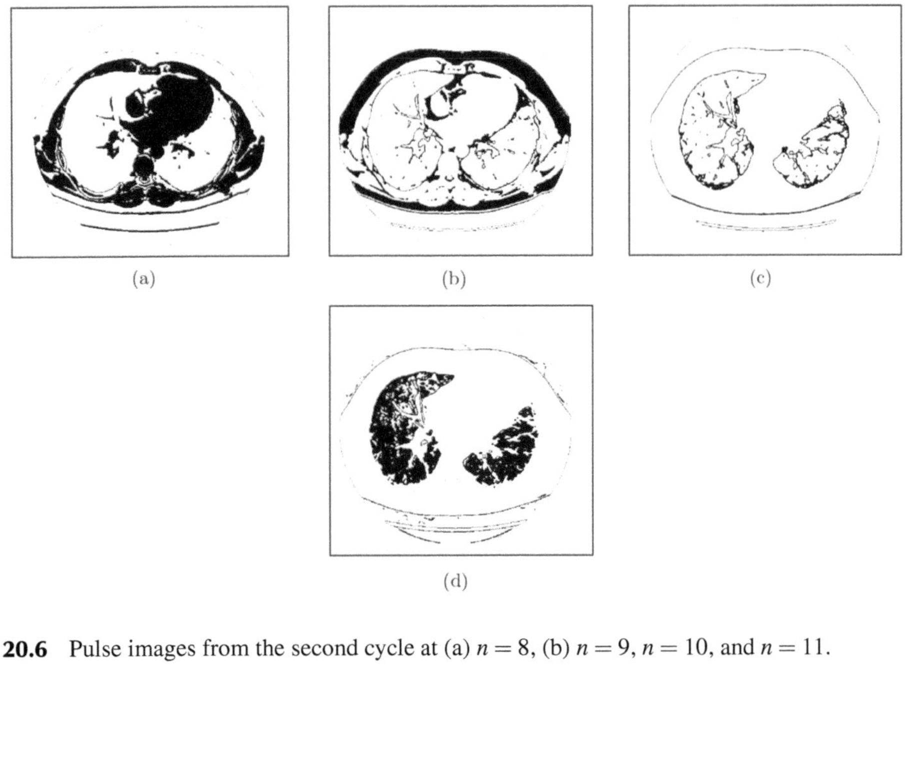

Figure 20.6 Pulse images from the second cycle at (a) $n = 8$, (b) $n = 9$, $n = 10$, and $n = 11$.

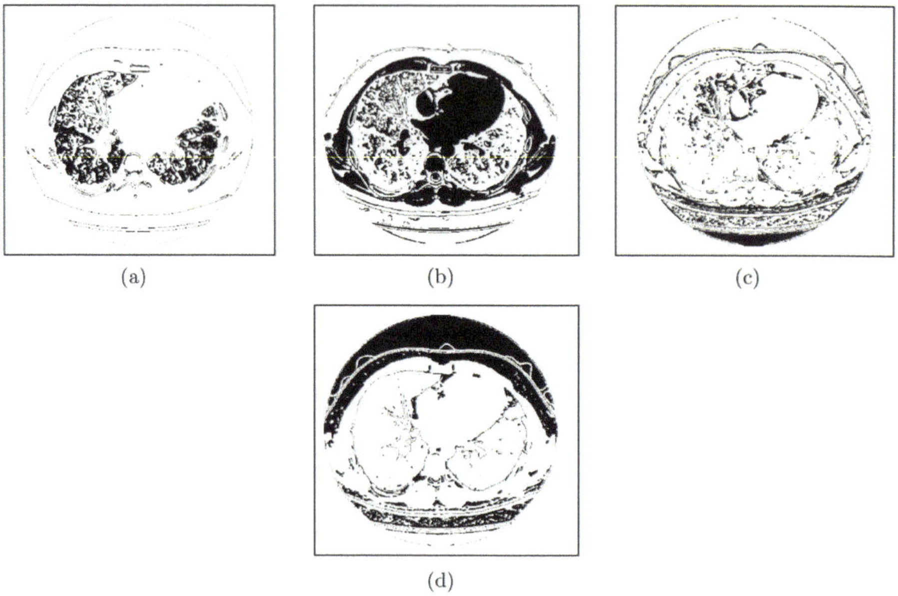

Figure 20.7 Pulse images from the third cycle at (a) $n = 19$, (b) $n = 20$, (c) $n = 21$, and (d) $n = 22$.

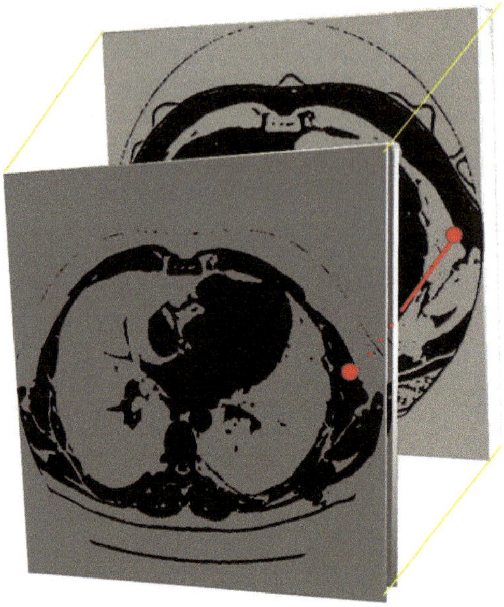

Figure 20.8 Only two slices are shown and the red line extracts a drill from the cube.

20.4 SUMMARY

The PCNN and ICM are digital models based on analog models of the mammalian visual cortex. Popularity in these models were gained from their applications to several image analysis processes. Both models expand the data space from the original $V \times H$ to $N \times V \times H$, where N is the number of iterations. From this expanded space, subsets of data were extracted and used for various image

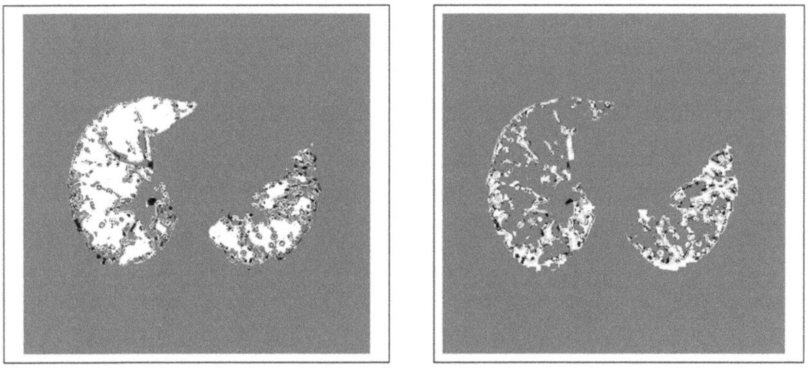

(a) (b)

Figure 20.9 Classification of the pixels as (a) healthy and (b) fibrotic.

processing applications. The output of the systems are pulse images, with binary-valued elements, which are distinctly advantageous when pursuing edge and corner information.

PROBLEMS

1. Create $\mathbf{a}[\vec{x}] = \mathbf{o}_{\vec{w};\vec{v},r}[\vec{x}]$, where $\vec{w} = (128,128)$, $\vec{v} = (64,64)$ and $r = 20$. Capture the first 12 PCNN pulse images and create a new image by adding these images together. Show the result.
2. Create $\mathbf{a}[\vec{x}] = \mathbf{r}_{\vec{w};\vec{v}_1,\vec{v}_2}[\vec{x}]$, where $\vec{w} = (128,128)$, $\vec{v}_1 = (32,32)$, and $\vec{v}_2 = (96,96)$. Compute $\mathbf{b}[\vec{x}] = \sum_{i=1}^{12} J_{12}\mathbf{a}[\vec{x}]$. Show the result.
3. Create $\mathbf{a}[\vec{x}] = \mathbf{r}_{\vec{w};\vec{v}_1,\vec{v}_2}[\vec{x}] + \mathbf{r}_{\vec{w};\vec{v}_3,\vec{v}_4}[\vec{x}]$, where $\vec{w} = (128,128)$, $\vec{v}_1 = (32,32)$, $\vec{v}_2 = (48,48)$, $\vec{v}_3 = (80,80)$, and $\vec{v}_4 = (96.96)$. Compute $\mathbf{b}[\vec{x}] = \sum_{i=1}^{11} J_{11}\mathbf{a}[\vec{x}]$. Show the result.
4. Create $\mathbf{a}[\vec{x}] = \mathbf{r}_{\vec{w};\vec{v}_1,\vec{v}_2}[\vec{x}] - \mathbf{r}_{\vec{w};\vec{v}_3,\vec{v}_4}[\vec{x}]$, where $\vec{w} = (128,128)$, $\vec{v}_1 = (32,32)$, $\vec{v}_2 = (96,96)$, $\vec{v}_3 = (48,48)$, and $\vec{v}_4 = (80,80)$. Compute $\mathbf{b}[\vec{x}] = \sum_{i=1}^{13} J_{13}\mathbf{a}[\vec{x}]$. Show the result.
5. Create $\mathbf{a}[\vec{x}] = \mathbf{o}_{\vec{w};\vec{v},r}[\vec{x}]$, where $\vec{w} = (128,128)$, $\vec{v} = (64,64)$ and $r = 20$. Compute the pulse signature $\vec{z} = \sum_{\vec{x}} J_{16}\mathbf{a}[\vec{x}]$. Plot the result. What does the linear increase in the plot from $n = 1$ to $n = 6$ represent?
6. Create $\mathbf{a}[\vec{x}] = \mathbf{o}_{\vec{w};\vec{v},r}[\vec{x}]$, where $\vec{w} = (128,128)$, $\vec{v} = (64,64)$ and $r = 36.18$. Compute the pulse signature $\vec{z} = \sum_{\vec{x}} J_{16}\mathbf{a}[\vec{x}]$. Plot the result.
7. This problem will compute two pulse signatures and plot them on the same graph. The frame size for all images is 128×128. The first is $\vec{z}_1 = \sum_{\vec{x}} J_{16}\mathbf{a}[\vec{x}]$ with $\mathbf{a}[\vec{x}] = \mathbf{r}_{\vec{v}_1,\vec{v}_2}[\vec{x}]$, where $\vec{v}_1 = (32,32)$ and $\vec{v}_2 = (96,96)$. The second is $\vec{z}_2 = \sum_{\vec{x}} J_{16}\mathbf{b}[\vec{x}]$. The input is $\mathbf{b}[\vec{x}] = \mathbf{r}_{\vec{v}_3,\vec{v}_4}[\vec{x}] + \mathbf{r}_{\vec{v}_5,\vec{v}_6}[\vec{x}] + \mathbf{r}_{\vec{v}_7,\vec{v}_8}[\vec{x}] + \mathbf{r}_{\vec{v}_9,\vec{v}_{10}}[\vec{x}]$. Use $\vec{v}_3 = (0,0)$, $\vec{v}_4 = (32,32)$, $\vec{v}_5 = (0,96)$ $\vec{v}_6 = (32,128)$, $\vec{v}_7 = (96,0)$, $\vec{v}_8 = (128,32)$, $\vec{v}_9 = (96,96)$, and $\vec{v}_{10} = (128,128)$. Plot $\vec{z}_1$ and $\vec{z}_2$ on the same graph.
8. Create $\mathbf{a}[\vec{x}] = \mathcal{S}_{10}\mathbf{o}_{\vec{w};\vec{v},r}[\vec{x}]$, where $\vec{w} = (128,128)$, $\vec{v} = (64,64)$ and $r = 10$. Compute the pulse signature $\mathbf{b}[\vec{x}] = \sum_{12} J_{12}\mathbf{a}[\vec{x}]$. Show the result.
9. Create $\mathbf{a}[\vec{x}] = 1 - Y(\texttt{geos.png})/255$. Compute the pulse signature $\mathbf{b}[\vec{x}] = \sum_{12} J_{12}\mathbf{a}[\vec{x}]$. Show the result.
10. Create $\mathbf{a}[\vec{x}] = 1 - Y(\texttt{geos.png})/255$. Before running the PCNN, initialize the value of t_2 to 150. Compute the pulse signature $\mathbf{b}[\vec{x}] = \sum_{12} J_{12}\mathbf{a}[\vec{x}]$. Show the result and compare to that of the previous problem.

11. Create $\mathbf{a}[\vec{x}] = Y(\texttt{geos.png})/255$. This is the negative of the image used in Problem 9. Compute the pulse signature $\mathbf{c}[\vec{x}] = \sum_{12} J_{12}\mathbf{a}[\vec{x}]$. Create $\mathbf{d}[\vec{x}] = \mathbf{b}[\vec{x}] + \mathbf{c}[\vec{x}]$, where $\mathbf{b}[\vec{x}]$ is the result from Problem 9. Show $\mathbf{d}[\vec{x}]$.
12. Create $\mathbf{a}[\vec{x}] = \mathcal{S}_1\mathbf{b}[\vec{x}]/\bigvee\mathbf{b}[\vec{x}]$, where $\mathbf{b}[\vec{x}] = \mathcal{L}_L Y(\texttt{bird.jpg})$. Compute $\mathbf{f}[\vec{x}] = \sum_{12} I\mathbf{a}[\vec{x}]$. Show the result.
13. Create $\mathbf{a}[\vec{x}] = \mathcal{S}_1\mathbf{b}[\vec{x}]/\bigvee\mathbf{b}[\vec{x}]$, where $\mathbf{b}[\vec{x}] = \mathcal{L}_L Y(\texttt{bird.jpg})$. Start the ICM threshold value for all elements to 1. Compute $\mathbf{f}[\vec{x}] = \sum_{12} I_{12}\mathbf{a}[\vec{x}]$. Show the result.
14. Create $\mathbf{a}[\vec{x}] = \mathcal{S}_1\mathbf{b}[\vec{x}]/\bigvee\mathbf{b}[\vec{x}]$, where $\mathbf{b}[\vec{x}] = \mathcal{L}_L Y(\texttt{bird.jpg})$. Start the ICM threshold value for all elements to 1. Collect the pulse images as a set $\{\mathbf{f}[\vec{x}]\} = I_{30}\mathbf{a}[\vec{x}]$. For each image in the set, isolate the contiguous regions, $\{\mathbf{g}[\vec{x}]\} = \{\mathcal{I}\mathbf{f}[\vec{x}]\}$. Keep only those that have more than 1000 pixels turned on. How many images are produced. Show the first three.
15. Create $\mathbf{a}[\vec{x}] = \mathcal{S}_1\mathbf{b}[\vec{x}]/\bigvee\mathbf{b}[\vec{x}]$, where $\mathbf{b}[\vec{x}] = \mathcal{L}_L Y(\texttt{bird.jpg})$. Start the ICM threshold value for all elements to 1. Collect the pulse images as a set $\{\mathbf{f}[\vec{x}]\} = I_{30}\mathbf{a}[\vec{x}]$. Keep only those images in which pixel (120,273) has a value of 1. This pixel is on the bird's beak. Show all pulse images in this final set.
16. Create $\mathbf{a}[\vec{x}] = \mathcal{S}_1\mathbf{b}[\vec{x}]/\bigvee\mathbf{b}[\vec{x}]$, where $\mathbf{b}[\vec{x}] = \mathcal{L}_L Y(\texttt{bird.jpg})$. Start the ICM threshold value for all elements to 1. Collect the pulse images as a set $\{\mathbf{f}[\vec{x}]\} = I_{30}\mathbf{a}[\vec{x}]$. Keep only those images in which pixel (120,273) has a value of 1. Compute $\mathbf{g}[\vec{x}] = \sum\mathbf{f}[\vec{x}]$. Show the result.
17. Create $\mathbf{a}[\vec{x}] = \mathcal{S}_1\mathbf{b}[\vec{x}]/\bigvee\mathbf{b}[\vec{x}]$, where $\mathbf{b}[\vec{x}] = \mathcal{L}_L Y(\texttt{bird.jpg})$. Start the ICM threshold value for all elements to 1. Collect the pulse images as a set $\{\mathbf{f}[\vec{x}]\} = I_{30}\mathbf{a}[\vec{x}]$. Keep only those images in which pixel (120,273) has a value of 1. Compute $\mathbf{g}[\vec{x}] = \sum\mathbf{f}[\vec{x}]$. Show the case of $\Gamma_{=4}\mathbf{g}[\vec{x}]$. These are the pixels that had the same pulsing behavior as the selected pixel at (120,273).

A Operators

This appendix presents the descriptions of each operator with accompanying Python script. The scripts will assume that the following modules have been loaded:

```
>>> import numpy as np
>>> import imageio
>>> import scipy.fftpack as ft
>>> import scipy.ndimage as nd
>>> import scipy.signal as ss
```

The descriptions will also assume that the following modules provided by the author are loaded:

```
>>> import correlate
>>> import dct2d
>>> import edge
>>> import eigenimage
>>> import fpf
>>> import geomops
>>> import hough
>>> import mgcreate
>>> import pca
>>> import rpolar
>>> import shape
>>> import warp
>>> import wavelet
```

Image variables are denoted as `amg`, `bmg`, `cmg`, etc. These have the form of $V \times H \times N$ were V and H are the vertical and horizontal dimensions, and N is the number of channels (if more than 1).

There are eight classes of operators. These are as follows:

- Creation operators create images or sets of images.
- Channel operators manage individual channels of images.
- Information operators extract information from images or sets of images.
- Intensity operators modify the intensity of the pixels but does not move the information.
- Geometric operators move the information to new locations but does not change the intensity of the objects in the image.
- Transformation operators create representations of the image in new coordinate spaces.
- Expansion operators exp dimensionality of the image.
- Contraction operators reduce the dimensional of the data, perhaps collapsing several images into a single image.

A.1 REPRESENTATIONS

A vector is represented with standard notation, $\vec{v}$. A set of vectors denoted by curly braces as in $\{\vec{v}_i\}$, where i is the index that spans the list of vectors. Selecting specific elements from a set of vectors is denoted as $\{\vec{v}\}$ or $\vec{v}_i$, where i is the index.

An image is represented by $\mathbf{a}[\vec{x}]$; $\vec{x} \in \mathbf{X}$, where $\vec{x}$ spans the spatial dimension of the image in domain $\mathbf{X}$. A set of images $\{\mathbf{a}[\vec{x}]\}$ or $\mathbf{a}_i[\vec{x}]$ $i = 1, \ldots, N$, where N is the number of images in the set.

A.2 CREATION OPERATORS

Creation operators create images or sets of images. They do not operate on other images.

A.2.1 LOAD OPERATOR

Purpose: Load an image from a file.
Notation: $\mathbf{a}[\vec{x}] = Y(\texttt{filename})$.
Input: File name.
Output: None.
Script:

```
>>> amg = imageio.imread( fileName )
```

A.2.2 ZERO OPERATOR

Purpose: Creates an image of a size specified by $\vec{w}$, in which all pixel values are set to 0.
Notation: $\mathbf{a}[\vec{x}] = \mathbf{z}_{\vec{w}}[\vec{x}]$.
Input: A vector of the frame size.
Output: An image.
Options: $\vec{w}$ may be omitted if implied in the protocol.
Script:

```
>>> amg = np.zeros( w )
```

A.2.3 RANDOM OPERATOR

Purpose: To create an image containing equally distributed random values between 0 and 1.
Notation: $\mathbf{a}[\vec{x}] = \mathbf{q}_{\vec{w}}[\vec{x}]$, where $\vec{w}$ is the frame size. The $\vec{w}$ may be omitted if implied in the protocol.
Input: `VH` is the size of the image.
Output: An image.
Script:

```
>>> amg = np.random.ranf( VH )
```

A.2.4 RECTANGLE OPERATOR

Purpose: To create an image containing a solid rectangle.
Notation: $\mathbf{a}[\vec{x}] = \mathbf{r}_{\vec{w};\vec{v}_1,\vec{v}_2}[\vec{x}]$. The $\vec{w}$ may be omitted if implied in the protocol.
Input: `VH` is the size of the image, `v1` and `v2` are the location of opposite corners of the rectangle.
Output: An image.
Script:

```
>>> amg = np.zeros( VH )
>>> amg[v1[0]:v2[0],v1[1]:v2[1]] = 1
```

A.2.5 CIRCLE OPERATOR

Purpose: To create an image containing a solid circle.
Notation: $\mathbf{a}[\vec{x}] = \mathbf{o}_{\vec{w};\vec{v},r}[\vec{x}]$, where $\vec{w}$ is the frame size, $\vec{v}$ is the location of the center, and r is the radius. The $\vec{w}$ may be omitted if implied in the protocol.
Input: `VH` is the size of the image, `ctr` is the location of the center of the circle, and `r` is the radius.

Output: An image.
Script:

```
>>> amg = mgcreate.Circle( VH, ctr, r )
```

A.2.6 ELLIPSE OPERATOR

Purpose: To create an image containing a solid ellipse.
Notation: $\mathbf{a}[\vec{x}] = \mathbf{p}_{\vec{w};\vec{v},a,b}[\vec{x}]$, where $\vec{w}$ is optional and the frame size, $\vec{v}$ is the center of the ellipse, and a and b are the parameters for the major and minor axes in the ellipse equation $\left(\frac{x}{a}\right)^2 + \left(\frac{y}{b}\right)^2 = 1$.
Input: `VH` is the size of the image, `ctr` is the location of the center of the circle, and the parameters `a` and `b` correspond to the major and minor axes.
Output: An image.
Script:

```
>>> amg = mgcreate.Ellipse(VH, ctr, a,b )
```

A.2.7 KAISER WINDOW

Purpose: To create an image with a Kaiser window.
Notation: $\mathbf{a}[\vec{x}] = k_{\vec{w};r_1,r_2}[\vec{x}]$, where $\vec{w}$ is the frame size, and r_1 and r_2 are the radii of the Kaiser window. The $\vec{w}$ may be omitted if implied in the protocol.
Input: `VH` is the size of the image, `r1` is the inner radius and `r2` is the outer radius.
Output: An image.
Script:

```
>>> amg = creation.Kaiser( VH, r1, r2 )
```

A.2.8 GABOR FILTERS

Purpose: To create a set of Gabor filters.
Notation: $\{\mathbf{a}[\vec{x}]\} = \mathfrak{G}_{\vec{w},f,t,\vec{\delta}}[\vec{x}]$, where $\vec{w}$ is the frame size, f is the frequency, t is the angle, and $\vec{\delta}$ is the extent.
Input: `VH` is the size of the image, `f` is the frequency, `theta` is the angle, and `deltax` and `deltay` are the extents.
Output: A set of images.
Script:

```
>>> amg = mgcreate.GaborCos( VH, f, theta, deltax, deltay )
```

A.2.9 ZERNIKE FILTERS

Purpose: To create a set of Zernike filters.
Notation: $\{\mathbf{a}_i[\vec{x}]\} = \mathcal{Z}_{\vec{w};r,m,n}[\vec{x}]$.
Input: `VH` is the size of the image, `r`, `m`, and `n` are Zernike parameters.
Output: A set of images.
Script:

```
>>> amg = mgcreate.Zernike( w, r, m,n )
```

A.3 CHANNEL OPERATORS

Channel operators manage the different channels in an image particularly in color images.

A.3.1 CHANNEL SEPARATOR

Purpose 1: To access individual channels.

Notation 1: $\begin{Bmatrix} \mathbf{a}[\vec{x}] \\ \mathbf{b}[\vec{x}] \\ \mathbf{c}[\vec{x}] \end{Bmatrix} = \mathbf{d}[\vec{x}]$

Input: None.
Output: Defined by use.
Script:

```
>>> d = imageio.imread( 'filename.png' )
>>> a = d[:,:,0]
>>> b = d[:,:,1]
>>> c = d[:,:,2]
```

Purpose 2: To modify channels.

Notation 2: $\mathbf{b}[\vec{x}] = \begin{Bmatrix} \alpha \\ \beta \\ \gamma \end{Bmatrix} \mathbf{a}[\vec{x}]$.

Script:

```
>>> a = imageio.imread( 'filename.png' )
>>> b[:,:,0] = alpha * a[:,:,0]
>>> b[:,:,1] = beta * a[:,:,1]
>>> b[:,:,2] = gamma * a[:,:,2]
```

Options:

- The symbol $\varnothing$ can be used to block a channel.
- The symbol $\bowtie$ can be used to pass a channel.

A.3.2 CHANNEL SUMMATION

Purpose: To combine channels through addition.

Notation: $\mathbf{b}[\vec{x}] = \sum_{\mathcal{L}} \begin{Bmatrix} \alpha \\ \beta \\ \gamma \end{Bmatrix} \mathbf{a}[\vec{x}]$.

Input: A multi-channel image.
Output: A grayscale image.
Script:

```
>>> amg = imageio.imread( 'filename.png' )
>>> bmg = alpha * amg[:,:,0] + beta * amg[:,:,1] + gamma * amg[:,:,2]
```

A.3.3 CHANNEL PRODUCT

Purpose: To combine channels through multiplication.

Notation: $\mathbf{b}[\vec{x}] = \prod_{\mathcal{L}} \begin{Bmatrix} 1 \\ 1 \\ 1 \end{Bmatrix} \mathbf{a}[\vec{x}]$.

Input: A multi-channel image.
Output: A grayscale image.
Script:

```
>>> amg = imageio.imread( 'filename.png' )
>>> bmg =  amg.prod(2)
```

A.4 INFORMATION OPERATORS

Information operators retrieve information about an image or set of images but do not change the input.

A.4.1 COMPARISON OPERATORS

Purpose 1: To compare variables.
Notation: To determine if a and b are equal $t = a \stackrel{?}{=} b$.
Input: Two scalars a and b.
Output: True or False.
Other comparisons replace $\stackrel{?}{=}$ with $>$, $<$, $\geq$, $\leq$
Script:

```
>>> t = a == b
>>> t = a > b # etc
```

Purpose 1: To compare two Python arrays of the same size.
Notation: $\vec{t} = \vec{a} \stackrel{?}{=} \vec{b}$.
Input: Two arrays of the same size a and b.
Output: An array with elements of `True` or `False`.
Other comparisons replace $\stackrel{?}{=}$ with $>$, $<$, $\geq$, $\leq$
Script:

```
>>> t = a == b
>>> t = a > b # etc
```

Purpose 3: To compare two images of the same size.
Notation: $\mathbf{t}[\vec{x}] = \mathbf{a}[\vec{x}] \stackrel{?}{=} \mathbf{b}[\vec{x}]$.
Other comparisons replace $\stackrel{?}{=}$ with $>$, $<$, $\geq$, $\leq$
Input: Two images of the same size $\mathbf{a}[\vec{x}]$ and $\mathbf{b}[\vec{x}]$.
Output: An image with elements of `True` or `False`.
Script:

```
>>> t = a == b
>>> t = a > b # etc
```

A.4.2 COMPLEX CONJUGATE

Purpose: To represent the complex conjugate of an image.
Notation: $\mathbf{b}[\vec{x}] = \mathbf{a}^{\dagger}[\vec{x}]$.
Input: A scalar or array.
Output: Same as the input.
Script:

```
>>> b = a.conjugate()
```

A.4.3 DIMENSION OPERATOR

Purpose: To retrieve the dimensions of an image.
Notation: $\vec{w} = Z\mathbf{a}[\vec{x}]$.
Input: An image.
Output: A vector.
Script:

```
>>> w = amg.shape
```

A.4.4 COUNT OPERATOR

Purpose: To retrieve the number of pixels in an image.
Notation: $d = N\mathbf{a}[\vec{x}]$.
Input: An image.
Output: An integer.
Script:

```
>>> d = np.prod( amg.shape )
```

A.4.5 GRADIENT

Purpose: Computes the gradient along specified axes.
Notation: $\{\mathbf{b}[\vec{x}]\} = \mathcal{G}_n\mathbf{a}[\vec{x}]$, where n may specify that only certain axes are used.
Input: An image.
Output: A set of images. The number of images is equal to the number of axes in the input. Each output image is the gradient along one of the axes.
Script:

```
>>> bmglist = np.gradient( amg )
```

A.4.6 ENERGY OPERATOR

Purpose: To retrieve the energy within an image.
Notation: $d = \mathcal{E}\mathbf{a}[\vec{x}]$.
Input: An image.
Output: An float.
Script:

```
>>> N = np.array( amg.shape ).prod() # number of pixels
>>> v = (amg**2).sum() / N
```

A.4.7 NONZERO OPERATOR

Purpose: Extracts the elements indexes that are not zero from either a vector or image.
Notation: The nonzero elements from a vector is $\vec{w} = \mathcal{N}\vec{v}$, and the script is shown in line 1.
The n-th nonzero element from a vector is $f = \mathcal{N}_1\vec{v}$, and the script is shown in line 2.
The nonzero elements in the vertical direction of an image produces a set of vectors $\{\vec{w}_i\} = \mathcal{N}_V\mathbf{a}[\vec{x}]$.
The selection of the first element that is nonzero in each column of an image is $\vec{w} = \mathcal{N}_{1V}\mathbf{a}[\vec{x}]$.
Input: An array.
Output: An array.
Script:

```
>>> w = v.nonzero()[0]
>>> f = v.nonzero()[0][0]
```

A.4.8 INNER PRODUCT OPERATOR

Purpose: To retrieve the inner product of two inputs.
Notation: $f = \mathbf{a}[\vec{x}] \cdot \mathbf{b}[\vec{x}]$.
Input: Two images.
Output: A scalar.
Script: For matrices, the **dot** function will perform matrix–matrix multiplication. Thus, it is not used to compute the dot product of two images.

```
>>> f = (amg * bmg).sum()
```

A.4.9 CENTER OF MASS OPERATOR

Purpose: To retrieve the center of mass of an image.
Theory: In two dimensions: $\vec{v} = \frac{\sum_{i,j} p_{i,j} x_{i,j}}{\sum_{i,j} p_{i,j}}$, where $p_{i,j}$ is the pixel intensity at location (i, j) and $x_{i,j}$ is the location (i, j).
Notation: $\vec{v} = \boxtimes \mathbf{a}[\vec{x}]$.
Input: A binary valued image.
Output: A vector.
Script:

```
>>> v = nd.center_of_mass( amg )
```

A.4.10 MAX AND MIN OPERATORS

Purpose: To retrieve the value of the max or min from an image.
Notation: Max: $f = \bigvee \mathbf{a}[\vec{x}]$ and Min: $f = \bigwedge \mathbf{a}[\vec{x}]$.
Input: An image.
Output: A scalar or complex value.
Script:

```
>>> mx = amg.max()
>>> mn = amg.min()
```

A.4.11 MAX AND MIN LOCATION OPERATORS

Purpose: To retrieve the location of the max or min from an image.
Notation: Max: $\vec{v} = A_{\vee} \mathbf{a}[\vec{x}]$ and Min: $\vec{v} = A_{\wedge} \mathbf{a}[\vec{x}]$.
Input: An image.
Output: A vector.
Script:

```
>>> v = divmod( amg.argmax(), H ) # H is horizontal dimension
>>> v = divmod( amg.argmin(), H )
```

A.4.12 AVERAGE OPERATOR

Purpose 1: To compute the average over an image.
Notation: $f = \mathcal{M} \mathbf{a}[\vec{x}]$.
Input: An image.
Output: A scalar or complex value.
Script:

```
>>> f = a.mean()
```

A.4.13 AVERAGE OPERATOR

Purpose 1: To compute the average over a defined region A in an image.
Notation: $f = \mathcal{M}_A \mathbf{a}[\vec{x}]$.
Input: An image.
Output: A scalar or complex value.
Script: Two options

```
# x ,y are lists of pixel locations or a bounding box
>>> f = amg[x,y].mean()
>>> b = nd.gaussian_filter(amg,r)
```

A.4.14 STANDARD DEVIATION OPERATOR

Purpose: To compute the standard deviation over an area A in an image.
Notation: $f = \mathcal{T}_A \mathbf{a}[\vec{x}]$.
Input: An image.
Output: An image.
Script: Two options

```
>>> f = amg[v1:h1,v2:h2].std()
>>> b = nd.standard_deviation( amg )
```

A.4.15 COVARIANCE MATRIX

Purpose: To compute the covariance matrix from a list of vectors.
Notation: $\mathbf{M} = V\{\vec{v}\}$.
Input: A list of vectors.
Output: A matrix.
Script:

```
>>> M = np.cov( vlist )
```

A.4.16 SORTING OPERATOR

Purpose: To sort a set of images according to user-defined criteria, C.
Notation: Sorting in descending order $\{\mathbf{b}[\vec{x}]\} = \mathcal{O}_{\downarrow,C}\{\mathbf{a}[\vec{x}]\}$.
Input: An image set.
Output: An image set.
Script: The criteria C is user defined and is applied to all of the images in the array `amgs`. The scores of the criteria are stored in `vec`. Line 1 determines the sort order from low to high. Line 2 will create an new array `bmgs` that is the images from `amgs` in the sort order.

```
>>> r = vec.argsort()
>>> bmgs = amgs[:,:,r]
```

A.4.17 DISTANCE OPERATOR

Purpose: To compute the Euclidean distances between two images.
Notation: $d = \mathfrak{D}(\mathbf{a}[\vec{x}], \mathbf{b}[\vec{x}])$.
Input: Two images.
Output: A scalar.
Script:

```
>>> d = np.sqrt(( (amg*bmg)**2).sum())
```

A.4.18 CO-OCCURRENCE MATRIX

Purpose: To compute the co-occurrence matrix for a given shift $\vec{\delta}$ from an image with N distinct gray levels.
Notation: $\mathbf{p} = \mathfrak{C}_{\vec{\delta},N}\mathbf{a}[\vec{x}]$.
Input: A grayscale image `amg`, a shift, and the total number of possible gray levels, `N`.
Output: A matrix.
Script:

```
>>> p = np.zeros((N,N))
>>> bmg = nd.shift( amg, shift )
>>> for i in range(N):
>>>     for j in range(N):
>>>         p[i,j] = ((amg==i)*(bmg==j)).sum()
```

A.4.19 FOURIER DESCRIPTORS

Purpose: To compute the Fourier descriptors of a given shape.
Notation: $\mathbf{p} = \mathcal{F}\mathbf{a}[\vec{x}]$.
Input: A binary valued image.
Output: A complex valued vector.
Script:

```
>>> vec = shape.FourierDescriptors( amg )
```

A.5 INTENSITY OPERATORS

Intensity operators modify the intensity of an image but does not move the information to new locations.

A.5.1 INTENSITY MODIFICATION

Purpose: To change the intensity by a simple scalar multiplication.
Notation: $\mathbf{b}[\vec{x}] = f\mathbf{a}[\vec{x}]$.
Script:

```
>>> bmg = f * amg
```

A.5.2 LOG

Purpose: To compute the log value of all elements.
Notation: $\mathbf{b}[\vec{x}] = L_m\mathbf{a}[\vec{x}]$, where m is an optional argument to indicate the base of the log operation. For example, a base L_{10} would compute $\log_{10}$ of each element.
Script:

```
>>> bmg = np.log( amg )
```

A.5.3 ELEMENTAL MATH

Purpose: To perform element-by-element mathematical operations.
Notation: Binary AND: $\mathbf{c}[\vec{x}] = \mathbf{a}[\vec{x}]\&\mathbf{b}[\vec{x}]$
Binary OR: $\mathbf{c}[\vec{x}] = \mathbf{a}[\vec{x}] \mid \mathbf{b}[\vec{x}]$
Addition: $\mathbf{c}[\vec{x}] = \mathbf{a}[\vec{x}] + \mathbf{b}[\vec{x}]$

Subtraction: $\mathbf{c}[\vec{x}] = \mathbf{a}[\vec{x}] - \mathbf{b}[\vec{x}]$
Multiplication: $\mathbf{c}[\vec{x}] = \mathbf{a}[\vec{x}] \times \mathbf{b}[\vec{x}]$
Input: Two images.
Output: An image.
Script:

```
>>> cmg = amg + bmg
>>> cmg = amg - bmg
>>> cmg = amg * bmg
```

A.5.4 BINARY SELECTION

Purpose: To select the lowest or highest bits in an integer-valued image.
Notation: Lo bits: $\mathbf{b}[\vec{x}] = \triangledown_n \mathbf{a}[\vec{x}]$, where n is the number of bits.
Hi bits: $\mathbf{b}[\vec{x}] = \triangle_n \mathbf{a}[\vec{x}]$
Input: An image.
Output: An image.
Script: The variable a is a matrix of unsigned bytes, and h is a hexadecimal value. Line 3 performs the operation for the lowest four bits. Use `h = 0xF0` for the highest four bits.

```
>>> amg = (255*np.random.ranf( (5,5) )).astype(np.ubyte)
>>> h = 0xF0
>>> bmg = h & amg
```

A.5.5 LOCATIONS

Purpose: To set pixel values to true of false based on a given condition. This is similar to the Threshold operator Γ except that f can describe operations other than simple thresholds.
Notation: Locations of value f: $\mathbf{b}[\vec{x}] = A_f \mathbf{a}[\vec{x}]$. Locations for values greater than f: $\mathbf{b}[\vec{x}] = A_{>f} \mathbf{a}[\vec{x}]$.
Input: An image.
Output: A binary valued image.
Script: The argument in the parenthesis is replaced by the conditional of the user's choice.

```
>>> locs = (amg == f)
```

A.5.6 BINARY FILL HOLES OPERATOR

Purpose: To fill the holes in a binary valued image.
Notation: $\mathbf{b}[\vec{x}] = \mathcal{B}\mathbf{a}[\vec{x}]$.
Input: A binary valued image.
Output: A binary valued image.
Script:

```
>>> bmg = nd.binary_fill_holes( amg )
```

A.5.7 EROSION

Purpose: To perform the erosion operator n times.
Notation: $\mathbf{b}[\vec{x}] = \triangleright_n \mathbf{a}[\vec{x}]$.
Input: An image.
Output: An image.
Script: Two options

```
>>> bmg = ndimage.binary_erosion( amg, iterations=3 )
>>> bmg = ndimage.grey_erosion( amg, iterations=3 )
```

A.5.8 DILATION OPERATOR

Purpose: To perform the dilation operator n times.
Notation: $\mathbf{b}[\vec{x}] = \lhd_n \mathbf{a}[\vec{x}]$.
Input: An image.
Output: An image.
Script: Two options

```
>>> bmg = nd.binary_dilation( amg, iteration=3)
>>> bmg = nd.grey_dilation( amg, iteration=3)
```

A.5.9 EDGE OPERATOR

Purpose: To enhance the edges. The default is a simple derivative method of enhancing edges. However, particular models can be specified with m.
Notation: $\mathbf{b}[\vec{x}] = E_m \mathbf{a}[\vec{x}]$, where m is an optional model specification.
Input: An image.
Output: An image.
Script: Three options listed

```
>>> bmg = edge.Edge( amg, n) # n is jump distance
>>> bmg = nd.filters.prewitt( amg )
>>> bmg = nd.filters.sobel( amg )
```

A.5.10 HARRIS OPERATOR

Purpose: To enhance corners using a Harris operator.
Notation: The α is a control factor, and the default is $\alpha = 0.2$.
$\mathbf{b}[\vec{x}] = \mathcal{H}_\alpha \mathbf{a}[\vec{x}]$.
Input: An image.
Output: An image.
Script: The default values is `alpha=0.2`.

```
>>> bmg = intensityops.Harris( amg, alpha )
```

A.5.11 THRESHOLD OPERATOR

Purpose: To apply a threshold to every pixel in an image.
Notation: $\mathbf{b}[\vec{x}] = \Gamma_n \mathbf{a}[\vec{x}]$, where n is a condition.
Greater than threshold with an binary output (type 1): $\mathbf{b}[\vec{x}] = \Gamma_{>\gamma} \mathbf{a}[\vec{x}]$
Equal to threshold with an binary output (type 2): $\mathbf{b}[\vec{x}] = \Gamma_{=\gamma} \mathbf{a}[\vec{x}]$
Between values (f and g) threshold with a binary output (type 3): $\mathbf{b}[\vec{x}] = \Gamma_{f<g} \mathbf{a}[\vec{x}]$
Passive threshold if greater than value γ (type 4): $\mathbf{b}[\vec{x}] = \Gamma_{p>\gamma} \mathbf{a}[\vec{x}]$
Input: An image.
Output: An image. Depending on the use the output can be a binary valued image or an image with the same data type as the input.
Script: Other types exist but have the same format as the following.

```
>>> bmg = amg > gamma # type 1
```

```
>>> bmg = amg == gamma # type 2
>>> bmg = (amg > f ) * ( amg < g ) # type 3
>>> bmg = amg*( amg > gamma )
```

A.5.12 SMOOTHING OPERATOR

Purpose: To create a smoothed version of the input image.
Notation: $\mathbf{b}[\vec{x}] = \mathcal{S}_n\mathbf{a}[\vec{x}]$, where n is a radius or algorithm selection.
Input: An image.
Output: An image.
Script: There are several algorithms that perform a smoothing operation. Below are only a few of the choices.

```
>>> bmg = ss.cspline2d( amg, 10*n)
>>> bmg = nd.gaussian_filter( amg, n )
# several choices for kernel
>>> kernel = np.ones( (m,m) ) # m is small integer
>>> bmg = scipy.signal.convolve2d( amg, kernel)
```

A.5.13 GAUSSIAN

Notation: $\mathbf{a}[\vec{x}] = G_{\vec{\sigma}}$.
Purpose: To correlate an image with a Gaussian surface.
Input: $\vec{\sigma}$ the standard deviations of the Gaussian surface for each dimension.
Output: An array.
Script: For an array amg of size $V \times H$, where s1 and s2 are the standard deviation values for each axis.

```
>>> bmg = nd.gaussian_filter( amg,(s1,s2))
```

A.6 GEOMETRIC OPERATORS

Geometric operators are designed to move pixel values from one location to another.

A.6.1 SCALING OPERATOR

Purpose: To change the spatial scale of an image.
Notation: There are two notations: $\mathbf{b}[\vec{x}] = \mathbf{a}[\alpha\vec{x}]$, where α is a scalar.
$\mathbf{b}[\vec{x}] = S_n\mathbf{a}[\vec{x}]$, where n as a float scales all dimensions and n as a vector scales each dimension with unique values.
Input: An image.
Output: An image.
Script: Use the function from Python Image Library.

```
>>> import Image
>>> # a is an Image
>>> H,V = a.size
>>> b = a.resize( (alpha*H, alpha*V) )
```

A.6.2 MAPPING OPERATOR

Purpose: To remap pixels in an image.
Notation: $\mathbf{b}[\vec{x}] = C_M\mathbf{a}[\vec{x}]$, where M is the new map.

Input: An image and an array of new locations.
Output: An image.
Script:

```
>>> bmg = nd.map_coordinates( amg, ndx )
```

A.6.3 AFFINE OPERATOR

Purpose: To perform an affine transformation.
Notation: $\mathbf{b}[\vec{x}] = \mathbf{a}[\mathbf{M}\vec{x}]$, where $\mathbf{M}$ is an $N \times N$ matrix, where N is the length of $\vec{x}$.
Input: An image.
Output: An image.
Script: Efficient scripts are application dependent.

A.6.4 WINDOW OPERATOR

Purpose: To isolate a subimage.
Notation: $\mathbf{b}[\vec{x}] = \square_{\vec{v}_1,\vec{v}_2}\mathbf{a}[\vec{x}]$, where $\vec{v}_1$ and $\vec{v}_2$ are the locations of opposing corners of the bounding rectangle.
Input: An image.
Output: An image.
Script: The script for a two-dimensional example is shown, where the upper left corner is (`v1`, `h1`), and the lower right corner is (`v2`, `h2`). Note that the (0,0) corner is in the upper left corner of the image.

```
>>> bmg = amg[v1:v2, h1:h2]
```

A.6.5 DOWNSAMPLE OPERATOR

Notation: $\mathbf{b}[\vec{x}] = \Downarrow_n \mathbf{a}[\vec{x}]$, where n is the instructions for downsampling. For example, $n = 2$ indicates that only those pixels on even numbered rows or columns are extracted. If n is a vector then it defines the sampling rate along each axis. If n is a letter, then it is referring to user-defined instructions.
Purpose: To extract a subimage with a smaller spatial resolution.
Input: An image.
Output: An image.
Script: For the case of $n = 2$

```
>>> V,H = a.shape
>>> bmg = amg[0:V:2, 0:H:2]
```

A.6.6 PLOP OPERATOR

Purpose: Places an image at the center of a larger frame.
Notation: $\mathbf{b}[\vec{x}] = U_{\vec{w}}\mathbf{a}[\vec{x}]$, where $\vec{w}$ is the size of the output frame.
Input: An image and frame size.
Output: An image.
Script:

```
>>> bmg = mgcreate.Plop( amg, w, background=0 )
```

A.6.7 CONCATENATE OPERATOR

Purpose: Builds an image from the concatenation of multiple images.
Notation: $\mathbf{b}[\vec{x}] = \mathcal{C}_d\{\mathbf{a}_i[\vec{x}]\}$, where d describes which axis the concatenation uses or is user defined. In a 2D application a vertical concatenation will stack the images in a vertical arrangement and so all of the horizontal dimensions are required to be the same.
Input: A list of images.
Output: An image.
Script: This function receives a list of the arrays that will be concatenated. The second optional argument indicates the axis in which the concatenation will occur.

```
>>> cmg = np.concatenate( (amg, bmg) ) # vertical
>>> cmg = np.concatenate( (amg ,bmg), 1 ) # horizontal
```

A.6.8 SWAP OPERATOR

Purpose: Swaps the quadrants of a two-dimensional image.
Notation: $\mathbf{b}[\vec{x}] = X\mathbf{a}[\vec{x}]$.
Input: An image.
Output: An image.
Script:

```
>>> bmg = ft.fftshift( amg )
```

A.6.9 FLIP OPERATOR

Purpose: Flips the image. The subscript m is user-defined that indicates how the image is being flipped. For example, $\mathfrak{L}_0$ indicates that the flip is in the first axis.
Notation: $\mathbf{b}[\vec{x}] = \mathfrak{L}_m\mathbf{a}[\vec{x}]$.
Input: An image.
Output: An image.
Script:

```
>>> bmg = np.flipud( amg ) # up and down, 0 axis
>>> bmg = np.fliplr( amg ) # left and right, 1 axis
```

A.6.10 RESHAPE OPERATOR

Purpose: Reshapes the input array, where the shape is defined by a vector $\vec{w}$.
Notation: $\mathbf{b}[\vec{x}] = \mathcal{V}_{\vec{w}}\mathbf{a}[\vec{x}]$, where $\vec{w}$ is the new window size.
Input: An array.
Output: An array.
Script:

```
>>> vec = amg.ravel() # to convert to a vector
>>> bmg = amg.reshape( (newV, newH) ) # generic
```

A.6.11 SHIFT OPERATOR

Purpose: Laterally shifts the input image.
Notation: $\mathbf{b}[\vec{x}] = D_{\vec{v}}\mathbf{a}[\vec{x}]$, where $\vec{v}$ indicates that number of pixels in each dimension of the shift.
Input: An image.
Output: An image.

Script: The shift vector is vec. In a 2D image positive values will move the image down and to the right. The optional cval is the fill value for pixels that are not defined after the shift.

```
>>> bmg = nd.shift( amg, vec )
>>> bmg = nd.shift( amg, vec, cval=5 )
```

A.6.12 ROTATION OPERATOR

Purpose: Rotations an image through an angle θ about a center $\vec{v}$.
Notation: $\mathbf{b}[\vec{x}] = \mathcal{R}_{\theta,\vec{v}}\mathbf{a}[\vec{x}]$. If $\vec{v}$ is not used then the center of rotation is the center of the frame.
Input: An image and the rotation angle.
Output: An image.
Script:

```
>>> bmg = nd.rotate( amg, theta ) # theta in degrees
```

A.6.13 BENDING OPERATOR

Purpose: Converts an image through a barrel or pincushion transformation.
Notation: $\mathbf{b}[\vec{x}] = B_\beta \mathbf{a}[\vec{x}]$, where β is the bending factor. If $\beta > 1$ then this becomes a barrel transformation. If $0 < \beta < 1$ then this is a pincushion transformation.
Input: An image.
Output: An image.
Script: The input cvh is the vector that is the user-defined center of transformation.

```
>>> bmg = geomops.Bend( amg, beta, cvh )
```

A.6.14 RADIAL COORDINATE OPERATOR

Purpose: Expands or contracts an image radially.
Notation: $\mathbf{b}[\vec{x}] = R_{\alpha,\vec{v}}\mathbf{a}[\vec{x}] = \mathcal{P}_{\vec{v}}^{-1} D_{(0,\alpha)} \mathcal{P}_{\vec{v}}\mathbf{a}[\vec{x}]$.
Input: An image, the location of the center of the operation, and a scaling factor.
Output: An image.
Script:

```
>>> temp1 = rpolar.RPolar( amg, v )
>>> temp2 = nd.shift( temp1, (0,alpha) )
>>> bmg = rpolar.IRPolar( temp2, v )
```

A.6.15 WARP OPERATOR

Purpose: Warps an image to a specified fiducial grid.
Notation: $\mathbf{b}[\vec{x}] = W_G \mathbf{a}[\vec{x}]$, where G is the output grid.
Input: An image, its fiducial grid and the output fiducial grid.
Output: An image.
Script: The input image data is amg and is fiducial grid is afid. The output fiducidal grid is bfid.

```
>>> rinf = warp.RInf( afid )
>>> b = warp.Warp( amg, afid, bfid, rinf )
```

A.6.16 MORPH OPERATOR

Purpose 1: Creates an image that is the morph between inputs $\mathbf{a}[\vec{x}]$ and $\mathbf{b}[\vec{x}]$.
Notation: $\mathbf{c}[\vec{x}] = M_\alpha (\mathbf{a}[\vec{x}], \mathbf{b}[\vec{x}])$.
Input: Two images, their respective fiducial grids and the morph factor `alpha`.
Output: An image.
Script: Usually, the morph operator is used to create several images. The following is used to create a single image with user-defined value for α. The inputs for $\mathbf{a}[\vec{x}]$ is `amg` for the image data and `afid` for the accompanying fiducial grid data. The image $\mathbf{b}[\vec{x}]$ has similar inputs.

```
>>> rinf1 = warp.RInf( afid )
>>> rinf2 = warp.RInf( bfid )
>>> fidm = i/NI * bfid + (NI-i)/NI*afid
>>> mata,d1 = warp.Warp( amg, afid, fidm, rinf2, 2 )
>>> matb,d2 = warp.Warp( bmg, bfid, fidm, rinf1, 2 )
>>> c = i/NI * matb + (NI-i)/NI*mata
```

Purpose 2: Creates a set of images that are the morphs between inputs $\mathbf{a}[\vec{x}]$ and $\mathbf{b}[\vec{x}]$. The total number of images equally spaced in the morphing process is N.
Notation: $\{\mathbf{c}[\vec{x}]\} = M_N (\mathbf{a}[\vec{x}], \mathbf{b}[\vec{x}])$.
Input: Two images, their respective fiducial grids and the number of images to create.
Output: A set of N output images.
Script: The inputs for $\mathbf{a}[\vec{x}]$ is `amg` for the image data and `afid` for the accompanying fiducial grid data. The image $\mathbf{a}[\vec{x}]$ has similar inputs. The input `NI` is the number of generated images.

```
>>> c = warp.Morph( amg, afid, bmg, bfid, NI=10)
```

A.7 TRANSFORMATION OPERATORS

The transformation operators convert an image from one coordinate system to another.

A.7.1 COLOR TRANSFORMATION

Purpose: Converts an image from one color space to another.
Notation: $\mathbf{b}[\vec{y}] = \mathcal{L}_M \mathbf{a}[\vec{x}]$, where M is the specified model.
Input: An image and a model.
Output: An image.
Script: There are many different color models. Python provides some of this in the *colorsys* module. To apply this to large arrays, see Code 5.3.

```
# for single pixels
>>> b = colorsys.rgb_to_hls( r, g, b )
>>> b = colorsys.rgb_to_hsv( r, g, b )
>>> b = colorsys.rgb_to_yiq( r, g, b )
>>> b = colorsys.hls_to_rgb(h, l, s )
>>> b = colorsys.hsv_to_rgb(h, s, v )
>>> b = colorsys.yiq_to_rgb( y, i, q)
```

A.7.2 DISTANCE TRANSFORMATION

Purpose: Returns the distance transformation image.
Notation: $\mathbf{b}[\vec{x}] = \mathcal{D}_n \mathbf{a}[\vec{x}]$, where n indicates the type of transform. The Euclidean transform uses $\mathcal{D}_E$.
Input: An image.

Output: An image.
Script:

```
>>> bmg = dt.distance_transform_edt( amg )
```

A.7.3 DERIVATIVE

Purpose: Returns the derivative of an array.
Notation: $\mathbf{b}[\vec{x}] = \partial_{\vec{x}}\mathbf{a}[\vec{x}]$.
Input: An array.
Output: An array.
Script:

```
>>> bmg = np.gradient( amg )
```

A.7.4 HOUGH TRANSFORMATION

Purpose: Returns the Hough transform of an image.
Notation: $\mathbf{b}[\vec{y}] = H\mathbf{a}[\vec{x}]$ for a linear Hough transform and H_0 for a circular Hough transform.
Input: An image.
Output: An image.
Script: The gamma is a passive threshold. Pixels below this value are not considered in the transform.

```
>>> b = hough.LineHough( a, gamma ) # linear Hough
>>> b = hough.CircleHough( a, radius, gamma ) # circle Hough.
```

A.7.5 POLAR COORDINATE TRANSFORMATION

Purpose: When applied to a vector, it converts the vector to polar coordinates. When applied to an array of complex values, it converts the complex representation from $x + \imath y$ to $re^{\imath\theta}$.
Notation: $\vec{\rho} = P\vec{x}$ or $\mathbf{b}[\vec{y}] = P\mathbf{a}[\vec{x}]$. The inverse transform is P^{-1}.
Input: A vector or an image.
Output: The same type as the input.
Script: For a vector:

```
>>> rho = np.zeros( 2 )
>>> rho[0] = np.sqrt( x[0]**2 + x[1]**2 )
>>> rho[1] = np.tan2( x[1], x[0] )
```

For an image:

```
>>> bmg = mgcreate.Rect2Polar( amg )
>>> cmg = mgcreate.Polar2Rect( bmg ) # inverse
```

A.7.6 RADIAL-POLAR TRANSFORMATION

Purpose: Converts a rectilinear image to radial-polar coordinates.
Notation: $\mathbf{b}[\vec{y}] = \mathcal{P}_{\vec{v}}\mathbf{a}[\vec{x}]$.
The inverse transformation is $\mathbf{c}[\vec{x}] = \mathcal{P}_{\vec{v}}^{-1}\mathbf{b}[\vec{y}]$.
Input: An image and the center of the transformation.
Output: An image.
Script:

```
>>> bmg = rpolar.RPolar( amg, v )
>>> cmg = rpolar.IRPolar( bmg, v )
```

A.7.7 WAVELET DECOMPOSITION OPERATOR

Purpose: Creates an image that is the wavelet decomposition of the input.
Notation: $\mathbf{b}[\vec{y}] = \mathcal{W}\mathbf{a}[\vec{x}]$.
Input: An image.
Output: An image.
Script: This operator works best if the dimensions of the input are factors of 2. Consider padding an image if the dimensions are else.

```
>>> bmg = wavelet.Divide( amg ) # for a single iteration
>>> bmg = wavelet.RepeatDivide( amg ) # for all iterations
```

A.7.8 DISCRETE COSINE TRANSFORM OPERATOR

Purpose: Applies the discrete cosine transform to the input signal.
Input: A vector or image.
Input: Vector or image (same as the input).
Notation: Forward transform: $\mathbf{b}[\vec{\omega}] = \mathcal{J}\mathbf{a}[\vec{x}]$.
Inverse transform: $\mathbf{b}[\vec{x}] = \mathcal{J}^{-1}\mathbf{a}[\vec{\omega}]$.
Script: There are four versions.

```
>>> bmg = ft.dct( amg ) # 1D
>>> bmg = ft.idct( amg ) # Inverse, 1D
>>> bmg = dct2d.dct2d( amg ) # 2D
>>> bmg = dct2d.idct2d( amg ) # Inverse, 2D
```

A.7.9 FOURIER TRANSFORM OPERATOR

Purpose: Transforms a data array into Fourier space.
Notation: The transformation of a vector is: $\vec{y} = \mathfrak{F}\vec{x}$.
The transformation of an image is: $\mathbf{b}[\vec{y}] = \mathfrak{F}\mathbf{a}[\vec{x}]$.
The inverse transformation is: $\mathbf{a}[\vec{x}] = \mathfrak{F}^{-1}\mathbf{b}[\vec{y}]$.
Input: The input can be a vector, matrix or tensor.
Output: The output will be the same shape as the input.
Script: There are three different functions depending on the shape of the data.

```
>>> y = ft.fft( x ) # vector
>>> b = ft.fft2( a ) # 2D array
>>> b = ft.fftn( a ) # N-D array
>>> x = ft.ifft( y ) # inverse
>>> a = ft.ifft2( b ) # inverse 2D array
>>> a = ft.ifftn( b ) # inverse N-D array
```

A.7.10 CORRLELATION OPERATOR

Purpose: Calculates the correlation of two inputs.
Notation: $\mathbf{c}[\vec{x}] = \mathbf{a}[\vec{x}] \otimes \mathbf{b}[\vec{x}]$.
Input: Two matrices.
Output: A matrix.
Script: Line 1 is a method to use if the kernel $\mathbf{b}[\vec{x}]$ is very small. Line 2 is used if both inputs are the same size.

```
>>> c = ss.correlate2d( a, b )
>>> c = correlate.Correlate( a, b )
```

A.8 EXPANSION OPERATORS

Expansion operators increase the dimensionality of the data in an attempt to lower the complexity of the pertinent information.

A.8.1 ISOLATION OPERATOR

Purpose: Isolates the noncontiguous binary shapes.
Notation: For the case of returning a set of binary valued images: $\{\mathbf{b}_i[\vec{x}]\} = \mathcal{I}\mathbf{a}[\vec{x}]$.
Input: A binary-valued image.
Output: Either a set of binary-valued images or a single integer-valued image.
For the case of returning a single integer-valued image: $\mathbf{b}[\vec{x}] = \mathcal{I}\mathbf{a}[\vec{x}]$.
Script: The Python function returns a single matrix in which each integer represents a different segment. Line 1 returns the integer-valued matrix, and line 2 separates it into individual binary valued images. The output b is the matrix and n is the number of different segments.

```
>>> b, n = nd.label( amg )
>>> clist = list(map( lambda x: b==x, range(n) ))
```

A.8.2 EIGENIMAGES OPERATOR

Purpose: Creates a list of eigenimages from a list of images.
Notation: $\{\mathbf{b}_i[\vec{x}]\} = T_n\{\mathbf{a}_j[\vec{x}]\}$, where n is the number of eigenimages, $i = 1, \ldots, N$ and j is the index over all of the input images.
Input: A list of images and the number of eigenimages to create.
Output: A list of images.
Script: Returns a list of matrices, emgs, and the associated eigenvalues, evls. Line 2 can be used to project other images into the eigenspace.

```
>>> emgs, evls = eigenimage.EigenImages( amg )
>>> c = eigenimage.ProjectEigen( emgs, indata )
```

A.8.3 PCNN OPERATOR

Purpose: Creates a list of binary valued images through the PCNN process.
Notation: $\{\mathbf{b}[\vec{x}]\} = J_n\mathbf{a}[\vec{x}]$, where n is the number of pulse images and $i = 1, \ldots, N$.
Input: An image and the number of iterations. PCNN parameters can also be adjusted.
Output: A list of images.
Script:

```
>>> net = pcnn.Net( amg.shape )
>>> Y = []
>>> for i in range( n ):
>>>     net.Iterate( amg ) # other iteration functions exist
>>>     Y.append( net.Y )
```

A.8.4 ICM OPERATOR

Purpose: Creates a list of binary valued images through the ICM process.
Notation: $\{\mathbf{b}[\vec{x}]\} = I_n\mathbf{a}[\vec{x}]$, where n is the number of pulse images and $i = 1, \ldots, N$.
Input: An image and the number of iterations. ICM parameters can also be adjusted.
Output: A list of images.
Script:

```
>>> net = icm.Net( amg.shape )
>>> Y = []
>>> for i in range( n ):
>>>     net.Iterate( amg ) # other iteration functions exist
>>>     Y.append( net.Y )
```

A.8.5 EMD OPERATOR

Purpose: Creates a list of binary valued images through the EMD process.
Notation: For the case of a vector input: $\{\vec{v}\} = \mathfrak{E}\vec{w}$.
For the case of an image input: $\{\mathbf{b}[\vec{x}]\} = \mathfrak{E}\mathbf{a}[\vec{x}]$.
Input: An image.
Output: A list of images.
Script:

```
>>> imfs, resid = emdspl.EMD2d( mat, 5 )
>>> imfs, resid = emdspl.EMD2d( amg, 6, 5 )
```

A.9 CONTRACTION OPERATORS

Contraction operators reduce the dimensionality of the data by combining the inputs.

A.9.1 SUMMATION OPERATOR

Purpose: Creates the sum of several input images.
Notation: $\mathbf{b}[\vec{x}] = \sum_i \mathbf{a}_i[\vec{x}]$.
Input: A list of images.
Output: An image.
Script: Line 1 is used if the data is collected in a tensor, and the first axis represents the channels. Lines 2 through 4 are used if the data is a list of images.

```
>>> b = a.sum(0)
>>> b = []
>>> for i in range( len( a ) ):
>>>     b += a[i]
```

A.9.2 PRODUCT OPERATOR

Purpose: Creates the product of several input images.
Notation: $\mathbf{b}[\vec{x}] = \prod_i \mathbf{a}_i[\vec{x}]$.
Input: A list of images.
Output: An image.
Script: Line 1 is used if the data is collected in a tensor, and the first axis represents the channels. Lines 2 through 4 are used if the data is a list of images.

```
>>> bmg = amg.prod(0)
>>> bmg = np.zeros( amg.shape[:2] )
>>> for i in range( len( amg ) ):
>>>     bmg *= amg[i]
```

A.9.3 FPF OPERATOR

Purpose: Creates a filter using the FPF algorithm.
Notation: $\mathbf{b}[\vec{x}] = Q_{\vec{c},\alpha}\{\mathbf{a}[\vec{x}]\}$, where $\vec{c}$ is the constraint vector, and α is the fractional power term.
Input: A matrix `X` in which the columns are the raveled, Fourier transforms of the images, a vector `cst` which is the constraint vector, and `alpha` which is the fractional power term.
Output: An image.
Script:

```
>>> bmg = fpf.FPF( X, cst, alpha )
```

A.10 MACHINE LEARNING

These algorithms perform supervised and unsupervised clustering.

A.10.1 PCA OPERATOR

Purpose: Transforms a set of vectors into a new PCA space.
Notation: $\{\mathbf{b}_i[\vec{y}]\} = \mathfrak{P}_n\{\mathbf{a}[\vec{x}]\}$, where n is the optional argument to indicate how many and/or which eigenvectors to use.
Input: A list of vectors.
Output: A list of vectors.
Script: The argument `D` is the number of eigenvectors to use. This function returns the coefficients and the eigenvectors. Line shows the function that will map other data into this same PCA space.

```
>>> cffs, evecs = pca.PCA( a, D=2 )
>>> cffs = pca.Project( evecs, datavecs )
```

B Operators in Symbolic Order

Table B.1
List of Defined Operators

Symbol	Name	Type
$\varnothing$	Block a channel	Channel
$\bowtie$	Pass a channel	Channel
$\mathbf{a}^{\dagger}[\vec{x}]$	Complex conjugate	Information
$\stackrel{?}{=}, >, <, \geq, \leq$	Comparisons	Information
$\mathbf{a}[\vec{x}] \cdot \mathbf{b}[\vec{x}]$	Inner product	Information
$\boxtimes$	Center of mass	Information
$\bigvee_m, \bigwedge_m$	Max & Min	Information
$f\mathbf{a}[\vec{x}]$	Intensity modification	Intensity
$\mathbf{a}[\alpha\vec{x}]$	Scaling	Geometric
$\mathbf{a}[\mathbf{M}\vec{x}]$	Affine	Geometric
$\mathbf{a}[\vec{x}] \& \mathbf{b}[\vec{x}]$	Elemental AND	Intensity
$\mathbf{a}[\vec{x}] \mid \mathbf{b}[\vec{x}]$	Elemental OR	Intensity
$\mathbf{a}[\vec{x}] + \mathbf{b}[\vec{x}]$	Elemental Addition	Intensity
$\mathbf{a}[\vec{x}] - \mathbf{b}[\vec{x}]$	Elemental Subtraction	Intensity
$\mathbf{a}[\vec{x}] \times \mathbf{b}[\vec{x}]$	Elemental Multiplication	Intensity
$\bigtriangledown_n$	Lo bits	Intensity
$\bigtriangleup_n$	Hi bits	Intensity
$\square_{\vec{v}_1,\vec{v}_2}$	Window	Geometric
$\Downarrow_n$	Downsample	Geometric
$\mathbf{a}[\vec{x}] \otimes \mathbf{b}[\vec{x}]$	Correlation	Transformation
$\rhd_n$	Erosion	Intensity
$\lhd_n$	Dilation	Intensity
$\{\vec{w}_i\}$	Vector set	
$\{\vec{w}_i\}[n]$	n-th element of Vector set	
$\{\mathbf{a}_i\}$	Image set	
$F^n\mathbf{a}[\vec{x}]$	Repeat function F n times	
$\left\{\begin{matrix}\mathbf{a}[\vec{x}]\\ \mathbf{b}[\vec{x}]\\ \mathbf{c}[\vec{x}]\end{matrix}\right\}$	Channel separator	Channel
$\Sigma\left\{\begin{matrix}\alpha\\ \beta\\ \gamma\end{matrix}\right\}$	Channel summation	Channel
$\Pi\left\{\begin{matrix}\alpha\\ \beta\\ \gamma\end{matrix}\right\}$	Channel product	Channel
$A_f\mathbf{a}[\vec{x}]$	Locations of value f	Intensity
$A_\vee, A_\wedge$	Max & Min locators	Information
$\mathcal{A}_n$	Append with opt. condition n	Creation
$\mathfrak{A}$	Genetic algorithm	Machine Learning
B_β	Bending (pincushion, barrel)	Geometric
$\mathcal{B}$	Binary Fill Holes	Intensity
C	Coordinate map	Geometric
$\mathcal{C}_a$	Concatenation	Geometric
$\mathfrak{C}$	Co-occurrence matrix	Information
$D_{\vec{v}}$	Shift	Geometric
$\mathcal{D}_n\mathbf{a}[\vec{x}]$	Euclidean Distance Transform	Transformation
$\mathfrak{D}(\{\mathbf{a}_i[\vec{x}]\}, \{\mathbf{a}_j[\vec{x}]\})$	Euclidean Distance	Information
$\partial_x\mathbf{a}[\vec{x}]$	Derivative	Transformation
E_m	Edge enhance, model m	Intensity

(Continued)

Table B.1 (*Continued*)
List of Defined Operators

Symbol	Name	Type
$\mathcal{E}$	Energy	Information
$\mathfrak{E}$	EMD	Expansion
F	User defined function	
$\mathcal{F}$	Fourier descriptors	Information
$\mathfrak{F}$	Fourier	Transformation
G	Gaussian window	Creation
$\mathcal{G}$	Gradient	Intensity
$\mathfrak{G}_{\vec{w},\vec{f},\vec{t},sw}[\vec{x}]$	Gabor	Creation
Γ_n	Threshold	Intensity
H	Hough	Transformation
$\mathcal{H}_\alpha$	Harris operator	Intensity
I_n	ICM	Expansion
$\mathcal{I}$	Isolation	Expansion
$\mathfrak{I}$	Imaginary component	Information
J_n	PCNN	Expansion
$\mathcal{J}$	Discrete Cosine Transform	Transformation
$k_{\vec{w};r_1,r_2}[\vec{x}]$	Kaiser	Creation
$\mathfrak{K}$	k-means clustering	Machine Learning
L	Log	Intensity
$\mathcal{L}_G$	Color model transformation	Transformation
$\mathfrak{L}_m$	Flip	Geometric
M_α	Morph	Geometric
$\mathcal{M}$	Average	Information
$\mathfrak{M}$	Sammon map	Machine Learning
N	Count	Information
$\mathcal{N}_m$	Nonzero	Information
$\mathfrak{N}$	Neural network	Machine Learning
$\mathcal{O}_{\downarrow,C}$	Sort	Information
$\mathbf{o}_{\vec{w};\vec{v},r}[\vec{x}]$	Circle	Creation
P	Polar coordinate transformation	Transformation
$\mathcal{P}_{\vec{v}}$	Radial polar	Transformation
$\mathfrak{P}_n$	PCA	Machine learning
$\prod_i\{\mathbf{a}_i[\vec{x}]\}$	Product over pixels	Contraction
$\mathbf{p}_{\vec{w};\vec{v},a,b}[\vec{x}]$	Ellipse	Creation
$\mathbf{q}_{\vec{w}}$	Random	Creation
$Q_{\vec{c},\alpha}$	FPF	Contraction
$\mathbf{r}_{\vec{w};\vec{v}_1,\vec{v}_2}[\vec{x}]$	Rectangle	Creation
$R_{\alpha,\vec{v}}$	Radial coordinate	Geometric
$\mathcal{R}_{\theta,\vec{v}}$	Rotation	Geometric
$\mathfrak{R}$	Real component	Information
S	Scaling	Geometric
$\mathcal{S}_n$	Smoothing	Intensity
$\mathfrak{S}$	SOM	Machine Learning
$\sum_i\{\mathbf{a}_i[\vec{x}]\}$	Summation over pixels	Contraction
T_n	Eigenimages	Expansion
$\mathcal{T}_A$	Regional Standard Deviation	Information
$U_{\vec{v}}$	Plop	Geometric
V	Covariance	Information
$\mathcal{V}_{\vec{w}}$	Reshape	Geometric
$\mathbf{w}_{\vec{w},\theta_1,\theta_2}$	Wedge	Creation
W_G	Warp	Geometric
$\mathcal{W}$	Wavelet decomposition	Transformation
X	Quadrant Swap	Geometric
Y	Load from File	Creation
Z	Dimension	Information
$\mathcal{Z}_{\vec{w};r,m,n}[\vec{x}]$	Zernike	Creation
$z_{\vec{w}}[\vec{x}]$	Zeros	Creation

C Lengthy Codes

Some Python scripts are too lengthy to embed into a chapter, and so they were placed in this chapter.

Code C.1 Programs to convert RGB to XYZ and then to CIE L*a*b*

```
# color.py
def RGB2XYZ( r,g,b ):
    V, H = r.shape
    mat = array( [[0.412453, 0.35758, 0.180423],\
                  [0.212671, 0.71516, 0.072169],\
                  [0.019334, 0.119193, 0.950227]] )
    c = zeros( (V,H,3),float )
    c[:,:,0] = r+0
    c[:,:,1] = g+0
    c[:,:,2] = b+0
    f = dot( c,transpose(mat))
    answ = zeros( (3,V,H), float )
    answ[0] = f[:,:,0]
    answ[1] = f[:,:,1]
    answ[2] = f[:,:,2]
    return answ

def LabHelp( d ):
    mask = greater( d, 0.008856 ).astype(int)
    answ = mask * (d ** (1./3) )
    answ = answ + (1-mask)*(7.787*d + 16./116)
    return answ

def XYZ2LAB( xyz ):
   xn=1003.061052; yn=1056.521491; zn =1148.503991
   t = xyz[0]/xn
   mask = greater( t, 0.0088556 ).astype(int)
   l = mask*(116* (xyz[1]/yn)**(1./3) -16)
   l = l + (1-mask)*( 903.3*xyz[1]/yn )
   a = 500 *( LabHelp( xyz[0]/xn ) - LabHelp(xyz[1]/yn))
   b = 200 *( LabHelp( xyz[1]/yn ) - LabHelp(xyz[2]/zn))
   return l,a,b
```

Code C.2 The **Zernike** function

```
# mgcreate.py
def Zernike( VH, rad, m,n ):
        rp = np.zeros( VH )
        horz = np.ones(VH[1])
        rr = (np.arange(rad)/float(rad))
        if m==0 and n==0: r = 1
        if abs(m)==1 and n==1:
                r = rr
        if m==0 and n==2:
                r = 2*(rr**2) -1
        if abs(m)==2 and n==2:
                r = rr**2
        if abs(m)==1 and n==3:
                r = 3*rr**3 - 2*rr
        if abs(m)==3 and n==3:
                r = rr**3
        if m==0 and n==4:
                r = 6 *rr**4 - 6*rr**2 +1
        if abs(m)==2 and n==4:
                r = 4*rr**4 - 3*rr**2
        if abs(m)==4 and n==4:
                r = rr**4
        if abs(m)==1 and n==5:
                r = 10 *rr**5 - 12* rr**3 + 3*rr
        if abs(m)==3 and n==5:
                r = 5 * rr**5 - 4 * rr**3
        if abs(m)==5 and n==5:
                r = rr**5
        if abs(m)==0 and n==6:
                r = 20*rr**6 - 30*rr**4 + 12*rr**2 - 1
        if abs(m)==2 and n==6:
                r = 15 * rr**6 - 20*rr**4 + 6*rr**2
        if abs(m)==4 and n==6:
                r = 6 *rr**6 - 5*  rr**4
        if abs(m)==6 and n==6:
                r = rr**6
        rp[:rad] = np.outer(r,horz)
        if m <0:
          rp *= np.sin( m * np.arange(VH[1]) /float(VH[1]) * 2 *np.pi )
        else:
          rp *= np.cos( m * np.arange(VH[1]) /float(VH[1]) * 2 *np.pi )
        ctr = int(VH[0]/2),int(VH[1]/2)
        answ = rpolar.IRPolar( rp, ctr )
        return answ
```

Code C.3 The **Plop** function

```
def Plop( data, VH, back=0):
    ans = np.zeros( VH, float ) + back
    vmax, hmax = VH
    V,H = data.shape
    vctr, hctr = V//2, H//2
    vactr, hactr = vmax//2, hmax//2
    valo = vactr - vctr
    if valo<0: valo = 0
    vahi = vactr + vctr
    if vahi>=vmax: vahi = vmax
    halo = hactr - hctr
    if halo<0: halo = 0
    hahi = hactr + hctr
    if hahi>=hmax: hahi = hmax
    vblo = vctr - vactr
    if vblo<=0: vblo = 0
    vbhi = vctr + vactr
    if vbhi>=V: vbhi= V
    hblo = hctr - hactr
    if hblo<=0: hblo = 0
    hbhi = hctr + hactr
    if hbhi>=H: hbhi = H
    if vahi-valo != vbhi-vblo:
        vbhi = vblo+vahi-valo
    if hahi-halo != hbhi-hblo:
        hbhi = hblo+hahi-halo
    ans[valo:vahi, halo:hahi] = data[vblo:vbhi, hblo:hbhi] + 0
    return ans
```

Code C.4 The **Warp** function

```
# delaunaywarp.py
def Warp( dela, fid, mat1 ):
        # dela from scipy.spatial.Delauany
        # fid is the target grid
        # mat1 is the image data
    V,H = mat1.shape
    mat2 = np.zeros( (V,H) )
    for i in range( V ):
        if i%25==0: print (i,end='')
        for j in range( H ):
            me = np.array( (i,j) )
            tnum = dela.find_simplex( me ) # which triangle is it in
            v1, v2, v3 = dela.vertices[tnum] # corners
            x1,y1 = dela.points[v1]
            x2, y2 = dela.points[v2]
            r3 = dela.points[v3]
            x3,y3 = r3
            T = np.array( ((x1-x3, x2-x3),((y1-y3),(y2-y3))))
            Ti = np.linalg.inv( T )
            l1, l2 = np.dot( Ti, me-r3)
            l3 = 1 - l1 - l2
            xx,yy = l1 * fid[v1] + l2*fid[v2] + l3*fid[v3]
            if 0 <= xx < V and 0 <= yy < H:
                mat2[i,j] = mat1[xx,yy]
    return mat2
```

Code C.5 The **KaiserMask** function

```
# mgcreate.py
def KaiserMask( shape, center, r1, r2 ):
    di, dj = center        # location of the center of the window
    v,h = shape
    theta = 2. * np.pi
    Iot = 1.0 + theta/4. + 2.*theta/64. + 3.*theta/2304.
    vindex = np.multiply.outer( np.arange(v), np.ones(h) )
    hindex = np.multiply.outer( np.ones(v), np.arange(h) )
    a = (di-vindex).astype(float)
    b = dj-hindex
    r = np.sqrt( a*a + b*b)
    del a,b
    mask = np.zeros( shape, float )
    mask = ( r<r1 ).astype(int)
    b = np.logical_and( (r<r2), (r>r1) )
    m = (r-r1)/(r2-r1)
    m = m*b
    a = theta * np.sqrt( 1.-m*m)
    a = 1.0 + a/4.0 + 2.0*a/64.0 + 3.0*a/2304.0
    a = a / Iot
    a = a * ( r< r2)
    a = a * (r>=r1).astype(int)
    mask = mask + a
    return mask
```

Bibliography

1 M. Acharya, J. Kinser, S. Nathan, M. C. Albano and L. Schlegel. An image analysis method for quantification of idiopathic pulmonary fibrosis. 2011 *IEEE Applied Imagery Pattern Recognition Workshop (AIPR)*, Washington, DC, 2011, pp. 1–5. doi: 10.1109/AIPR.2011.6176357.
2 A. Busch and W. W. Boles. Texture classification using multiple wavelet analysis. In *DICTA2002: Digital Image Computing Techniques and Applications*, January 2002.
3 A. M. Bronstein, M. M. Bronstein, A. M. Bruckstein, and R. Kimmel. Analysis of two-dimensional non-rigid shapes. *International Journal of Computer Vision (IJCV)*, 78/1:67–88, June 2008.
4 Y. I. Balkarey, M. G. Evtikhov, and M. I. Elinson. Autowave media and neural networks. In *Proceedings of SPIE*, volume 1621, pages 238–249, 1991.
5 J. D. Brasher and J. M. Kinser. Fractional-power synthetic discriminant functions. *Pattern Recognition*, 27(4):577–585, 1994.
6 P. Brodatz. *Textures: A Photographic Album for Artists and Designers*. Dover, New York, 1966.
7 J. W. Cooley and J. W. Tukey. An algorithm for the machine calculation of complex Fourier series. *Mathematics of Computation*, 19:297–301, 1965.
8 D. Cheng, X. Tang, J. Lu, and X. Liu. Multi-object segmentation based on improved pulse coupled neural network. *Computer and Information Science*, 1(4):91–98, 2008.
9 X. Feng, S. Dagou, and Y. Hongchen. Image edge detection based on improved pcnn. In *Information Science and Engineering (ICISE), 2010 2nd International Conference on*, pages 3757–3760, Hangzhou, China, Dec 2010.
10 P. J. B. Hancock, R. J. Baddeley, and L. S. Smith. The principal components of natural images. *Network: Computation in Neural Systems*, 3(1):61–70, 1992.
11 R. M. Haralick, K. Shanmugan, and I. Dinstein. Texture features for image classification. *IEEE Transactions on Systems, Man and Cybernetics*, SMC-3(6):610–621, 1973.
12 N. E. Huang, Z. Shen, S. R. Long, M. C. Wu, H. H. Shih, Q. Zheng, N-C. Yen, C. C. Tung, and H. H. Liu. The empirical mode decomposition and the hilbert spectrum for nonlinear and non-stationary time series analysis. *Proceedings of the Royal Society of London A*, 454:903–995, 1998.
13 Anaconda Inc. https://www.anaconda.com, December 2017.
14 J. L. Johnson and M. L. Padgett. PCNN models and applications. *IEEE Transactions on Neural Networks*, 10(3):480–498, 1999.
15 J. L. Johnson and D. Ritter. Observation of periodic waves in a pulse-coupled neural network. *Optics Letters*, 18(15):1253–1255, 1993.
16 J. M. Kinser and H. J. Caulfield. O(no) pulse-coupled neural network performing humanlike logic. *Proceedings of SPIE–the International Society for Optical Engineering*, 2760:555–562, 1996.
17 J. M. Kinser. Foveation by a pulse-coupled neural network. *IEEE Transactions on Neural Networks*, 10(3):621–626, 1999.
18 J. M. Kinser. *Kinematic Labs with Mobile Devices*. Morgan & Claypool, San Rafael, CA, 2015.
19 J. M. Kinser and C. Nguyen. Image object signatures from centripetal autowaves. *Pattern Recognition Letters*, 21(3):221–225, 2000.
20 J. M. Kinser, C. L. Wyman, and B. L. Kerstiens. Spiral image fusion: a 30 parallel channel case. *Optical Engineering*, 37(2):492–498, 1998.
21 K. I. Laws. Texture energy measures. In *Proceedings of Image Understanding Workshop*, pages 47–51, 1979.
22 O. A. Mornev. *Elements of the Optics of Autowaves in Self-Organization Autowaves and Structures far from Equilibrium*, pages 111–118. Springer-Verlag, 1984.
23 K. Mikolajczyk and C. Schmid. Scale & affine invariant interest point detectors. *International Journal of Computer Vision*, 60(1):63–86, 2004.
24 J. C. Nunes, Y. Bouaoune, E. Delechelle, O. Niang, and Ph. Bunel. Image analysis by bidimensional empirical mode decomposition. *Image and Vision Computing*, 21:1019–1026, 2003.

25 H. S. Own and A. E. Hassanien. Rough wavelet hybrid image classification scheme. *Journal of Convergence Information Technology*, 3(4):65–75, 2008.

26 Trygve Randen. `http://www.ux.uis.no/~tranden/brodatz.html` or `http://www.cipr.rpi.edu/resource/stills/brodatz.html`, 2011. Accessed 18 April 2011.

27 A. Rosenfeld and M. Thurston. Edge and curve detection for visual scene analysis. *IEEE Transactions on Computers*, C-20(5):562–569, may 1971.

28 scipy. `http://docs.scipy.org/doc/numpy/reference/generated/numpy.rollaxis.html`, 2014.

29 P. Simmons, H. J. Caulfield, J. L. Johnson, M. P. Schamschula, F. T. Allen, and J. M. Kinser. Hearing shapes: Auditory recognition of two-dimensional spatial patterns. *Proceedings of the SPIE*, 2824:84–99, 1996.

30 R. Srinivasan and J. Kinser. A foveating-fuzzy scoring target recognition system. *Pattern Recognition*, 31(8):1149–1158, 1998.

31 A. N. Skourikhine, L. Prasad, and B. R. Schlei. Neural network for image segmentation. In J. C. Bes-dek B. Bosacchi, D. B. Fogel, editor, *Applications and Science of Neural Networks, Fuzzy Systems, and Evolutionary Computation III, Proceedings of SPIE*, volume 4120, pages 28–35, 2000.

32 M. Turk and A. Pentland. Eigenfaces for recognition. *Journal of Cognitive Neuroscience*, 3(1):71–86, January 1991.

33 w3techs.com. The png image file format is now more popular than gif. http://w3techs.com/blog/entry/the_png_image_file_format_is_now_more_popular_than_gif, New York, May 2015.

34 J. N. Wilson and G. X. Ritter. *Handbook of Computer Vision Algorithms in Image Algebra.* Technology & Engineering. CRC Press, 2000.

35 X. Wang and X. Tang. Bayesian face recognition using gabor features. In *Proceedings of the WBMA03*, pages 70–73, 2003.

36 F. Xue, K. Zhan, Y-D. Ma, and W. Wang. The model of numerals recognition based on PCNN and FPF. In *Proceedings of the 2008 International Conference on Wavelet Analysis and Pattern Recognition*, Hong Kong, 2008.

37 D. Zhang and Y. Y. Tang. Extraction of illumination-invariant features in face recognition by empirical mode decomposition. In M. Tistarelli and M.S. Nixon, editors, *ICB 2009, LNCS 5558*, 2009.

Index